FRACTIONAL KINETICS IN SPACE

Anomalous Transport Models

FRACTIONAL KINETICS IN SPACE

Anomalous Transport Models

Vladimir Uchaikin
Renat Sibatov
Ulyanovsk State University, Russia

World Scientific

NEW JERSEY · LONDON · SINGAPORE · BEIJING · SHANGHAI · HONG KONG · TAIPEI · CHENNAI · TOKYO

Published by

World Scientific Publishing Co. Pte. Ltd.

5 Toh Tuck Link, Singapore 596224

USA office: 27 Warren Street, Suite 401-402, Hackensack, NJ 07601

UK office: 57 Shelton Street, Covent Garden, London WC2H 9HE

Library of Congress Cataloging-in-Publication Data
Names: Uchaikin, V. V. (Vladimir Vasilevich), author. | Sibatov, Renat, author.
Title: Fractional kinetics in space : anomalous transport models / by Vladimir V. Uchaikin
    (Ulyanovsk State University, Russia), Renat Sibatov (Ulyanovsk State University, Russia).
Description: New Jersey : World Scientific, 2018. | Includes bibliographical references.
Identifiers: LCCN 2017041018 | ISBN 9789813225428 (hardcover : alk. paper)
Subjects: LCSH: Fractional calculus. | Cosmic physics.
Classification: LCC QC20.7.F75 U34 2018 | DDC 523.01--dc23
LC record available at https://lccn.loc.gov/2017041018

British Library Cataloguing-in-Publication Data
A catalogue record for this book is available from the British Library.

For any available supplementary material, please visit
http://www.worldscientific.com/worldscibooks/10.1142/10581#t=suppl

"The Cosmos is all that is or was or ever will be. Our feeblest contemplations of the Cosmos stir us – there is a tingling in the spine, a catch in the voice, a faint sensation, as if a distant memory, of falling from a height. We know we are approaching the greatest of mysteries."

Carl Sagan

Contents

Overview

It is amazing how wildly different-sized objects exist in the Universe! The red giant Betelgeuse is so monstrously larger than our Sun that it could fill the orbit of Mars. But compared to our home Galaxy, the Milky Way, Betelgeuse is just a grain of sand on a beach. Clearly, these sizes are relative, and to distribute various objects over their sizes, we use a special term *size scale*. The size scale of an object is a region large enough to include the entire object but not so large that the object becomes insignificant.

Hermes Trismegistus, the founder of the philosophical school of Hermeticism, is famous for saying: ***"Know then the Greatest Secret of the Universe: As Above, So Below As Within, So Without"***. Nowadays, this idea is expressed by the term *scaling*. Scaling (or *self-similarity*, being a synonym of this term) of a dynamical process $x(t)$ is a special kind of its symmetry such that a change in scale of some variables can be compensated by a corresponding rescaling of others. Dealing with phenomena relating to different scales, we often meet the problems with graphical representation of dependencies under investigation for comparing them. To put the graphs with different scales on the same plot, the logarithmic transformation is often used

$$y = f(x) \quad \mapsto \quad \log y = \phi(\log x).$$

One of the specific properties of this transformation is its ability to straighten power function graphs:

$$y = x^\alpha \quad \mapsto \quad \log y = \alpha \log x.$$

This is an important sign of the property called the ***self-similarity***. We can change the units of both x and y, but the slope of the log-log plot remains the same. Functions of such kind characterize self-similar structures called ***fractals*** [Mandelbrot (1983)]. Many astrophysicists believe

1

that interplanetary medium, interstellar medium, galaxy distributions possess fractal structure signs. The fractal concept lies in the base of modern description of turbulence in hydrodynamics and plasma.

The concept of self-similarity plays a leading role in the probability theory: you just recall the central limit theorem. Its extension to continuous time provides us such universal stochastic model as the *Brownian motion*. The first who saw the permanent chaotic motion of tiny pollen grains suspended in water was the Scotland botanist Robert Brown (1827). He was very surprised with this discovery and thought that this motion had the living origin. However, he would be much more surprised if he knew that his name will be associated with the movement of matter on giant cosmic scales! Namely this image had served as a basis for interpretation of many astrophysical phenomena and especially for description of cosmic rays propagation in the Galaxy [Berezinskii *et al.* (1984)].

Because of its chaotic character, Brownian motion does not look like Newtonian motion of planets along Keplerian orbits, and its mathematical description was not an easy task in those days. Only many decades later, Albert Einstein and independently of him Marian von Smoluchowski solved this problem on the base of the random process theory. The physical explanation of the phenomenon was based on assumption that atoms and molecules actually exist, and was later verified experimentally by Jean Perrin in 1908. Perrin was awarded the Nobel Prize in Physics in 1926 "for his work on the discontinuous structure of matter". But 19th centuries earlier Roman poet Titus Lucretius Carus ("On the Nature of Things") gave an instructive descripton of this phenomenon: "*You will see a multitude of tiny particles mingling in a multitude of ways... their dancing is an actual indication of underlying movements of matter that are hidden from our sight... It originates with the atoms which move of themselves.*"

In one-dimensional case, the Brownian *propagator* (probability density function for one tracer with a fixed initial point) has the self-similar form

$$p(x,t) = t^{-1/2}g(xt^{-1/2}), \quad -\infty < x < \infty,$$

where $g(x)$ is the Gaussian density. Brownian trajectories are continuous but nowhere differentiable curves (Fig. 0.1). The length of its segment between any two points of such curve is infinite, which is a sign of its fractality. Observe that giving t natural values 1,2,3,..., we bridge to the *random walk model* underlying such fundamental results of the probability theory as the **Large Numbers Law** and the **Central Limit Theorem**.

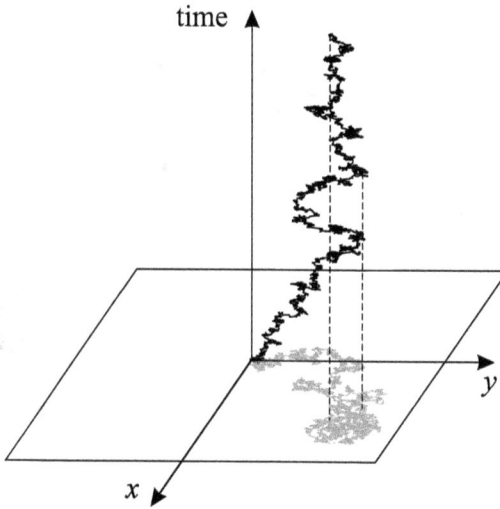

Fig. 0.1 A sample trajectory of the Brownian motion.

According to the latter, the size of a single walker probability cloud after N steps is

$$R(N) = R(1)N^{1/2}.$$

Here $R(1)$ is the size of one step. Associating $R(1)$ with the Compton length $h/m_n c$, we recognize the *Eddington-Weinberg formula* for interpretation of the aggregation process for structures observed in the Universe. Estimating the number of nucleons in a galaxy as 10^{68}, we arrive at the galaxy radius $R \simeq 1-10$ kpc and plausible interrelations for some other cosmic structures (see Table 0.1).

Table 0.1. $N - R$ interrelations for some structures.

Structure	Number of nucleons N	Evaluated size R
Galaxies	10^{68}	1–10 kpc
Clusters of galaxies	10^{72}	1–10 Mpc
Super-clusters of galaxies	10^{73}	10–100 Mpc

Perhaps, this is the most impressive case of scaling relations covering all from vanishingly small particles to unimaginably huge cosmical systems. To express his feelings generated by all-swallowing self-similarity of cosmic

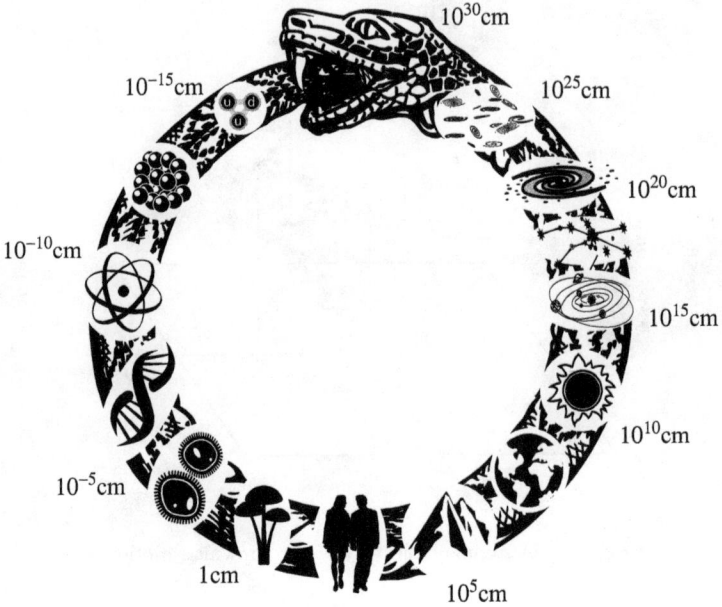

Fig. 0.2   Cosmic Uroboros.

structures, Sheldon Glashow, 1979 Nobel Laureate in Physics, resorted to the image of the *Uroboros*, laid on him some sort of dial with time-scale replaced by length-scale numbered from 10^{-30} cm to 10^{30} cm and named this image the *Cosmic Uroboros*[1] (Fig. 0.2).

Following the self-similarity idea, the outstanding French mathematician Paul Lévy powerfully pushed limits of probability theory by discovering a new class of laws and processes bearing now his name. The ***Lévy motion*** propagator has the form

$$p(x,t;\alpha) = t^{-1/\alpha} g(xt^{-1/\alpha};\alpha), \quad -\infty < x < \infty,$$

with positive constant $\alpha \in (0,2]$ called the *Lévy exponent*. The case $\alpha = 2$ recovers the Brownian motion with Gaussian propagator, but when $\alpha < 2$ we have the whole family of propagators – ***stable Lévy distributions*** – with infinite variances and long tails of inverse power type. The corresponding probability packets expand more rapidly than in the Brownian motion;

---

[1]The image of a serpent has led many cultures to associate it symbolically with the creation of the world and the unity of all things, especially when the serpent is represented as swallowing its own tail.

such processes are often classified as *superdiffusion*. Term **Lévy-flights** is also used for modeling the processes by broken lines composed of random segments with independent lengths (distributed according to inverse power laws) and isotropic direction.

In contrast to the Brownian trajectory, a Lévy-flight trajectory is not continuous and its trace in (2d- or 3d-space) looks like a set of clusters randomly scattered in the space. Mandelbrot noticed the striking similarity of these node distributions with distributions of galaxies in the Universe and used it for statistical simulations [Mandelbrot (1977)].

Finally, let us come back for a moment to the Lévy propagator and express its characteristic function (the Fourier image)

$$\widetilde{p}(k,t;\alpha) = \int\limits_{-\infty}^{\infty} e^{ikx} p(x,t;\alpha) dx = \exp(-|k|^{\alpha} t).$$

Clearly, it obeys the equation

$$\frac{\partial \widetilde{p}}{\partial t} = -|k|^{\alpha} \widetilde{p}(k,t;\alpha),$$

whose right-hand side contains multiplier $|k|^{\alpha}$ which is none other than the Fourier transform of the differential operator $-d^2/dx^2$ (or $-\Delta$ in a multidimensional case) raised to the non-integer power $\alpha/2$ ($\alpha \in (0,2]$). Thus, the self-similarity idea, symbolically represented by the Uroboros image, directly led us to such a fascinating area of modern analysis as the *fractional calculus*.

However, everything in the world has its limits, and self-similarity is no exception. As there are no infinite straight lines and planes in nature, so no endless fractals, distributions with infinite variance. The power functions reflecting these properties quickly descend to zero outside their relevance. Truncated distributions, meso-fractals and tempered fractional operators replace ideal self-similar structures and operations.

The central object of our attention in this book is the transport phenomena in interplanetary and interstellar media. They are very specific media uniting properties of gases, fluids, plasmas, subjected actions of gravity, magnetic fields, and cosmic radiations. In spite of extreme diversity in behavior of these components, reveal a common fundamental property: the *turbulence*. In fluid dynamics, turbulence or turbulent flow is a fluid regime characterized by chaotic, stochastic changes of dynamical variables, demonstrating low momentum diffusion, high momentum convection, and rapid variation of pressure and velocity in space and time. Nobel Laureate

Richard Feynman claimed the turbulence as "the most important unsolved problem of classical physics".

Turbulence is ubiquitous in the interstellar medium. It plays a major role in the formation of dense structures and stars, the stability of molecular clouds, the amplification of magnetic fields, the propagation of cosmic rays in the Galaxy. Notice that despite its importance, this phenomenon cannot be considered as being fully understood. Literature on the turbulence is extremely wide and the presentation of the theory of turbulence is not included in our plans. We just slightly touch on the problem only to the extent in which it reveals the need for fractional calculus.

In principle, the equations that govern turbulent flows are the same as that describing laminar flows, i.e., the Navier–Stokes equations. Given an initial state of a fluid, such equation suggests that the evolving velocity field $u(\mathbf{x}, t)$ is deterministic. However, we are uncertain about the uniqueness of the solution and therefore cannot characterize the phenomenon as a deterministic process. Moreover, the number of degrees of freedom characterizing a turbulent flow is extremely large, and therefore, the statistical approach is more appropriate than deterministic description. For this reason, the turbulent media dynamics is usually described in terms of *random fields*.

There exists a special parameter which is of the utmost importance for understanding conditions for emergence of turbulence. This is the *Reynolds number*, a dimensionless number that gives a measure of the ratio of *inertial forces* $\rho v^2 / L$ to *viscous forces* $\eta v^2 / L^2$. Although laminar-turbulent transition is not governed by Reynolds number, the same transition occurs if the size of the object is gradually increased, or the viscosity of the fluid is decreased, or if the density of the fluid is increased.

The basis for the statistical description is formed by three *Kolmogorov's hypotheses*, relating to statistically homogeneous and isotropic small-scale turbulence under large Reynolds numbers [Kolmogorov (1941)].

This book[2] will allow you to look at various aspects of cosmic kinetics through these "mathematical spectacles" which are still treated by many theorists as too exotic tools.

The first chapter introduces the reader to the scope of such concepts as self-similar structure, self-similar processes, stochastic fractals, Markov processes and Lévy motion. The latter is closely connected with the derivatives of fractional order. There are given examples and the physical sense of sim-

---

[2]We thank the Russian Foundation for Basic Research (grants 16-01-00556, 15-01-99674) for financial support.

ple partial differential equations of fractional order is discussed. The links between solutions of fractional-order equations and those of integer-order ones are shown, and fractional stable distribution family is represented. The chapter ends with demonstration of generalized versions of the Poisson process and Brownian motion.

The second chapter contains a summary of nonlocal diffusion models in turbulent environments, which include in particular the interplanetary, interstellar and intergalactic media. Comparing the Einstein's derivation of Brownian motion equation with Taylor's approach, Fock's one-dimensional walk model and linearized Boltzmann's equation, we demonstrate the difference between kinetic models and diffusion approximation. The physical nature of this distinction lies in the laws of turbulence, leading eventually to a nonlocal transport equations. Due to the power kind of these laws, the nonlocal operators turn out to be differ-integral operators of fractional orders. Some particular cases are discussed at the end of the chapter.

Chapter 3 begins consecutive exposition of fractional models applications to various transport problem with processes in interstellar medium (diffusion in interstellar plasma, walking magnetic field lines, diffusion in molecular clouds, and light propagation through turbulent interstellar medium).

Chapter 4 considers processes on the solar system scales (subdiffusion in photosphere, diffusion in a granular surface, solar flares, solar cosmic rays).

Chapters 5, 6 and 7 describe fractional models of propagation and acceleration of cosmic rays in galaxy. In Chapter 5, first models of anomalous diffusion of cosmic ray transport are reviewed and some inconsistencies of these models are discussed. Chapter 6 considers effects of anomalous transport in acceleration mechanisms: sub- and superdiffusive shock acceleration and fractional reacceleration of cosmic rays on supernova remnants. Chapter 7 contains validation of the new approach to cosmic ray propagation in the Galaxy (Nonlocal Relativistic Diffusion model – NoRD model) developed by the authors. We discuss some results obtained in its framework.

In Chapters 8 and 9, the reader will find the discussion of the large scale statistics and some aspects of the fractional cosmology. We consider spatial distribution of galaxies and its modeling on the base of Mandelbrot-Lévy walk scheme, fractional generalization of the Ornstein-Zernike phenomenology, and fractional generalization of the excursion set formalism in application to the large-scale structure formation.

Chapter 1

Mathematical prelude

The central theme of this chapter is the demonstration of how self-similarity being the universally observed property of material systems and processes on the cosmic scales of the universe inevitably involves fractional calculus in their description. The most visual example of such a property is given by the spatial distribution of galaxies. Inverse power type correlations observed in these distributions gave birth to a new trend in the science of space – fractal cosmology. A special type of self-similarity that relates spatio-temporal variables to one scaling variable forms a family of self-similar random processes, to which the Brownian motion (Bm) and other known processes belong (section 1.2). Section 1.3 contains some information about Markovian subfamily of the self-similar processes family, the main property of which is the independence of the future development from the past under a fixed presence state. Such properties are inherent in the Bm and Lévy motion. There are considered probability densities of these processes known as Lévy-stable distributions. At the end of this section, the reader can see how naturally fractional operators (Laplacians of fractional orders) appear in description of the space-generalized Bm. Section 1.4 considers an alternative case: the time-fractional generalization of Bm, involving Riemann-Liouville fractional derivatives with respect to time. In section 1.5, we discuss non-Markovian self-similar process governing by bifractional differential equation and its decomposition into two independent random walk processes with discrete time. The reader will become acquainted with fractional-stable and fractional Poisson distributions.

1.1 Self-similar structures

Geometrical models of objects, self-similarity of which ranges indefinitely were called *fractals* (see examples in Fig. 1.1). Apparently, the most impressive feature of a fractal system is represented by spatial galaxy distribution. Already in the 18th century, Immanuel Kant and Johann Lambert conceived a hierarchical structure of the Universe, in which stars clustered into larger systems, called nowadays galaxies, and which, in turn, were clustered into larger systems and so on. Many astronomers (John Herschel and Richard Proctor, Carl Charlier and Fournier d'Albe etc.) shared the idea of such constitution of the Universe. Two centures later, Gerard de Vaucouleurs found observational evidence supporting this idea in the distribution of galaxies in clusters and superclusters. In 1982, Benoit Mandelbrot reinterpreted this hierarchical structures in terms of infinite hierarchical self-similar structures called *fractals* [Mandelbrot (1983)].

Fig. 1.1 The first iterations of the formation of deterministic fractals. If iterated infinitely, each would be self-similar at any level and each would have a fractional dimension.

In Fig. 1.2 we can see an elementary fractal distribution of points in space whose construction is evident. Starting from a point occupied by an object and counting how many objects are present within a volume characterized by a certain length scale, we get $\overset{\circ}{N}_1$ point objects within a radius r_1, $\overset{\circ}{N}_2 = q\overset{\circ}{N}_1$ objects within a radius $r_2 = pr_1$, $\overset{\circ}{N}_3 = q\overset{\circ}{N}_2 = q^2\overset{\circ}{N}_1$ objects within $r_3 = pr_2 = p^2 r_1$ and so on. In general we have

$$\overset{\circ}{N}_n / \overset{\circ}{N}_1 = q^{n-1}$$

and

$$r_n / r_1 = p^{n-1},$$

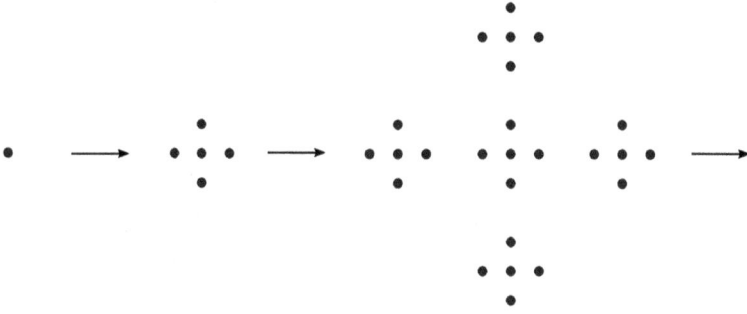

Fig. 1.2   A simple deterministic fractal with fractal dimension $D = 1.2$.

where p and q are constants. By taking the logarithm of the equations and dividing one by the other we get

$$\overset{\circ}{N}_n = Br_n^D \tag{1.1}$$

with

$$B = \overset{\circ}{N}_1 \, r_1^{-D}, \quad D = \frac{\ln q}{\ln p},$$

where B is a prefactor related to the lower cutoffs $\overset{\circ}{N}_1$ and r_1 of the fractal system, that is, the inner limit where the fractal system ends, and D is a fractal dimension $(D < 3)$. If we smooth out the point structure we get the continuum limit of equation (1.1)

$$\overset{\circ}{N}(r) = Br^D.$$

Further generalization of the fractal idea has led to the concept of *random* or *stochastic fractals*. In the frame of this approach,

(1) $\overset{\circ}{N}(r)$ is a random variable with the mean value

$$\langle \overset{\circ}{N}(r) \rangle = Br^D,$$

(2) the normalized random variable $Z = \overset{\circ}{N}(r)/\langle \overset{\circ}{N}(r) \rangle$ has the same distribution at all radii, i.e., a fractal structure is always connected with large fluctuations and clustering at all scales, and

(3) all its points are statistically equivalent with respect to their environments, there is no privileged points among them.

The density averaged over a sphere $V(r)$

$$\langle \overset{\circ}{n} \rangle_V \equiv \langle \overset{\circ}{N}(r) \rangle / V(r) = (3/4\pi)Br^{-\gamma}, \qquad \gamma = 3 - D$$

tends to zero as $r \to \infty$ (global number density) and diverges as $r \to 0$ (local number density). Therefore, one can use neither the densities nor correlation functions in the fractal case. Instead of these functions, the conditional density is used

$$\overset{\circ}{n}(r) = [d\langle \overset{\circ}{N}(r)\rangle/dr]/S(r) = (BD/4\pi)r^{-\gamma}, \qquad (1.2)$$

where $S(r)$ is the area of a spherical shell of radius r. Usually, the exponent $\gamma = 3 - D$ that defines the decay of the conditional density is called the *codimension*.

It is worth to keep in mind that fractal is not a real object but just a model which approximates its property in some limited region of space, $\lambda < r < \Lambda$. Most of astronomers and cosmologists believe that spatial distribution of galaxies described by power kind density (1.2) up to some distance Λ becomes uniformly distributed beyond this limit. On large scales, galaxies form a kind of cellular structure with a characteristic size of cells about 100 Mpc (≈ 0.33 billion light years). Inside the cells there is a deficiency of galaxies (so-called voids). Considering the scales of hundreds of megaparsecs, we observe that density fluctuations are smoothed out and the distribution of the visible substance becomes more homogeneous. Such turnover of fractal structure to homogeneous distribution can be covered by term *mesofractal* [Uchaikin (2004a)].

1.2   Self-similar processes

Let us now pass to dynamic processes describing moving structures. As a simplest case, we consider a single particle with $x(t)$ being one of its coordinates at time t. Assume that the function $x(t)$ obeys the scaling relation

$$x(at) = cx(t)$$

for any $a > 0$ and $t > 0$. A positive constant c is usually represented in the form

$$c = a^H,$$

where the *Hurst exponent* $H > 0$ determines the order of self-similarity. The process possessing the property

$$x(at) = a^H x(t)$$

is called the *self-similar process with index* $H > 0$ (in short H-ss process). Observe that the motion of a free particle $x(t) = vt$ is a 1-ss process.

The motion of a dynamical system with a known Hamiltonian including determined external forces is totally determined by its initial condition. The behavior of such system with time can be pictured by a sole smooth curve in the proper phase space. However, if the external forces are random, we have to take into account the whole family of possible trajectories called the *statistical ensemble*. Normalized to 1, their distribution over the phase space at any time t can be interpreted as a probability distribution. For detailed description of such motion, joint probability distributions $p_n(x_1, t_1; \ldots; x_n, t_n)$ are involved:

$$p_n(x_1, t_1; \ldots; x_n, t_n) dx_1 \ldots dx_n$$

$$= \mathrm{Prob}\left(X(t_1) \in dx_1, \ldots, X(t_n) \in dx_n\right), \quad n = 1, 2, 3, \ldots .$$

According to Kolmogorov axiomatics, the infinite set of such multivariate joint pdf's determines a stochastic (random) process.

A stochastic (random) process $\{X(t)\}$ is said to be *self-similar with index* $H > 0$ if for any $a > 0$ and any $n \geq 1$, $t_1, t_2, \ldots, t_n$ the joint distribution of the random variables $X(at_1), \ldots, X(at_n)$ is identical with the joint distribution of $a^H X(t_1), \ldots, a^H X(t_n)$:

$$\{X(at_1), \ldots, X(at_n)\} \stackrel{d}{=} \{a^H X(t_1), \ldots, a^H X(t_n)\},$$

where superscript over the equality sign means *equality in distribution*.

Having the family of joint pdf's, one can build the set of conditional pdf's:

$$p_n(x_n, t_n | x_1, t_1; \ldots; x_{n-1}, t_{n-1}) dx_n$$

$$= \frac{\mathrm{Prob}\left(X(t_1) \in dx_1, X(t_2 \in dx_2, \ldots, X(t_n) \in dx_n\right)}{\mathrm{Prob}\left(X(t_1) \in dx_1, \ldots, X(t_{n-1}) \in dx_{n-1}\right)}, \quad t_1 < t_2 < \cdots < t_n.$$

Observe that in case of deterministic process, additional information about the prehistory of the system, i.e., $x(t_{n-2})$, $x(t_{n-3})$, $\ldots$ to the known $x(t_{n-1})$ doesn't change prediction for the future motion $x(t_n)$. In case of stochastic process, we should specially postulate such property:

$$p_n(x_n, t_n | x_1, t_1; \ldots; x_{n-1}, t_{n-1}) = p_n(x_n, t_n | x_{n-1}, t_{n-1}).$$

Processes possessing such property are called *Markovian processes*. A Markov process homogeneous with respect to space and time is called the *Lèvy process* (L-*process*, see the Lévy and Bertoin monographs [Lévy (1965); Bertoin (1998)]), its transition pdf depends only on differences of arguments

$$x_k - x_{k-1} \equiv x_{k-1,k}, \quad t_k - t_{k-1} \equiv t_{k-1,k} \quad k = 1, \ldots, n:$$

$$p(x_{k-1}, t_{k-1} \to x_k, t_k) = p(0, 0 \to x_k - x_{k-1}, t_k - t_{k-1}) \equiv p(x_{k-1,k}, t_{k-1,k}).$$

In other words, the L-process is the *homogeneous process with independent increments*. Because

$$p_n(x_{01}, t_{01}; \ldots; x_{n-1,n}, t_{n-1,n}) = p(x_{01}, t_{01}) \ldots p(x_{n-1,n}, t_{n-1,n}),$$

the only self-similarity of $p(x, t)$,

$$p(x, t) = t^{-H} p(xt^{-H}, 1), \tag{1.3}$$

is enough for self-similarity of the whole process. Equation (1.3) can be rewritten in the dimensionless form

$$p(x, t)dx = g(\xi)d\xi, \quad \xi = xt^{-H}. \tag{1.4}$$

Certainly, inserting xt^{-H} instead of ξ into any probability distribution, one can give it the *form* of a random process, saying nothing about its random trajectories (*realizations*).

However, if we supplement the self-similarity condition with the Markovian condition, we arrive at a wide class of random processes called the *Lévy motion* (Fig. 1.3). The most known of them is the *Brownian motion*, for which

$$H = 1/2, \quad g(\xi) = \frac{1}{2\sqrt{\pi}} \exp\left(-\xi^2/4\right), \quad -\infty < \xi = xt^{-1/2} < \infty.$$

Examples of Gaussian and Lévy ($\alpha = 1.5$) noises and corresponding trajectories of Brownian and Lévy motion are presented in Fig. 1.3. For $\alpha = 1$, we have the Cauchy process (Fig. 1.4). Trajectories of Lévy motion are self-similar in the stochastic sense (Fig. 1.5).

1.3 Generalized (Lévy) statistics

The consequence of Markovian property

One can show that Brownian motion is the only ss-L-process, having a finite variance. Let's now consider an arbitrary L-process $\{X(t), \ t > 0\}$. The random coordinates of a particle performing motion of such kind at times t_1 and $t_1 + t_2$ are linked via the relation

$$X(t_1 + t_2) \overset{d}{=} X(t_1) + X(t_2).$$

Provided that

$$X(0) = 0,$$

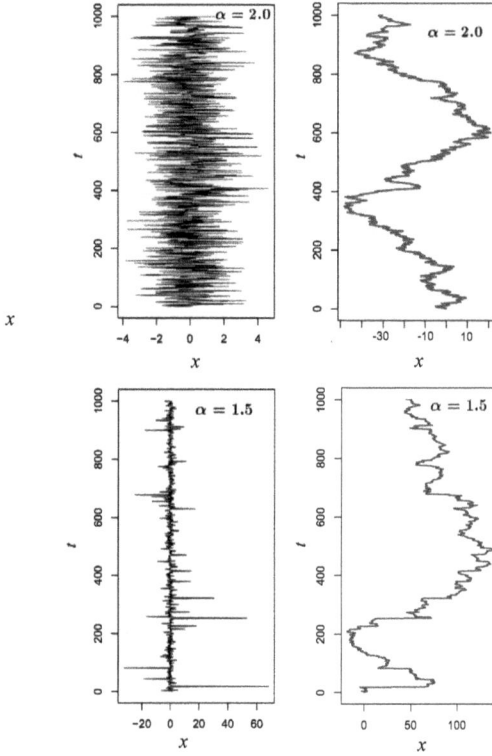

Fig. 1.3 Standard Gaussian and Lévy ($\alpha = 1.5$) noises (left panels) and corresponding trajectories of Brownian and Lévy motion (right panels).

the random variables $X(t_1)$ and $X(t_2)$ are increments of process in non-intersecting intervals $(0, t_1)$ and $(t_1, t_1 + t_2)$ and, hence, are independent. The probability density function of their sum is given by convolution of densities:

$$p(x, t_1 + t_2) = \int_{-\infty}^{\infty} p(x - x', t_1) \, p(x', t_2) \, dx'.$$

In terms of characteristic functions

$$\tilde{p}(k, t) = \langle e^{ikX(t)} \rangle = \int_{-\infty}^{\infty} e^{ikx} p(x, t) \, dx$$

we have

$$\tilde{p}(k, t_1 + t_2) = \tilde{p}(k, t_1) \, \tilde{p}(k, t_2).$$

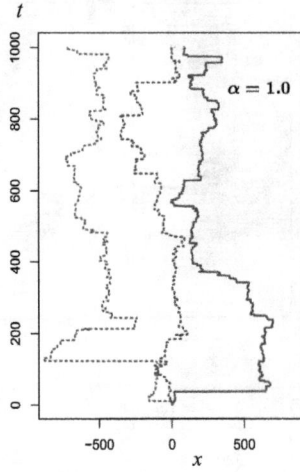

Fig. 1.4   Trajectories of Lévy motion for $\alpha = 1.0$ (Cauchy process).

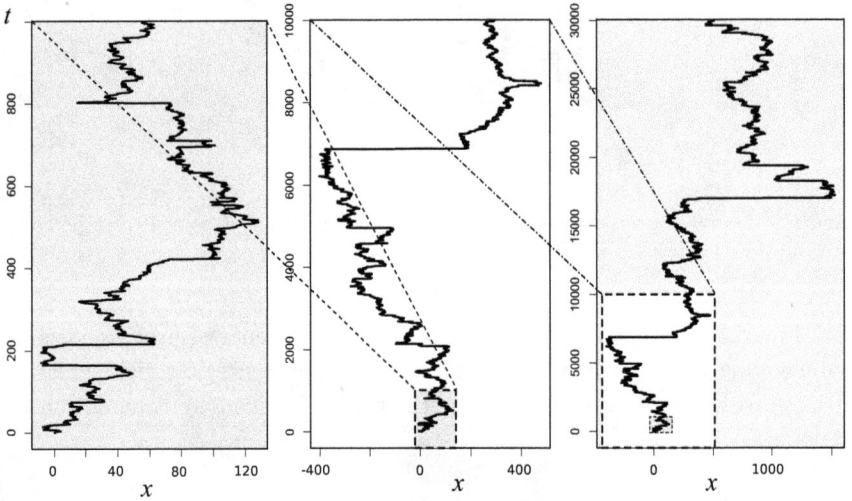

Fig. 1.5   Self-similarity of the trajectory of Lévy motion ($\alpha = 1.5$).

Taking the initial condition $\tilde{p}(k, 0) = 1$ and the property $|\tilde{p}(k, t)| \leq 1$, we find that

$$\tilde{p}(k, t) = e^{-\psi(k)t}, \qquad (1.5)$$

where $\mathrm{Re}\,\psi > 0$.

The consequence of self-similarity

Using the condition of self-similarity (1.3) yields

$$\tilde{p}(k,t) = \int_{-\infty}^{\infty} e^{ikx} p(xt^{-H},1)t^{-H} dx = \tilde{g}(kt^H).$$ (1.6)

Comparison of the right side of Eqs. (1.5) and (1.6) yields

$$\psi(k) = k^\alpha = |k|^\alpha \exp[-i(\theta\alpha\pi/2)\text{sign}k].$$

Here $\alpha = 1/H \in (0,2]$ is the *Lévy-exponent*, and θ is the *skew parameter*, $|\theta| \leq \theta_\alpha = \min\{1, 2/\alpha - 1\}$. So, the characteristic function of the Lévy stable distribution has the form (so-called form C)

$$\tilde{g}(k;\alpha,\theta) = \exp\left\{-|k|^\alpha \exp[-i\theta(\alpha\pi/2) \text{ sign}(k)]\right\}.$$

Another representation (form A) is

$$\tilde{g}^{(\alpha,\beta)}(k) = \exp\left\{-|k|^\alpha[1 - i\beta\tan(\alpha\pi/2) \text{ sign}(k)]\right\},$$

where β is linked to θ via relation

$$\beta = \tan(\theta\alpha\pi/2)/\tan(\alpha\pi/2).$$

The details of computing stable pdf's by inverting Fourier transform can be found in books [Lukacs (1970); Feller (1971); Zolotarev (1986); Samorodnitsky and Taqqu (1994); Uchaikin and Zolotarev (1999)].

Main properties of stable densities

All stable densities are unimodal, variances of them except Gaussian ($\alpha = 2$) are infinite (one or both of its tails in this case fall in a power type, $\propto |x|^{-\alpha-1}$), the mean values of stable distributions with exponents $\alpha < 1$ do not exist.

The densities with $\theta = 0$ are symmetrical with respect to 0. We will omit in this case the zero-parameter in arguments:

$$g(x;\alpha,0) = g(x;\alpha).$$

If the skew parameter $\theta \neq 0$, the most part of probability is distributed over the positive semiaxis in case $\theta > 0$ and over the negative semiaxis in case $\theta < 0$. When $\alpha < 1$, the distributions with $\theta = \pm 1$ become one-sided, occupying only one half of the axis. We will mark these densities by subindexes $\pm$:

$$g(x;\alpha,\pm 1) = g_\pm(x;\alpha), \quad \alpha \leq 1.$$

When $\alpha = 1$, the probability concentrate at two points:

$$g_{\pm}(x; 1) = \delta(x \mp 1).$$

The following stable densities can be represented in terms of elementary functions: the Gaussian distribution

$$g(x; 2) = \frac{1}{2\sqrt{\pi}} \exp\left\{-\frac{x^2}{4}\right\}, \quad -\infty < x < \infty,$$

the Cauchy distribution

$$g(x; 1) = \frac{1}{\pi(1 + x^2)}, \quad -\infty < x < \infty,$$

and the Lèvy-Smirnov distribution

$$g_{+}(x; 1/2) = \frac{1}{2\sqrt{\pi x^3}} \exp\left\{-\frac{1}{4x}\right\}, \quad 0 < x < \infty.$$

Some of the stable pdf's are pictured in Figs. 1.6, 1.7 and 1.8.

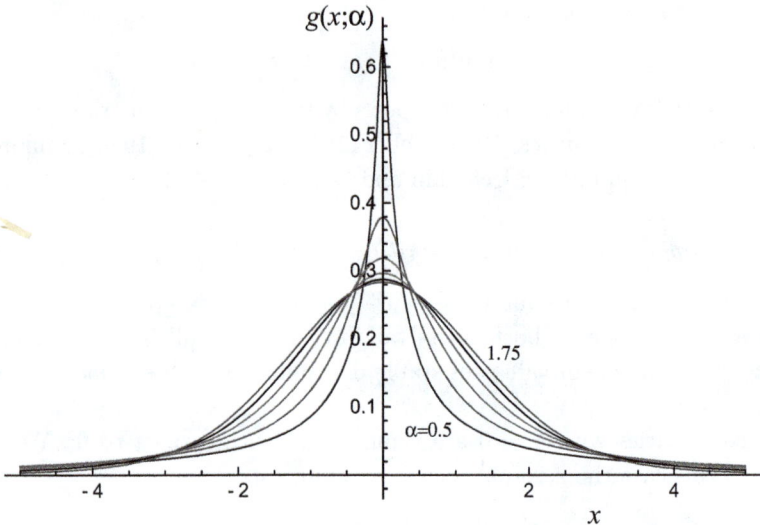

Fig. 1.6   Symmetrical Lévy stable densities.

Multivariate Lévy-distributions

Considering a set of $d > 1$ random variables, we can interpret this set as some random d-vector, given by these components.

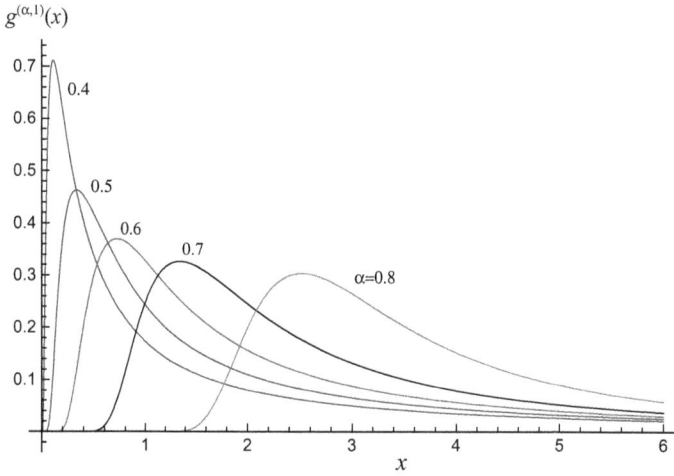

Fig. 1.7 One-sided Lévy stable densities.

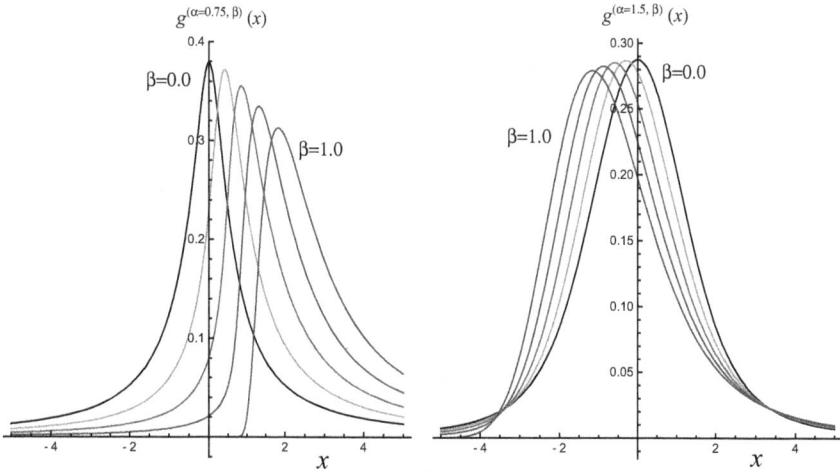

Fig. 1.8 Lévy stable densities for different values of asymmetry parameter ($\beta = 0.0$, 0.25, 0.5, 0.75, 1.0).

Let $\mathbf{G}$ be a two-dimensional Gaussian isotropic vector with variance equals 2, so its characteristic function is spherical symmetric and reads

$$\tilde{g}_2(\mathbf{k}; 2) = \mathsf{E} e^{i\mathbf{k}\cdot\mathbf{G}} = e^{-k^2}.$$

But the random vector possesses one more important property: its components are mutually independent:

$$\tilde{g}_2(\mathbf{k}; 2) = e^{-(k_1^2 + k_2^2)} = e^{-k_1^2} e^{-k_2^2} = g_1(k_1) g_1(k_2).$$

Both representations can be written in a common form,

$$\tilde{g}_2(\mathbf{k}; 2, \mu(\cdot)) = \exp\left\{ -\int_{S_2} |\mathbf{u} \cdot \mathbf{k}|^2 \mu_2(d\mathbf{u}) \right\},$$

where $\mu_2(d\mathbf{u})$ is a spectral measure given on the unite 2d-sphere S_2, i.e., on the circle of the unit radius $|\mathbf{u}| = 1$. In the first case the measure is uniformly distributed over the circle,

$$\mu_2^0(d\mathbf{u}) = \frac{d\varphi}{\pi}, \quad \varphi \in (1, 2\pi),$$

whereas in the second one it is equally concentrated on four points $(1, 0)$, $(0, 1)$, $(-1, 0)$ and $(0, -1)$:

$$\mu_2^1(d\mathbf{u}) \equiv \mu_2^1(d\varphi) = \frac{1}{2}[\delta(\varphi) + \delta(\varphi - \pi/2) + \delta(\varphi - \pi) + \delta(\varphi - 3\pi/2)]d\varphi.$$

Both the spectral measure distributions are depicted in Fig. 1.9. Each of these different forms produces the same isotropic normal density.

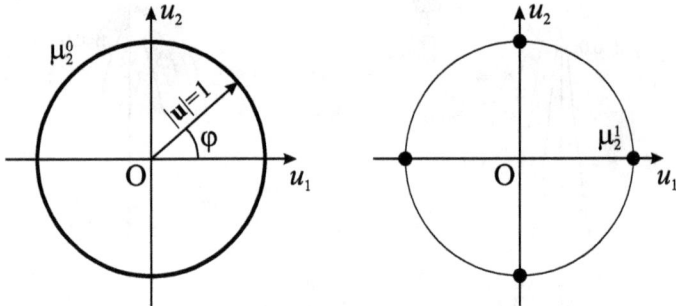

Fig. 1.9　Spectral measures of two-dimensional isotropic (left panel) and with independent projections (right panel) random central symmetrical vector.

The property takes place in Cartesian space of any dimension higher than 1, but only for Gaussian distributions. For other central symmetric stable distributions (i.e., for distributions satisfying the condition $g(\mathbf{x}) = g(-\mathbf{x})$), there exists a whole family of such (recall, the *central symmetrical*) distributions described by common form of the characteristic function with the same α but different μ.

$$\tilde{g}_d(\mathbf{k}; \alpha, \mu(\cdot)) = \exp\left\{ -\int_{S_d} |\mathbf{u} \cdot \mathbf{k}|^\alpha \mu_d(d\mathbf{u}) \right\}. \tag{1.7}$$

Denoting as before, the spectral measure for isotropic distribution by μ^0, and for the distribution with independent components by μ^1, we obtain

$$\tilde{g}_d(\mathbf{k}; \alpha, \mu^0(\cdot)) = \exp\left\{-\left(\sum_{j=1}^{d} k_j^2\right)^{\alpha/2}\right\} \tag{1.8}$$

and

$$\tilde{g}_d(\mathbf{k}; \alpha, \mu^1(\cdot)) = \exp\left\{-\sum_{j=1}^{d} |k_j|^\alpha\right\}. \tag{1.9}$$

Examples of two such densities (the 2d-Cauchy distributions with μ^0 and μ^1) are plotted in Fig. 1.10.

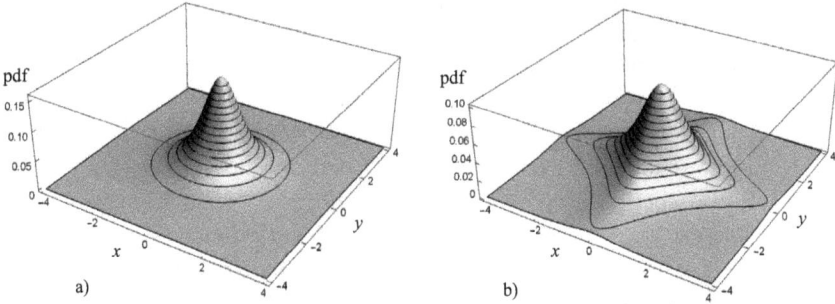

Fig. 1.10   Densities (a) $g_2(x_1, x_2; 1, \mu_2^0)$, and (b) $g_2(x_1, x_2; 1, \mu_2^1)$.

More detailed information can be found in [Samorodnitsky and Taqqu (1994)] and [Uchaikin and Zolotarev (1999)].

Fractional Laplacian

Let us return to fractals. As is said above, the main attribute of an isotropic (in average) embedded in three-dimensional space stochastic fractal is its power kind of the mean density

$$n(\mathbf{x}) = (B\alpha/4\pi)r^{\alpha-3}, \quad r = |\mathbf{x}|.$$

This allows us to reveal a direct link between such a property of matter as the fractality and such a special mathematical operation as the fractional Laplacian operator.

Taking into account application to the Universe, we will consider $1 < \alpha < 2$. Denote by $\mathcal{F}$ the operator of Fourier transformation

$$\tilde{n}(\mathbf{k}) = (\mathcal{F}n)(\mathbf{k}) \equiv \int \exp(i\mathbf{k} \cdot \mathbf{x})n(\mathbf{x})d\mathbf{x},$$

$$n(\mathbf{x}) = (\mathcal{F}^{-1}\widetilde{n})(\mathbf{x}) \equiv (2\pi)^{-3} \int \exp(-i\mathbf{k} \cdot \mathbf{x})\widetilde{n}(\mathbf{k})d\mathbf{k}.$$

Computing the first of them,

$$\widetilde{n}(\mathbf{k}) = B\alpha k^{-\alpha} \int_0^\infty z^{\alpha-2} \sin z \, dz = Ak^{-\alpha}, \quad A = B\alpha\Gamma(\alpha - 1)|\cos(\alpha\pi/2)|,$$

and representing result in the form

$$(k^2)^{\alpha/2}\widetilde{n}(\mathbf{k}) = A,$$

we recognize the Fourier image of the equation with fractional Laplacian[1]

$$(-\Delta)^{\alpha/2}n(\mathbf{x}) = A\delta(\mathbf{x}). \tag{1.10}$$

The corresponding evolution equation

$$\frac{\partial p(\mathbf{x}, t)}{\partial t} = -K_\alpha(-\Delta)^{\alpha/2}p(\mathbf{x}, t) + S(t)\delta(\mathbf{x}), \tag{1.11}$$

describes the d-dimensional ($\mathbf{x} \in \mathbb{R}^d$, $d = 1, 2, 3, \ldots$) Markovian self-similar process with start point at the origin of coordinates also called the *d-dimensional isotropic Lévy motion*, or *Lévy-Feldheim motion*. Parameter K_α is an anomalous diffusion coefficient (when $\alpha = 2$, K_2 represents ordinary diffusion coefficient $K_2 \equiv D$). If $S(t) = S_0 = $ const, then we arrive at Eq. (1.10) with $A = S_0/K$. In the case of instantaneous point source ($S(t) = \delta(t)$), solution of Eq. (1.11) is the probability density function (pdf) defined by its *characteristic function*

$$\widetilde{p}(\mathbf{k}, t) = e^{-K_\alpha|\mathbf{k}|^\alpha t}, \quad t \geq 0. \tag{1.12}$$

Trajectories of symmetric two-dimensional Lévy-Feldheim motion are presented in Fig. 1.11.

For central symmetric but anisotropic stable distributions considered in **1.4**, the following generalization of Eq. (1.12) results:

$$\widetilde{p}(\mathbf{k}, t; \mu(\cdot)) = \exp\left\{-K_\alpha t \int_{S_d} |\mathbf{u} \cdot \mathbf{k}|^\alpha \mu(d\mathbf{u})\right\} = \tilde{g}_d\left(\mathbf{k}(K_\alpha t)^{1/\alpha}; \alpha, \mu(\cdot)\right).$$

Its particular forms corresponding to two extreme cases (isotropic (1.8) and with independent components (1.9)) are obtained from the corresponding expressions for characteristic function replacing $|\mathbf{k}|^\alpha$ by $-(-\Delta)^{\alpha/2}$. The first of them is already represented by Eq. (1.11), and the second reads

$$\frac{\partial p(\mathbf{x}, t; \mu^1)}{\partial t} = -K_\alpha \sum_{j=1}^d \left(-\frac{\partial^2}{\partial x_j^2}\right)^{\alpha/2} p(\mathbf{x}, t; \mu^1) + \delta(\mathbf{x})\delta(t).$$

Here, we assume $K_{\alpha,j} = K_\alpha$ for any j.

---

[1] In case of a homogeneous boundless space.

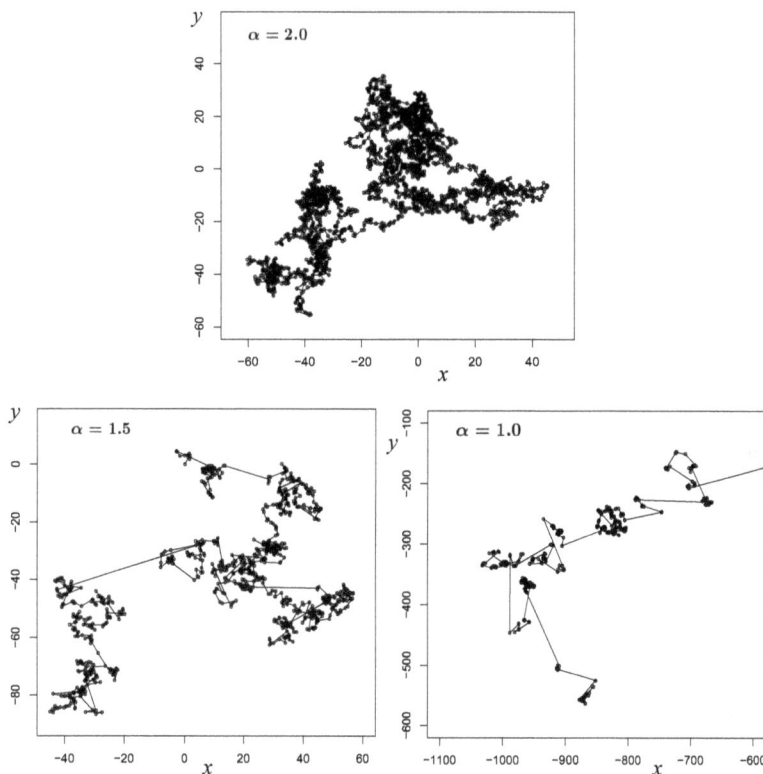

Fig. 1.11 Trajectories of isotropic two-dimensional Brownian (upper panel) and Lévy-Feldheim (lower panels) motion.

1.4 Fractional Brownian motion

The Lévy-motion described in the previous section could be named the space-fractional generalization of the Brownian motion. Here, we introduce the time-fractional generalization of the Brownian motion. However, before doing this, we consider transformation of the Bm under action of ordinary differential and integral operators.

Differential and integral Bm process

As is known, correlations of Bm coordinates at an arbitrary pair of times t_1, t_2 are described by the covariance function

$$\langle B(t_1)B(t_2)\rangle = [\sigma_0^2/2](|t_1| + |t_2| - |t_1 - t_2|).$$

Consider the *differential Bm process* (*dBm*), i.e., the process of Bm increments

$$dB(t) \equiv B(t + dt) - B(t), \quad dt = \text{const.}$$

Evidently,

$$\langle dB(t) \rangle = 0, \quad \sigma_{dB}^2 = \sigma_0^2 dt.$$

Autocorrelations in $dB(t)$ are described by the covariance function $\text{Cov}(dB(t_1), dB(t_2))$ which can easily be calculated from correspondent expression for Bm:

$$\langle dB(t_1)dB(t_2) \rangle = \frac{\partial^2 \langle B(t_1)B(t_2) \rangle}{\partial t_1 \partial_2} dt_1 dt_2 = \sigma_0^2 \delta(t_1 - t_2) dt_1 dt_2.$$

The differential Bm process $dB(t)$ is an example of *stochastic differentials* $dX(t)$. Many authors prefer to write

$$dB(t) = \xi(t)dt$$

or even

$$\frac{dB(t)}{dt} = \xi(t) \tag{1.13}$$

and call equations of such kind *stochastic equations*, and the functions $\xi(t)$ *random noises*. In this special case, when $B(t)$ represents Brownian motion, the noise $\xi(t)$ is called the *white noise* and expressed as

$$\xi(t) = B^{(1)}(t).$$

The stochastic process

$$B^{(-1)}(t) = {}_0 I_t B(t) \equiv \int_0^t B(t')dt' \tag{1.14}$$

is called the *integral Bm* (*iBm*). The iBm process is also a Gaussian process. One can easily verify it by representing the integral as a limit of approximation sums

$$S_n(t) = \sum_{j=1}^{n} B(t_j)\Delta t_j = \sum_{j=1}^{n} \left[\sum_{k=1}^{j} \Delta B(t_k) \right] \Delta t_j = \sum_{k=1}^{n} (t_n - t_{k-1})\Delta B(t_k) \tag{1.15}$$

and taking into account that any set of linear superpositions of independent normally distributed random variables $\Delta B(t_k) \equiv B(t_k) - B(t_{k-1})$, $k = 1, 2, 3, \ldots$, is jointly normal.

Since $\{B^{(-1)}(t),\ t \geq 0\}$ is Gaussian, its distribution is completely determined by its mean value and covariance function. They are easily computed and have the form:

$$\langle B^{(-1)}(t) \rangle = \left\langle \int_0^t B(t')dt' \right\rangle = \int_0^t \langle B(t') \rangle dt' = 0;$$

and, for $t_1 < t_2$,

$$\langle B^{(-1)}(t_1)B^{(-1)}(t_2) \rangle = \int_0^{t_1} dt' \int_0^{t_2} dt'' \langle B(t')B(t'') \rangle = \sigma_0^2 t_1^2 \left(\frac{t_2}{2} - \frac{t_1}{6} \right).$$

Note, that $\{B^{(-1)}(t), t \geq 0\}$ is not a Markov process, however, the vector process $\{(B^{(-1)}(t), B(t)), t \geq 0\}$ is again a Markov process.

Fractional Brownian motion

Three kinds of stochastic processes considered above and written in terms of derivatives,

$$B^{(-1)}(t) = {}_0\mathsf{D}_t^{-1} B(t),$$

$$B(t) = {}_0\mathsf{D}_t^0 B(t),$$

$$B^{(1)}(t) = {}_0\mathsf{D}_t^1 B(t),$$

provoke to introduce the *fractional Brownian motion (fBm)* determined as a fractional time-derivative of the ordinary Bm

$$_0 B_H(t) \equiv {}_0\mathsf{D}_t^\nu B(t) = \frac{1}{\Gamma(H + 1/2)} \int_0^t (t - t')^{H-1/2} dB(t'), \quad t > 0. \quad (1.16)$$

Though the process is self-similar, its increments are stationary only in case the Hurst exponent $H = 1/2$ when it becomes the ordinary Bm:

$$_0 B_{1/2}(t) = \int_0^t dB(t') = B(t).$$

Trajectories of fBm are demonstrated in Fig. 1.12.

In contrast with the Bm, the fBm possesses a memory about its prehistory. For this reason, restriction of the low limit of time-integration with zero doesn't look natural, except the case, when the fBm-tracer didn't exist

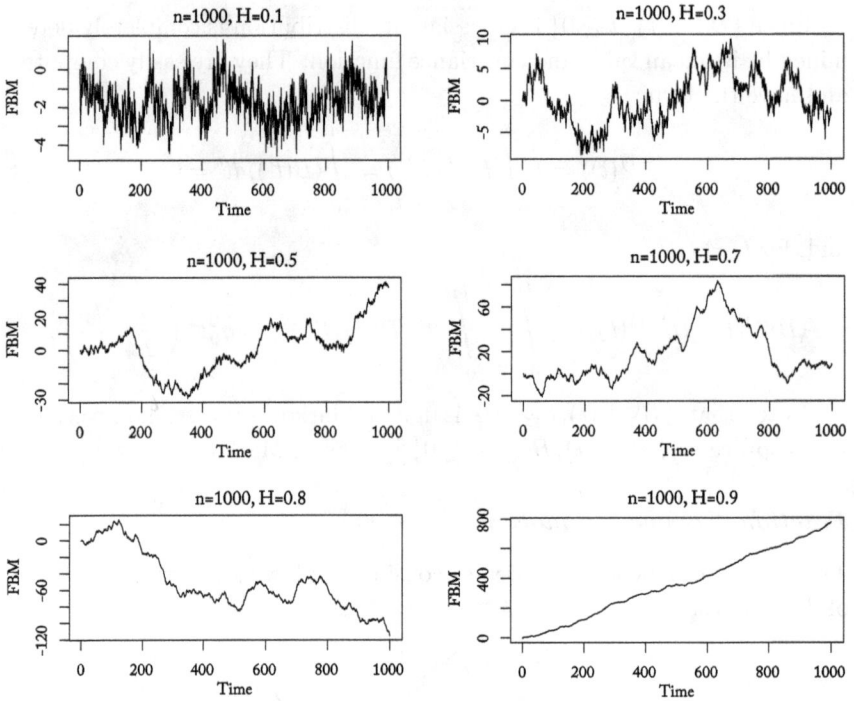

Fig. 1.12   Trajectories of fBm.

before $t = 0$. To generalize this model Mandelbrot and van Ness [Mandelbrot and Van Ness (1968)] extended the integral to the whole time-axis:

$$B_H(t) = \frac{1}{\Gamma(H + 1/2)} \int\limits_{-\infty}^{\infty} \left[(t - t')_+^{H-1/2} - (-t')_+^{H-1/2} \right] dB(t').$$

Direct computation yields the autocovariance function:

$$\langle B_H(t_1) B_H(t_2) \rangle = \frac{1}{[\Gamma(H + 1/2)]^2} = \frac{\sigma_H^2}{2} \left[|t_1|^{2H} + |t_2|^{2H} - |t_1 - t_2|^{2H} \right],$$

where

$$\sigma_H^2 = \langle B_H^2(1) \rangle = \frac{\Gamma(1 - 2H) \cos(H\pi)}{H\pi} \sigma_0^2.$$

By definition, the Hurst exponent is a self-similarity index which should be positive. On the other hand, if $H < 1$, the fBm is the only self-similar Gaussian process with stationary increments [Samorodnitsky and Taqqu (1994)].

The case $1/2 < H < 1$ relates to *persistent* or *fractional superdiffusion* (enhanced diffusion), the process with $H < 1/2$ describes *antipersistent* or *fractional subdiffusion*. Note that all these processes are characterized by Gaussian one-dimensional distribution:

$$p(x,t) = \frac{1}{2\sqrt{\pi\sigma}t^H} \exp\left\{-\frac{x^2}{4\sigma^2 t^{2H}}\right\}.$$

Riemann-Liouville fractional derivatives

The basic disadvantage of this definition is its difficulty for interpretation, especially when ν is not an integer. Another form of fractional operator called the Riemann-Liouville form seems to be more handy for statement of a problem originated from physical investigations. First of all, it is not hard to prove that determined by Eq. (1.16) operator for $\nu = 0$ becomes the identity operator and for $\nu = -m < 0$ ($m \in \mathbb{N}$) takes an m-integral form which can be transformed into the Cauchy formula

$$\frac{d^{-m}f(x)}{dx^{-m}} = \frac{1}{(m-1)!} \int_0^x (x-\xi)^{m-1} f(\xi)d\xi.$$

In its turn, this formula can be extended to non-integer orders $\mu > 0$,

$$\frac{d^{-\mu}f(x)}{dx^{-\mu}} = \frac{1}{\Gamma(\mu)} \int_0^x (x-\xi)^{\mu-1} f(\xi)d\xi.$$

Combining the operators of natural (n) and fractional negative $(-\mu \in (-1,0))$ orders, we obtain the *Riemann-Liouville fractional derivative* of order $\nu = n - \mu \in (n, n+1)$:

$$_0D_x^\nu f(x) = \frac{1}{\Gamma(\mu)} \frac{d^n}{dx^n} \int_0^x (x-\xi)^{\mu-1} f(\xi)d\xi.$$

Changing order of action of the operators, we obtain another form — the *Caputo fractional derivative*

$$_0^\nu D_x f(x) = \frac{1}{\Gamma(\mu)} \int_0^x (x-\xi)^{\mu-1} \frac{d^n f(\xi)}{d\xi^n} d\xi.$$

Although the differ-integral representations are more instructive for application to physical problems under investigation than the difference ones, their direct use for deriving fractional differential equations often remains not obvious. In many cases, applying the Laplace transform makes the problem easier for consideration. It is not hard to make sure that for $\nu \in (0,1)$ and $f(0) = 0$

$$\{\mathcal{L} \; {}_0D_x^\nu f(x)\}(\lambda) = \lambda^\nu \{\mathcal{L}f(x)\}(\lambda),$$

or

$$f^{(\nu)}(x) \overset{\mathcal{L}}{\leftrightarrow} \lambda^\nu \widehat{f}(\lambda).$$

1.5   Bifractional equation and random walks

Time-fractional generalization of space-fractional equation

Equation (1.11) is a space-fractional generalization of the normal diffusion equation. Continuing the generalization process, we replace the partial time-derivative by its fractional counterpart. Here, we consider isotropic case and omit μ^0 in $p(\mathbf{x},t;\mu^0)$. If the Brownian tracer starts its motion from the origin of coordinates, $p(\mathbf{x},0) = \delta(\mathbf{x})$. Observe that this condition admits three representations:

$$\frac{\partial p}{\partial t} = -K_\alpha(-\Delta)^{\alpha/2}p(\mathbf{x},t), \quad t \in (0,\infty), \quad p(\mathbf{x},0+) = \delta(x),$$

$$\frac{\partial p}{\partial t} = -K_\alpha(-\Delta)^{\alpha/2}p(\mathbf{x},t) + \delta(\mathbf{x})\delta(t), \quad t \in [0,\infty), \quad p(\mathbf{x},0-) = 0,$$

and

$$\frac{\partial[p(\mathbf{x},t) - \delta(\mathbf{x})1_+(t)]}{\partial t} = -K_\alpha(-\Delta)^{\alpha/2}p(\mathbf{x},t), \quad t \in [0,\infty), \quad p(\mathbf{x},0-) = 0,$$

$$(1.17)$$

where

$$1_+(t) = \begin{cases} 1, t > 0; \\ 0, t < 0. \end{cases}$$

The third of these forms is the most convenient for passage to time-fractional operator because

$${}_0D_t^\beta[p(\mathbf{x},t) - \delta(\mathbf{x})1_+(t)] = {}_0D_t^\beta p(\mathbf{x},t) - \delta(\mathbf{x})\delta_\beta(t),$$

where

$$\delta_\beta(t) = {}_0D_t^\beta \, 1_+(t) = \frac{t_+^{-\beta}}{\Gamma(1-\beta)}.$$

As a result, we arrive at the bifractional equation

$${}_0D_t^\beta p(\mathbf{x},t) = -K(-\Delta)^{\alpha/2}p(\mathbf{x},t) + \delta(\mathbf{x})\delta_\beta(t), \qquad (1.18)$$

where $K \equiv K_{\alpha,\beta}$.

In the one-dimensional case, the fractional Laplacian could be expressed through the left and right Riemann-Liouville fractional derivatives as

$$-(-\Delta)^{\nu/2} = \left(2\cos\frac{\pi\nu}{2}\right)^{-1}({}_{-\infty}D_x^\nu + {}_xD_\infty^\nu). \qquad (1.19)$$

Stochastic interpretation of bifractional equation

For adequate understanding of this equation, we pass to Fourier-Laplace variables,

$$p(\mathbf{x}, t) \mapsto \tilde{p}(\mathbf{k}, \lambda) = \int_0^\infty dt\, e^{-\lambda t} \int dx\, e^{i\mathbf{kx}} p(\mathbf{x}, t).$$

The equation becomes

$$[\lambda^\beta + K|\mathbf{k}|^\alpha]\, \tilde{p}(\mathbf{k}, \lambda) = \lambda^{\beta-1}. \tag{1.20}$$

Representing K as c_α/c_β, rewrite the equation in the following asymptotically ($\lambda \to 0$, $|\mathbf{k}| \to 0$) equivalent form:

$$[1 - (1 - c_\beta \lambda^\beta)(1 - c_\alpha|\mathbf{k}|^\alpha)]\, \tilde{p}(\mathbf{k}, \lambda) = \frac{1 - (1 - c_\beta \lambda^\beta)}{\lambda}. \tag{1.21}$$

The terms $1 - c_\beta \lambda^\beta$ and $1 - c_\alpha|\mathbf{k}|^\alpha$ can be considered as main asymptotical terms in integral transforms of some time and space probability densities $q(t)$ and $p(\mathbf{x})$

$$\tilde{q}(\lambda) = \int_0^\infty e^{-\lambda t} q(t) dt \sim 1 - c_\beta \lambda^\beta, \quad \lambda \to 0 \tag{1.22}$$

and

$$\tilde{p}(\mathbf{k}) = \int_{\mathbb{R}^d} e^{-i\mathbf{k} \cdot \mathbf{x}} p(\mathbf{x}) d\mathbf{x} \sim 1 - c_\alpha|\mathbf{k}|^\alpha, \quad |\mathbf{k}| \to 0. \tag{1.23}$$

Thus, Eq. (1.21) can be considered as the asymptotical representation of some 'exact' equation

$$[1 - \tilde{q}(\lambda)\tilde{p}(\mathbf{k})]\, \tilde{p}(\mathbf{k}, \lambda) = \frac{1 - \tilde{q}(\lambda)}{\lambda},$$

looking in time-space variables as

$$p(\mathbf{x}, t) = \int_0^t dt' \int d\mathbf{x}'\, q(t') p(\mathbf{x}') p(\mathbf{x} - \mathbf{x}', t - t') + \delta(\mathbf{x})\bar{Q}(t), \tag{1.24}$$

where

$$\bar{Q}(t) = \mathsf{P}(T > t) = \int_t^\infty q(t') dt'$$

(the bar-sign distinguishes the complement distribution function from the ordinary cumulative distribution function $Q(t) = \mathsf{P}(T < t)$).

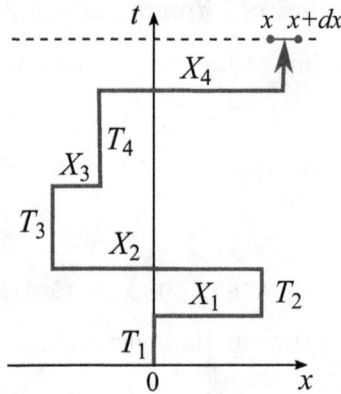

Fig. 1.13   One-dimensional CTLJ process.

Equation (1.24) being a prelimit image of the bifractional differential equation describes the following jump process. A particle placed at origin $\mathbf{x} = 0$ at time $t = 0$, stays there during a random time T distributed with probability density $q(t)$. At the moment T it performs a jump on a random vector $\mathbf{x}$ distributed with 3-dimensional density $p(\mathbf{x})$ and stays in this new position during a random time T' with the same distribution as T. Then the particle repeats such jumps again and again till the observation time t and solution of Eq. (1.24) gives pdf for random position of the particle at observation time t.

Processes with random instantaneous jumps between which the particle is immobile during random waiting time are known as CTRW (*Continuous Time Random Walk*) processes. For the sake of convenient, we will use for subfamily of CTRW satisfying conditions (1.22)-(1.23) abbreviation CTLJ (*Continuous Time Lévy Jumps*). Schematic of one-dimensional CTLJ is shown in Fig. 1.13.

Similarly to the use of difference approximations for numerical solving of integer-order differential equations, the CTLJ-approximation is a power tool for solving fractional-order differential equations. In addition, the CTLJ-model has the clear probabilistic sense and for this reason it often occurs closer to the real process under numerical investigation, for which the fractional equation itself serves as asymptotical long-time and long-scale approximation. Nevertheless, the fractional calculus is a very effective tool for analytical investigation of non-Markovian processes running in nonhomogeneous (i.e., porous, turbulent, fractal) media.

Mixing-time interrelations

By means of Laplace transform with respect to time, one can ascertain that solution $f_{1,\beta}(\mathbf{x}, t)$ of time-fractional diffusion equation

$$_0D_t^\beta f_{1,\beta} = D\Delta f_{1,\beta}(\mathbf{x}, t) + \delta_\beta(t)\delta(\mathbf{x}), \quad \delta_\beta(t) = \frac{t_+^{-\beta}}{\Gamma(1 - \beta)}, \quad \beta \in (0, 1],$$

is expressed through the solution $f_{1,1}(\mathbf{x}, t)$ of the standard diffusion equation

$$\frac{\partial f_{1,1}}{\partial t} = D\Delta f_{1,1}(\mathbf{x}, t) + \delta(t)\delta(\mathbf{x}),$$

via integral relation

$$f_{1,\beta}(\mathbf{x}, t) = \int_0^\infty w_{1,\beta}(\tau|t) f_{1,1}(\mathbf{x}, \tau) d\tau, \tag{1.25}$$

whereas the solution of equation with the fractional Laplacian

$$\frac{\partial f_{\gamma,1}}{\partial t} = -(-\Delta)^\gamma f_{\gamma,1}(\mathbf{x}, t) + \delta(t)\delta(\mathbf{x}), \quad \gamma \in (0, 1]$$

is related to the solution of the basic equation via the *Bochner formula* [Bochner (1949)]

$$f_{\gamma,1}(\mathbf{x}, t) = \int_0^\infty w_{\gamma,1}(\tau|t) f_{1,1}(\mathbf{x}, \tau) d\tau. \tag{1.26}$$

These formulas differ from each other only by functions $w(\tau|t)$, which are expressed, however, through the same one-sided stable density $g_+(x; \alpha)$ but in different ways:

$$w_{1,\beta}(\tau|t) = (t/\beta)g_+(t\tau^{-1/\beta}; \beta)\tau^{-1/\beta-1},$$

$$w_{\gamma,1}(\tau|t) = t^{-1/\gamma}g_+(\tau t^{-1/\gamma}; \gamma).$$

The function has a sense of *operational time distribution* at fixed *observational time* t, if we interpret these expressions in terms of conception of *subordinated processes* [Feller (1971)], or of distribution of travel time along different paths if we think in terms of *mixing-length theory*.[2] Let us underline that when $\alpha = 1$ then $g_+(x; 1) = \delta(x - 1)$ and both operational and observational times coincide.

---

[2]From physical point of view, the latter interpretation is more visual, but applicable only in the case without trapping. The first one which can be called *mixing-length concept* is applicable in both cases.

For bifractional equation,

$$_0D_t^\beta f(\mathbf{x}, t) = -(-\Delta)^\gamma f(\mathbf{x}, t) + \delta(\mathbf{x})\delta^{(\beta)}(t), \quad \gamma \in (0, 2], \ \beta \in (0, 1], \quad (1.27)$$

the solution can be represented in the form [Uchaikin and Sibatov (2012b)]:

$$f_{\gamma, \beta}(\mathbf{x}, t) = \int_0^\infty f_{1,1}(\mathbf{x}, \tau) \, t^{-\beta/\gamma} q_+(\tau t^{-\beta/\gamma}; \gamma, \beta) \, d\tau, \qquad (1.28)$$

where

$$q_+(\tau; \gamma, \beta) = \int_0^\infty \xi^{\beta/\gamma} g_+(\tau \xi^{\beta/\gamma}; \gamma) g_+(\xi; \beta) \, d\xi \qquad (1.29)$$

is a one-sided fractional stable density [Kolokoltsov *et al.* (2001)]. This representation is convenient to compute the propagator by means of the Monte-Carlo algorithm,

$$f_{\gamma, \beta}(\mathbf{x}, t) = \left\langle f_{1,1}\left(\mathbf{x}, S_+(\gamma)/[S_+(\beta)]^{-\beta/\gamma}\right) \right\rangle. \qquad (1.30)$$

Here, $S_+(\gamma)$ is a one-sided Lévy stable variable [Uchaikin and Zolotarev (1999)].

Fractional stable distributions

Combining Eqs. (1.25) and (1.26), we write solution of Eq. (1.18) in the self-similar form

$$p(\mathbf{x}, t) = t^{-d\beta/\alpha} \Psi_d^{(\alpha, \beta)}(\mathbf{x} t^{-\beta/\alpha}; \mu_d^0),$$

where

$$\Psi_d^{(\alpha, \beta)}(\mathbf{x}; \mu_d^0) = \int_0^\infty g_d(\mathbf{x} \tau^{\beta/\alpha}; \alpha, \beta, \mu_d^0) \, g_+(\tau; \beta) \tau^{d\beta/\alpha} d\tau$$

is the *isotropic fractional stable distribution* (see for detail [Uchaikin (2002)]). One-dimensional fractional stable distributions are plotted in Fig. 1.14. The densities with fractional β (panels c and d) have a singularity at the origin, dependent on β. When $\beta = 1$, the singularity disappears (panels a and b) and $\Psi_d^{(\alpha, \beta)}$ becomes the isotropic Lévy-Feldheim density $g_d(\mathbf{x}; \alpha)$.

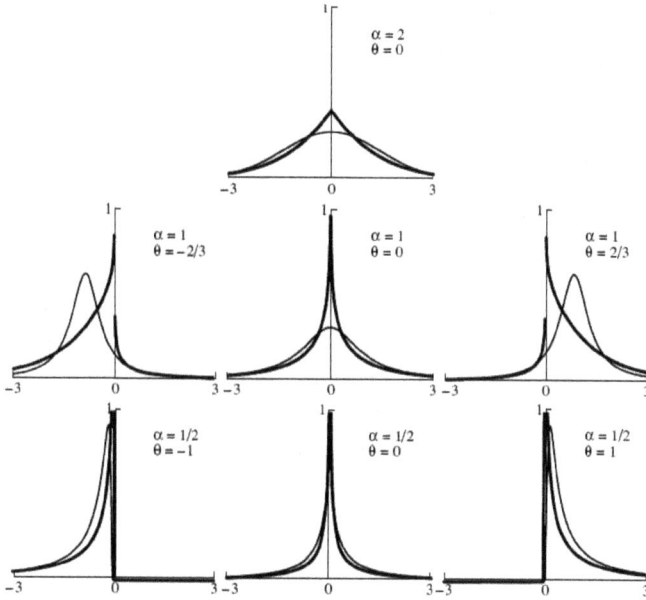

Fig. 1.14 Densities of one-dimensional stable distributions (smoothly peaked light curves) and of fractionally stable distributions with the exponent $\beta = 1/2$ (sharply peaked heavy curves).

Decomposition of CTLJ into DTLJ and FPP

Results of the above-stated discussion give the basis for decomposition CTLJ-process into two independent processes with discrete time $n = 0, 1, 2, \ldots$:

$$\{0, X(1), X(2), X(3), \ldots, X(n), \ldots\} \tag{1.31}$$

and

$$\{0, T(1), T(2), T(3), \ldots, T(n), \ldots\}, \tag{1.32}$$

where

$$X(n) = X_1 + X_2 + X_3 + \cdots + X_n$$

and

$$T(n) = T_1 + T_2 + T_3 + \cdots + T_n$$

respectively.

The sequence Eq. (1.31) calling *Discrete Time Lévy Jumps* (DTLJ) is completely determined by transition pdf $p(x) \equiv p(x, 1)$,

$$p(x|n) = p^{n\star}(x), \tag{1.33}$$

where sign $\star$ marks the Fourier-convolution:

$$p^{0\star}(x) = \delta(x), \; p^{1\star}(x) = p(x), \; p^{n\star}(x) = \int\limits_{-\infty}^{\infty} p^{(n-1)\star}(x - x')p(x')dx'.$$

In its turn, all properties of another discrete time process represented by nondecreasing random sequence $T(n)$ with independent increments T_1 are determined by giving pdf $q(t)$. This process can be interpreted as a time-flow of instantaneous events (say, random pulses generated in electric circuit by Geiger counter measuring flow of gamma-quants). The random function $N(t)$ is called the *counting function* and determined by definition

$$N(0) = 0, \quad N(t) = \max\{n : T(n) \le t\}.$$

Denoting by $\mathsf{P}(A)$ probability of an event A, one can represent the probability distribution of $N(t)$ in the form

$$p_n(t) \equiv \mathsf{P}(N(t) = n) = \mathsf{P}\left(\sum_{j=1}^{n} T_j > t\right) - \mathsf{P}\left(\sum_{j=1}^{n+1} T_j > t\right), \quad n = 0, 1, 2, \ldots$$

and get to the following system of integral equations for $p_n(t)$

$$p_n(t) = \delta_{n0} \int\limits_{t}^{\infty} q(\tau)d\tau + [1 - \delta_{n0}] \int\limits_{0}^{t} q(t - \tau)p_{n-1}(\tau)d\tau, \; n = 0, 1, 2, \ldots .$$

$$\tag{1.34}$$

Counting processes belong to the class of the renewal processes. Feller noted that considering renewal processes we deal merely with sums of independent identically distributed random variables, and the only reason for introducing a special term is using such a power analytic tool as the renewal equation. The mean number of events by time t

$$\langle N(t) \rangle = \sum_{n=0}^{\infty} np_n(t)$$

is called the *renewal function*. It obeys the *renewal equation*

$$\langle N(t) \rangle = \int\limits_{0}^{t} [1 + \langle N(t - t') \rangle]q(t')dt',$$

which means that the mean number of events within $(0, t)$ is equal to the contribution of the first event plus the mean number of the subsequent events. Differentiation of this equation yields equation for the rate of the event flow

$$f(t) = \frac{d\langle N(t)\rangle}{dt},$$

having the form:

$$f(t) = q(t) + \int_0^t f(t - t')q(t')dt'. \tag{1.35}$$

Let us find the kernel $q(t)$ for which the process $N(t)$ is self-similar in average. In other words, we want to find such $q(t) \equiv \psi_\nu(t)$ that

$$\langle N_\nu(t)\rangle = \mu_\nu t^\nu, \quad f_\nu(t) = \mu_\nu \nu t^{\nu-1}, \quad \mu_\nu > 0, \quad 0 < \nu \leqslant 1. \tag{1.36}$$

Following Mandelbrot [Mandelbrot (1983)], we will call the ensemble of random points on t-axis the *fractal dust* and the pdf $\psi_\nu(x)$ the *fractal dust generator*. As follows from above, it is linked with the mean fractal dust density $f_\nu(x)$ via equation

$$\psi_\nu(t) = f_\nu(t) - \int_0^t f_\nu(t - t')\psi_\nu(t')dt'. \tag{1.37}$$

Applying the Laplace transform

$$f(t) \mapsto \widehat{f}(\lambda) = \mathcal{L}\{f(t)\}(\lambda) \equiv \int_0^\infty e^{-\lambda t} f(t)dt$$

yields the expression

$$\widehat{\psi}_\nu(\lambda) = \frac{\widehat{f}_\nu(\lambda)}{1 + \widehat{f}_\nu(\lambda)} = \frac{\mu}{\mu + \lambda^\nu}, \quad \mu = \mu_\nu \nu \Gamma(\nu), \quad \nu \leq 1, \tag{1.38}$$

which for $\nu = 1$ coincides with the corresponding expression for the ordinary Poisson process:

$$\{\mathcal{L}\psi_1(t)\}(\lambda) = \frac{\mu}{\mu + \lambda}, \quad \psi_1(t) = \mu e^{-\mu t}.$$

For $\nu < 1$ it is expressed through two-parameter Mittag-Leffler function:

$$\psi_\nu(t) = \mu t^{\nu-1} \sum_{j=0}^\infty \frac{(-\mu t^\nu)^j}{\Gamma(\nu j + \nu)} = \mu t^{\nu-1} E_{\nu, \nu}(-\mu t^\nu) \tag{1.39}$$

having for small and large time algebraic asymptotics.

The renewal process with waiting time $\psi_\nu(t)$ is called *fractional Poisson process*. Performing the Laplace transform of Eq. (1.34) and inserting there (1.38) lead us to the system of equations

$$\lambda^\nu \widetilde{p}_n(\lambda) = -\mu \widetilde{p}_n(\lambda) + \mu \widetilde{p}_{n-1}(\lambda) + \lambda^{\nu-1} \delta_{n0}, \ n = 0, 1, 2, \ldots, \ \widetilde{p}_{-1} = 0.$$

This is a Laplace image of the fractional equation systems describing the fractional Poisson process:

$$_0\mathsf{D}_t^\nu p_n(t) = \mu[p_{n-1}(t) - p_n(t)] + \frac{t^{-\nu}}{\Gamma(1-\nu)} \delta_{n0}, \ 0 < \nu \le 1. \qquad (1.40)$$

When $\nu \to 1$ it becomes the well-known system for the standard Poisson process:

$$\frac{dp_n(t)}{dt} = \mu[p_{n-1}(t) - p_n(t)] + \delta(t)\delta_{n0}.$$

1.6 Tempered fractional processes

Dealing with long-tailed distributions under consideration we admitted that the related physical variables can take any large values. But we always find arguments to claim principally boundedness of physical variables X_j and others connected to them. As noticed in [Mantegna and Stanley (1994)], "in real physical systems, an unavoidable cutoff is always present".

Note that using the Central Limit Theorem, we ignore the cutoff idea although the Gaussian distribution has an infinite support as well. This is because the tails of an original distribution fall very quickly in this case and from practical point of view may be considered as if they were truncated. Tracing the development of the sum Σ_n in the case $\langle X_j^2 \rangle < \infty$ one can observe that for small n the Σ_n pdf depends on individual pdf of X_j, while for large n it has a normal (Gaussian) shape independently of $p_{X_j}(x)$. Thus, the n-axis is divided into two domains: initial domain (small n) and asymptotic domain (large n) separated by transition interval.

Let us now imagine that $p(x)$ meets the demands of the generalized limit theorem (they say that $p(x)$ or X_j belongs to the domain of attraction of a correspondent Lévy stable law). In the region of small n, the Σ_n distribution will depend on original distribution as in the previous case, but at large n it takes the Lévy stable form. But what happens when long tailed original pdf will be truncated by cutting a very small tip? [Mantegna and Stanley (1994)] proposed the model of truncated Lévy flights, a process showing a slow convergence to a Gaussian. The distribution of jump length has a power law behavior up to some large scale, at which it has a cutoff and thus have finite moments of any order. Smoothly (exponentially) truncated Lévy flights, introduced by Koponen [Koponen (1995)], provided a convenient analytic representation of results. Authors [Cartea and del Castillo-Negrete (2007)] derived the diffusion equation with the tempered fractional derivative describing the exponentially tempered Lévy flight. In papers [Stanislavsky *et al.* (2008); Gajda and Magdziarz (2010)], the approach based on the tempered fractional calculus [Baeumer and Meerschaert (2010)] was developed to describe crossover from subdiffusive to Gaussian transport. Tempered subdiffusion model assumes exponentially truncated power law distribution of waiting times [Sabzikar *et al.* (2015)].

[Cartea and del Castillo-Negrete (2007)] used a special form of infinitely divisible characteristic function for justifying free path pdf in the form

$$p_{\alpha,L}(r) = p_\alpha(r)e^{-r/L}, \ r > 0, \ L > 0,$$

where

$$p_\alpha(r) \propto r^{-\alpha-1}, \ r \to \infty.$$

They arrived at the tempered superdiffusion equation

$$ {}_0^\beta D_t P = -V\partial_x P \frac{\sigma^2}{2}\partial_x^2 P + c\mathcal{D}_x^{\alpha,\gamma}P - \nu P, \tag{1.41}$$

where

$$\mathcal{D}_x^{\alpha,\gamma} = le^{-\gamma x} {}_{-\infty}\mathsf{D}_x^\alpha e^{\gamma x} + re^{\gamma x} {}_x\mathsf{D}_\infty^\alpha e^{-\gamma x}, \tag{1.42}$$

and V, c, l and r are constant parameters (see details in [Cartea and del Castillo-Negrete (2007)]).

Authors [Sabzikar *et al.* (2015)] introduced the tempered fractional derivative defined by the following Fourier transform

$$\left[(\gamma + ik)^\alpha - \gamma^\alpha - ik\alpha\lambda^{\alpha-1}\right]\hat{f}(k)$$

$$= \frac{\alpha(\alpha-1)}{\Gamma(2-\alpha)}\hat{f}(k)\int_0^\infty (e^{-iky} - 1 + iky)e^{-\gamma y}y^{-\alpha-1}dy.$$

Its inverse has the form

$$\mathsf{D}_x^{\alpha,\gamma}f(x) = \frac{\alpha(\alpha-1)}{\Gamma(2-\alpha)}\int_0^\infty (f(x-y) - f(x) + yf'(x))e^{-\gamma y}y^{-\alpha-1}dy. \tag{1.43}$$

One may call this expression the left-side tempered fractional derivative. The right-side derivative has the transform $(\gamma + ik)^\alpha - \gamma^\alpha - ik$, it is defined by

$$\mathsf{D}_{-x}^{\alpha,\gamma}f(x) = \frac{\alpha(\alpha-1)}{\Gamma(2-\alpha)}\int_0^\infty (f(x+y) - f(x) - yf'(x))e^{-\gamma y}y^{-\alpha-1}dy. \tag{1.44}$$

The tempered fractional integrals are defined in [Sabzikar *et al.* (2015)] by relations

$$_{-\infty}I_x^{\alpha,\gamma}f(x) = \frac{1}{\Gamma(\alpha)}\int_{-\infty}^x f(u)(x-u)^{\alpha-1}e^{-\gamma(x-u)}du, \qquad (1.45)$$

$$_xI_\infty^{\alpha,\gamma}f(x) = \frac{1}{\Gamma(\alpha)}\int_{-\infty}^x f(u)(u-x)^{\alpha-1}e^{-\gamma(u-x)}du. \qquad (1.46)$$

The tempered fractional derivatives of the Riemann-Liouville type have the following form,

$$_{-\infty}D_x^{\alpha,\gamma}f(x) = e^{-\gamma x}\,_{-\infty}D_x^\alpha\left[e^{\gamma x}f(x)\right] + \gamma^\alpha f(x) + \alpha\gamma^{\alpha-1}f'(x). \qquad (1.47)$$

The tempered power law form of free path length distribution results in transition from linear dependence to a constant value for diffusion coefficient $D(t)$ determined through the mean squared displacement. Such behavior is often observed in simulation of particle transport in turbulent magnetic fields [Casse *et al.* (2001); Pucci *et al.* (2016)]. Interpretation of parallel diffusion in terms of tempered Lévy walk can be very helpful in analysis of experimental and numerical results. [Perri and Zimbardo (2007)] reported about superdiffusion of solar electrons and normal diffusion of protons at distance $\simeq 5$ AU. On the other hand, [Perri and Zimbardo (2009b)] have shown that "proton transport upstream of the solar wind termination shock, at roughly 84 AU, is superdiffusive". As is known, the turbulence level δb decreases with the distance from the Sun, so truncation length γ^{-1} increases, and exponent α decreases. This provides superdiffusive type of parallel motion at larger scales, but in any case this scale is limited and use of normal diffusive transport can be justified, for example in calculations of Solar modulation of Galactic CRs. In any case, corresponding quantitative estimations for different energies have to be made and more general model of parallel walk (tempered superdiffusive model) has to be used.

Chapter 2

Nonlocal diffusion models in hydrodynamics

When observing the chaotic motion of a mote in a turbulent flow, we can easily see that its small consecutive displacements are not independent but statistically connected by motion of the turbulent vortex as a whole. These long-range correlations indicate an essentially nonlocal character of the turbulent diffusion considered in this chapter. Section 2.1 critically analyzes the Einstein derivation of his diffusion equation. Section 2.2 describes some modifications of the model by involving ballistic regime with a finite velocity. They relate to different levels of descriptions, but none of them reflects an essential feature of this diffusion – its nonlocality. The principal step in understanding this nonlocality is implication of the turbulent power spectrum as a result of averaging over statistical ensemble of turbulent realizations of the medium. The nonlocal equation obtained under Kolmogorov's axioms is discussed in Section 2.3. Sections 2.4 and 2.5 complete constructing mathematical model of nonrelativistic turbulent diffusion on the base of statistical hydrodynamics.

2.1 The Navier-Stokes equation

We begin discussion of the diffusion concept with a plane hydrodynamic phenomenon described by Navier-Stokes equation. For an incompressible fluid, it reads (in standard notation)

$$\frac{\partial \mathbf{u}}{\partial t} + (\mathbf{u} \cdot \nabla)\mathbf{u} = -\frac{1}{\rho}\nabla p + \nu \Delta \mathbf{u}.$$

Let us imagine an infinite boundless motionless liquid (so that $p = \text{const}$) with an absolutely rigid large-size infinitely thin plate immersed. Introduce an inertial coordinate system with the origin in the plane of this plate and with the z-axis directed along its normal. Now force this plate to move along the x-axis with a varying velocity $V(t)$. The moving plate will drive the fluid, and in a region far from boundaries of this plate, its velocity field will have only one component $u_y \equiv f(x,t)$ obeying the equation

$$\frac{\partial f}{\partial t} = \nu \frac{\partial^2 f}{\partial x^2},$$

because under these conditions

$$(\mathbf{u} \cdot \nabla)u_x = u_x \frac{\partial u_x}{\partial x} + u_y \frac{\partial u_x}{\partial y} + u_z \frac{\partial u_x}{\partial z} = 0.$$

If suppose, that the driving velocity has an impulsive character, say $V(t) = \delta(t)$, then

$$f(x,0) = \delta(x),$$

and we meet the mathematical image of the diffusion problem. We express ourselves so cautiously because in reality f is the velocity and not the probability density as one should expect in the diffusion case. Nevertheless, both these problems are identical in form and have identical solutions. It is important that both physical processes are linked with the same property of the liquid, with the viscosity. And the second noteworthy fact: there exists an explicit analogy between a real liquid in hydrodynamics and imaginary liquid called probability in the stochastic theory. We may talk about evolution of spatial distribution of probability, of its density and of its current as if it is a real physical substance.

2.2 Einstein's diffusion equation

In spite of the wide popularity of the Brownian motion process in modeling many physical phenomena, it possesses some peculiarities which are often ignored. To understand them more clearly, let us discuss the Einstein article [Einstein (1905)] and trace the derivation of the Brownian diffusion equation. The author described behavior of a particle bombarded by surrounding atoms in different directions forcing the particle to move in a seemingly chaotic way under action of random imbalance of obtained momenta. Einstein wrote: "Obviously, it must be assumed that *each individual particle moves independently of other particles; in addition, motions of the same particle at different time intervals should be regarded as independent of each other until these intervals become too small.*"

By introducing the probability density function (pdf) $f(x,t)$ for the coordinate of a particle randomly walking along the x-axis and the symmetric pdf $\varphi(\Delta, \tau)$ for a random displacement Δ of the particle during time

τ, Einstein derived the equation[1]

$$f(x, t + \tau) = \int_{-\infty}^{\infty} f(x + \Delta, t)\varphi(\Delta, \tau)d\Delta. \tag{2.1}$$

Let us look at how Einstein explains derivation of this equation [Einstein (1956)] (pp. 14-16).

"Now, since τ is very small, we can put

$$f(x, t + \tau) = f(x, t) + \tau\frac{\partial f}{\partial t}.$$

Further, we can expand $f(x + \Delta, t)$ in powers of Δ:

$$f(x + \Delta, t) = f(x, t) + \Delta\frac{\partial f(x, t)}{\partial x} + \frac{\Delta^2}{2!}\frac{\partial^2 f(x, t)}{\partial x^2} \cdots .$$

We can bring this expansion under the integral sign, since only very small values of Δ contribute anything to the latter. We obtain

$$f + \frac{\partial f}{\partial t}\tau = f\int_{-\infty}^{\infty} \varphi(\Delta, \tau)d\Delta + \frac{\partial f}{\partial x}\int_{-\infty}^{\infty} \Delta\varphi(\Delta, \tau)d\Delta + \frac{\partial^2 f}{\partial x^2}\int_{-\infty}^{\infty} \frac{\Delta^2}{2}\varphi(\Delta, \tau)d\Delta \cdots .$$

On the right-hand side the second, fourth, etc., terms vanish since $\varphi(\Delta, \tau) = \varphi(-\Delta, \tau)$; whilst of the first, third, fifth, etc., terms, every succeeding term is very small compared with the preceding. Bearing in mind that

$$\int_{-\infty}^{\infty} \varphi(\Delta, \tau)d\Delta = 1,$$

and putting[2]

$$\frac{1}{\tau}\int_{-\infty}^{\infty} \frac{\Delta^2}{2}\varphi(\Delta, \tau)d\Delta = D, \tag{2.2}$$

and taking into consideration only the first and third terms on the right-hand side, we get from this equation

$$\frac{\partial f(x, t)}{\partial t} = D\frac{\partial^2 f(x, t)}{\partial x^2}, \quad -\infty < x < \infty. \tag{2.3}$$

This is the known differential equation for diffusion, and D is the diffusion coefficient."

---

[1] Instead of $x + \Delta$ in the argument of f, it should be $x - \Delta$, but the symmetry of pdf $\varphi(\Delta)$ neutralizes Einstein's slip of the pen.

[2] Evidently, the existence of this limit is implied.

Under initial condition

$$f(x,0) = \delta(x) \tag{2.4}$$

solution of Eq. (2.3) has the form of Gaussian density,

$$f(x;t) = \frac{1}{2\sqrt{\pi Dt}} e^{-x^2/4Dt} \tag{2.5}$$

with the variance equal to $2Dt$ and the diffusion coefficient

$$D = \lim_{\tau \to 0} \frac{1}{\tau} \int_{-\infty}^{\infty} x^2 \varphi(x,\tau) dx. \tag{2.6}$$

We note that introducing the time interval τ, Einstein defined it as "a very small compared with the observed interval of time, but, nevertheless, of such a magnitude that the movements executed by a particle in two consecutive intervals of time τ are to be considered as mutually independent phenomena." During intervals much shorter than t, the particle can move without collisions with atoms of the medium; correlations are then strictly determined by Newton's law of motion. This is a reason that Einstein defines $\varphi(\Delta, \tau)$ in Eq. (2.1) separately, not identifying this pdf with $f(x, \tau)$. Assumption of self-similarity would have led him to the nonlinear Chapman-Kolmogorov equation for Markov processes,

$$f(x,t+\tau) = \int_{-\infty}^{\infty} f(x+\Delta,t)f(\Delta,\tau)d\Delta, \quad f(x,0) = \delta(x), \tag{2.7}$$

which underlies the theory of random processes with independent increments.

The information on short-time behavior of the particle disappears, but Eq. (2.7) takes the same long-time asymptotics as (2.3) with the same fundamental solution (2.5) being valid under condition (2.6) applied to solution $f(x,t)$ itself.

- The absence of ballistic regime can be explained by enormous difference in sizes of atoms and Brownian tracer, so it is more natural to consider the combined action of surrounded atoms on a slowly moving tracer as a continuous rapidly changing force.
- Right away after switching on a point source, the probability to find out the particle becomes *nonzero at all distances*. This contradicts not only to assumption on enormous difference in masses of atoms and Brownian tracer, but to the relativistic principle of the limited velocity

as well, being valid even the tracer mass is vanishingly small (as in case of elementary particle diffusion).[3]

- There are no space-time correlations between successive collisions of the particle with atoms and molecules. This assumes that the atoms acting on the Brownian tracer do not interact with each other, so their positions and momentum are independent. This is valid for short-acting interaction between neutral molecules (under condition that the medium is sufficiently rarified) but not for long-range Coulomb or gravitational Newton forces.

- There are excluded repeated collisions of the particle with the same atom, the particle doesn't have influence on its environment. The Basset force excluded, neither particle nor environment keep the memory about retarded prehistory.

- When considering a system of many tracers, they are assumed not to be interact between each other either directly or by means of their environment, otherwise we don't have the right to extent one-particle propagator obtained from a linear equation to many-particle system.

2.3 Ballistic correlations

Despite evident difference between the rate of the diffusion packet expansion and individual velocity of the particle, one should conclude that the maximum velocity should also be infinite (otherwise this packed should be limited in size). At the same time, the width of the packet, determined for instance through its mean square radius, grows with time as $\sqrt{Dt}$. The cause of this uncertainty in velocity roots in the limit character of the Brownian motion understandable as some abstract mathematical construction. In order to get rid of this uncertainty, one has to concretize pdf $\varphi(x,t)$ in Eq. (2.2). We indicate here two schemes of such kind. The first is the instantaneous jump (or CTLJ) model, which is often used for modeling of random motion of charged particles in disordered solids. In frame of this model

$$\varphi(x,t) \sim \bar{Q}(t)\delta(x) + Q(t)p(x), \ t \to 0, \tag{2.8}$$

[3]Einstein was not fully satisfied by the result he obtained, which clearly contradicted the most important principle of his theory of relativity: a diffusion packet, being concentrated at the initial instant at the coordinate origin, the next instant fills the entire space, including its remotest regions. Practically, this did not cause any inconveniences due to the vanishingly small probability of the residence of the diffusion packet there, but conceptually Einstein could not help feeling some discomfort, as has now become fashionable to say. However, he did not develop this topic any further.

where probabilities $Q(t) = \mathsf{P}(T < t)$ and $\bar{Q}(t) = \mathsf{P}(T > t)$ are linked with pdf $q(t)$ of random waiting time T between successive jumps:

$$Q(t) = \int_0^t q(\tau)d\tau, \quad \bar{Q}(t) = \int_t^\infty q(\tau)d\tau.$$

With (2.8) inserted in (2.6), we obtain

$$D = \lim_{\tau \to 0} \frac{1}{\tau} \int_{-\infty}^{\infty} x^2 \left(\bar{Q}(\tau)\delta(x) + Q(\tau)p(x) \right) dx = q(0) \int_{-\infty}^{\infty} x^2 p(x)dx. \quad (2.9)$$

This result demonstrates that the necessary conditions for the normal diffusion should include not only the finiteness of the mean square of elementary displacement but also nonzero value of $q(0)$.

The second way of introducing the correlations is given by the Boltzmann collision concept, when a particle having velocity $\mathbf{v}$ at time t, will have

$$\mathbf{V} = \begin{cases} \mathbf{v}, & 1 - \mu\Delta t; \\ \mathbf{V}', & \mu\Delta t, \end{cases}$$

at time $t + \Delta t$, where $\mu > 0$ is the rate of collisions, and $\mathbf{V}'$ is a random vector standing for the particle velocity after the collision. Observe that the correlations

$$\langle \mathbf{v}\mathbf{V} \rangle - \langle \mathbf{v} \rangle \langle \mathbf{V} \rangle = \mathbf{v}(1 - \mu\Delta t)\mathbf{v} - \mathbf{v}\mu\Delta t \langle \mathbf{V}' \rangle$$

are kept up even in case of isotropic distribution of $\mathbf{V}'$:

$$\langle \mathbf{v}\mathbf{V} \rangle - \langle \mathbf{v} \rangle \langle \mathbf{V} \rangle = \mathbf{v}(1 - \mu\Delta t)\mathbf{v}.$$

In the coordinate space, each such trajectory is represented by a broken line consisting of randomly directed segments of various and also random length. The direction distribution may be both isotropic and anisotropic, absolute value of the velocity may change or remain constant, but the constituents length is usually taken in the form of exponential function

$$P(\xi > x) = e^{-\sigma x}, \quad \sigma = \mu/v, \mu = \text{const.}$$

However, the first application of this idea to turbulent diffusion problem was undertaken by Taylor who used its oversimplified version [Taylor (1920)], Taylor noted that "in any continuous motion there is necessarily a correlation between the movement in any one short interval of time and the next. This correlation will evidently increase as the interval of time diminishes, till, when the time is short compared with the time during which a finite change in velocity takes place, the coefficient of correlation tends to the limiting unity". He introduced this idea into the process of a symmetric one-dimensional walk

$$X(n\tau) = X_1 + X_2 + \cdots + X_n,$$

where $X_k = V_k \tau$ are random displacements during intervals $((k-1)\tau, \ k\tau)$ of the tracer with the random velocity V_k. In the simplest theory of random walks, when the velocities V_k are mutually independent and equal to $+v$ or $-v$ $(v > 0)$ with equal probabilities $p = q = 1/2$, the mean square of the tracer coordinate X is proportional to time $t_n = n\tau$,

$$\langle X^2(t_n)\rangle = v^2 \tau t_n$$

and the diffusion coefficient is equal to the same value

$$D = \frac{\langle X^2(t_n)\rangle}{t_n} = v^2 \tau$$

for any time t_n. Considering the case when V_{k+1} is correlated with V_k by a correlation coefficient c (that is $p \neq q$, $c = p - q$) meanwhile the partial correlations of V_k with V_{k+2}, $V_{k+3}, \ldots$ are all zero, Taylor showed that if $\tau/(1-c)$ tends to the limit A when τ and $1-c$ tend to zero,

$$\langle X^2(t_n)\rangle = 2\tau^2 \langle V_1 V_2 + V_1 V_3 + \cdots + V_{n-1} V_n \rangle$$

$$\to 2v^2 A \left[t - A \left(1 - e^{-t/A} \right) \right] \sim \begin{cases} (vt)^2, & t \to 0; \\ 2v^2 At, & t \to \infty. \end{cases} \tag{2.10}$$

Now, we observe different behavior of the mean square of displacement at short and long times. The constant A measures the rate at which the correlation between the velocity directions of the successive jumps falls with increasing time interval between observations, the covariance reads

$$C_x(t) = \frac{1}{2}\frac{d^2 \langle X^2(t)\rangle}{dt^2} = \frac{d^2}{dt^2}\left\{ v^2 A \left[t - A \left(1 - e^{-t/A} \right) \right] \right\} = v^2 e^{-t/A}.$$

Five years later, Fock derived the equation for one-dimensional walk of particles with a finite speed v and exponential intercollision time distribution [Fock (1926)]. Separating the probability $p(x,t)dx$ for a particle to be in $(x, x+dx)$ at time t into two parts related to positive and negative directions of motion ($p_+(x,t)dx$ and $p_-(x,t)dx$ respectively), he wrote difference equations

$$p_+(x,t) = p_+(x - vdt, t - dt)(1 - \mu dt) + p_-(x + vdt, t - dt)\mu dt,$$

$$p_-(x,t) = p_-(x + vdt, t - dt)(1 - \mu dt) + p_+(x - vdt, t - dt)\mu dt,$$

which in the limit $dt \to 0$ led to a system of differential equations

$$\mu(p_+ - p_-) + v\frac{\partial p_+}{\partial x} + \frac{\partial p_+}{\partial t} = 0;$$

$$\mu(p_+ - p_-) + v\frac{\partial p_-}{\partial x} - \frac{\partial p_-}{\partial t} = 0.$$

Summing and subtracting these expressions, one obtains the following equations for concentration $n(x,t) = p_+(x,t) + p_-(x,t)$ and x-projection of flux $j(x,t) = v(p_+(x,t) - p_-(x,t))$:

$$\frac{\partial n}{\partial t} + \frac{1}{2\mu}\frac{\partial^2 n}{\partial t^2} = D\frac{\partial^2 n}{\partial x^2}, \qquad (2.11)$$

$$j(x,t) + \frac{1}{2\mu}\frac{\partial j}{\partial t} = -D\frac{\partial n}{\partial x}. \qquad (2.12)$$

Here $D = v^2/(2\mu)$ is the diffusion coefficient. Equation (2.11) called the telegraph equation has the Green function (the solution under conditions $n(x,0) = \delta(x,0)$ and $\partial n(x,t)/\partial t|_{t=0} = 0$) which consists of two parts,

$$G(x,t) = G^{(0)}(x,t) + G^{(s)}(x,t).$$

First of them (singular)

$$G^{(0)}(x,t) = \frac{1}{2}e^{-\mu t}[\delta(x - vt) + \delta(x + vt)]$$

describes distribution of unscattered particles, concentrated at points $x = \pm vt$, the second one (continuous on $(-vt, vt)$)

$$G^{(s)}(x,t) = e^{-\mu t}$$

$$\times \left[\frac{\mu}{2v}I_0\left(\mu\sqrt{t^2 - x^2/v^2}\right) + \frac{\mu t}{2v}I_1\left(\mu\sqrt{t^2 - x^2/v^2}\right)/\sqrt{t^2 - x^2/v^2}\right],$$

represents concentration of particles which changed their motion direction at least once. Using known properties of the modified Bessel functions, one can verify that the second moment of the distribution

$$\int\limits_{-vt}^{vt} x^2 G(x,t)dx = (v/\mu)^2[\mu t - (1/2)(1 - e^{-2\mu t})]$$

coincides with that in the Taylor model (2.10) when $A = 1/(2\mu)$. Evolution of the propagator of the telegraph equation (for $\mu = 1$, $v = 1$) is shown in Fig. 2.1.

Equation (2.12) known as the Maxwell-Cattaneo equation, was first introduced in the heat conduction and found experimental confirmation.

A.S. Monin applied Eqs. (2.11) and (2.12) to describe the turbulent diffusion in the atmosphere [Monin (1955)].

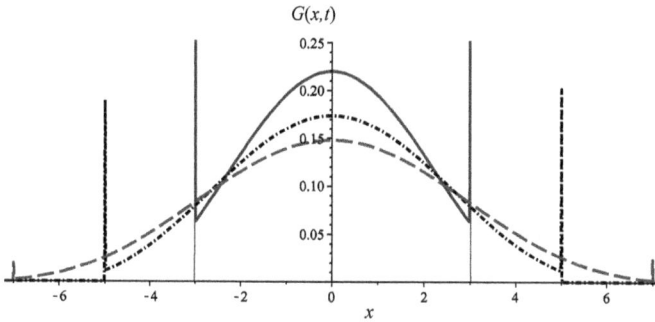

Fig. 2.1 Green function of the telegraph equation for $\mu = 1$, $v = 1$ and three values of time $t = 3$, 5 and 7. Delta-peaks formally reflect the impact of delta-fronts produced by unscattered particles from the instantaneous point source.

2.4 The Schönfeld eddy interpretation

Ballistic correlations, at least in the form discussed above, do not solve the problem of adequate description of the turbulent diffusion. The reason is that the exponential distribution of free paths (or the corresponding times) actually assumes the mutual independence of scattering centers (for example, of atoms or molecules). In other words, this model describes the diffusion if Brownian particle in an ideal gas rather than in a real gas and even more so in a viscous fluid. The following elementary reasoning clarifies this statement.

Let us direct the x axis in the direction of motion of the particle at the initial moment at the point 0, and construct on its forthcoming way the segment $[0, l]$. Then, we drop on this segment n independent random points ('atoms') uniformly distributed over this interval, and select a small interval $[0, x]$ at the beginning of the interval. Obviously, the random number of points in a small segment will be characterized by the binomial probability distribution with the expectation nx/l, so the probability that the small segment will be empty, that is, the probability that the particle's free path before the next collision will exceed x

$$P(\xi > x) = (1 - x/l)^n \cong e^{-\sigma x}, \quad \sigma = n/l.$$

Consequently, this model relates to the Brownian diffusion in a rarefied ideal gas.

The turbulent diffusion in fluids implies that mass is transferred by means of the mixing of turbulent eddies. This mechanism is sufficiently different from the processes determining molecular diffusion. The eddy diffusion coefficient depends mainly on the properties of the fluid flow and ratio of eddy and 'spot' scales. Figure 2.2 demonstrates three examples of the mixing of an isolated puff (or spot of something) in a turbulent flow according to [Seinfeld (1986)]. There are three cases of ratio of scales: the size of puff is large (a), very small (b), or comparable (c) to the size of the turbulent eddies. In reality, turbulent eddies of all distributed sizes are present simultaneously. An idealized trajectory of a tracer in a turbulent flow is shown in Fig. 2.3.

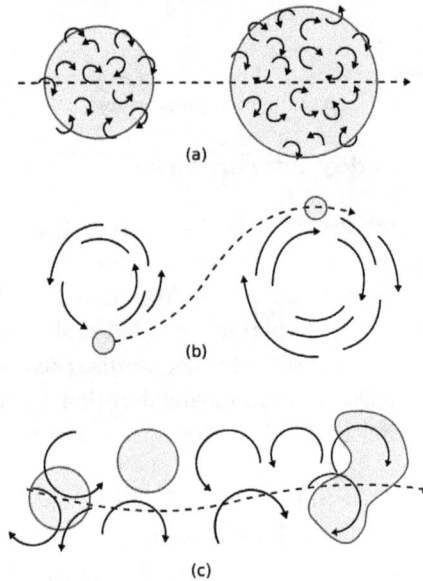

Fig. 2.2   Scales of turbulence according to [Seinfeld (1986)].

Like molecular diffusion, the turbulent diffusion representing migration of passive tracers as they are carried along with a turbulent flow is a linear process if the convected particles have no reaction on the flow. Observe that in the case when free path molecules are small compared with the characteristic length of concentration change, and for this reason the diffusion flux is determined by the local concentration gradient at the point under consideration. In turbulent movement, however, the paths may be on such a large scale that the transport should be supposed to depend not so much on the local gradient of the concentration as on the distribution of the concentration in a larger area. To amend the equation, we represent it

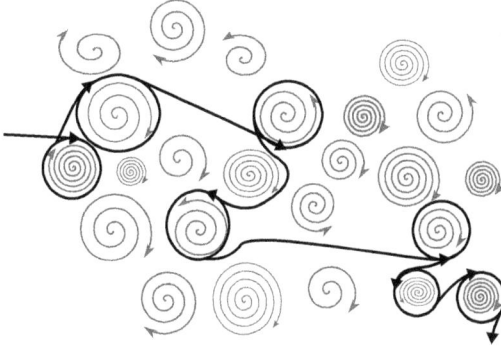

Fig. 2.3 Schematic of turbulent transport over distributed eddies.

as

$$\mathbf{j}(\mathbf{x}, t) = -\nabla \int d\mathbf{x}' D\delta(\mathbf{x} - \mathbf{x}') n(\mathbf{x}', t)$$

and make the replacement

$$D\delta(\mathbf{x} - \mathbf{x}') \mapsto K(\mathbf{x} - \mathbf{x}'),$$

so the flux vector takes the form

$$\mathbf{j}(\mathbf{x}, t) = -\nabla \int d\mathbf{x}' K(\mathbf{x} - \mathbf{x}') n(\mathbf{x}', t).$$

With this generalization of Fick's law inserted into the balance equation, we arrive at the integro-differential version of the diffusion equation:

$$\frac{\partial n(\mathbf{x}, t)}{\partial t} = \Delta \int d\mathbf{x}' K(\mathbf{x} - \mathbf{x}') n(\mathbf{x}', t) + S(\mathbf{x}, t). \qquad (2.13)$$

In a frame of reference where the drift is absent, the kernel $K(\mathbf{x}) = K(r)$, $\mathbf{x} = \mathbf{x} - \mathbf{x}'$ at large r falls that outlines some region having an influence upon motion of the tracer.

[Schönfeld (1962)] gave the following inductive explanation of this phenomenon. He considered an 'eddy' characterized by a length measure ϱ, engaged at the point (x, y). The eddy produces a velocity w in a direction at an angle χ with the x axis so the concentration at (x, y) incidentally deviates from the mean value $n(x, y)$. This deviating value was estimated by

$$n(x, y) \mapsto n(x - \varrho \cos \chi, y - \varrho \sin \chi),$$

so the contribution of this eddy into x-component of the diffusion flux was written as

$$\delta j_x = n(x - \varrho \cos \chi, y - \varrho \sin \chi) w \cos \chi.$$

The resulting flux is obtained via averaging the contribution of all possible position of the eddy and let $W d\varrho/\varrho$ be the probability that the distance falls into the interval $(\varrho, \varrho + d\varrho)$. Assuming that all directions χ are equiprobable and that

$$\omega \equiv W w$$

depends on ϱ only, one obtains:

$$j_x = \frac{1}{2\pi} \int\limits_0^\infty \frac{d\varrho}{\varrho} \, \omega(\varrho) n(x - \varrho \cos\chi, y - \varrho \sin\chi) \cos\chi.$$

Fourier transforming with respect to both spatial coordinates yields

$$\widetilde{j}_x = ik_x \widetilde{K}(k) \, \widetilde{n}(k_x, k_y),$$

and analogously

$$\widetilde{j}_y = ik_y \widetilde{K}(k) \, \widetilde{n}(k_x, k_y),$$

where k_x and k_y are projections of the wave vector $\mathbf{k}$,

$$\widetilde{K}(k) = \frac{1}{k} \int\limits_0^\infty \frac{d\varrho}{\varrho} \, \omega(\varrho) j_1(\varrho k)$$

and $k = |\mathbf{k}| = \sqrt{k_x^2 + k_y^2}$. Substitution of these functions into Fourier image of the continuity equation

$$\frac{\partial \widetilde{n}}{\partial t} = ik_x \widetilde{j}_x + ik_y \widetilde{j}_y + \widetilde{S}$$

yields

$$\frac{\partial \widetilde{n}}{\partial t} + k^2 \widetilde{K}(k) \widetilde{n}(\mathbf{k}, t) = \widetilde{S}(\mathbf{k}, t). \tag{2.14}$$

This is the Fourier image of the nonlocal diffusion equation (2.13).

[Chavanis (2002)] gave a more comprehensive description of this approach. Considering the random set of point vortices in two-dimensional incompressible flow in terms of statistical mechanics, and expressing (according to the Schwarz theorem) the fluid velocity through a stream function ψ

$$\mathbf{u} = -[\mathbf{z}, \nabla\psi]$$

with $\mathbf{z}$ denoting the unit vector normal to the flow plane xy, he represents a single vorticity

$$\omega = \operatorname{rot} \mathbf{u} = \omega\mathbf{z}, \quad \omega = \frac{\partial u_y}{\partial x} - \frac{\partial u_x}{\partial y}.$$

Thus, in the case of molecular diffusion in a dilute gas, a tracer interacts with almost independent molecules per collisions and this fact makes the diffusion equation local, but in the case of turbulent diffusion the motion of neighboring fluid elements are correlated, and the tracer motion continuously affected by the elements is described by the nonlocal equation,

$$\frac{\partial n(\mathbf{x}, t)}{\partial t} = \Delta \int d\mathbf{x}' K(\mathbf{x} - \mathbf{x}') n(\mathbf{x}', t) + S(\mathbf{x}, t). \qquad (2.15)$$

One cannot say that this rather formal result had attracted much attention of 'turbulent community', but with the course of time, the interest to nonlocal ideas in the turbulence phenomenology was growing and more physical arguments were found.

2.5 Monin's hypothesis

The first implementation of fractional concept into turbulent diffusion modeling was made by outstanding Soviet scientist, one of Kolmogorov's disciples A. S. Monin in his work [Monin (1955)] recommended by Kolmogorov himself. He discussed the process rather formally, nevertheless it would be very instructive to trace the way of his reasoning and explanation. Let us do it intermitting by our own remarks.

Monin begins his article with indication that peculiarity of the turbulent diffusion is caused by action of eddies of various size existing in a turbulent medium. The distance between neighboring tracers essentially changes only under action of eddies comparable in size with this distance: the diffusion packet expands. Because the velocity of this expansion grows with growing eddy size, the diffusion coefficient should increase with space scale l. This fact confirmed experimentally was used by Richardson who proposed the one-dimensional diffusion equation for pdf $g(l, t)$ of the distance between two tracers with the diffusion coefficient $D \propto l^{4/3}$ [Richardson (1926)]. This ansatz was interpreted by Kolmogorov and Obukhov as a consequence of the *similarity hypothesis* [Kolmogorov (1941); Obukhov (1941)] which states that in a wide scale intervals the locally isotropic turbulence is completely determined by the only dimensional characteristic ϵ (the rate of turbulent energy dissipation). It follows from the dimensional reasoning, that

$$D(l) = c \epsilon^{1/3} l^{4/3},$$

where c is a dimensionless constant value. But the relative distance pdf $g(l, t)$ was inconvenient for calculations of spatial distributions of admixtures, as Monin wrote in his article, and the procedure proposed by Richardson for passage from relative distance pdf $g(l, t)$ to actual concentration

$p(\mathbf{x}, t)$ in his opinion was not enough justified and effective. Monin wanted to derive an equation directly for the averaged concentration field in the framework of moving coordinate system associated with the mean flow (i.e., with the center of mass of diffusion packet), but solution of this equation should be in aggreement being in agreement with experimentally observed *enhanced diffusion* (*superdiffusion*) of admixture in a homogeneous uniform stationary 2D-turbulent medium. However, the diffusion coefficient in this case should not depend on coordinates or time. The only thing that remains for changing this equation is the Laplace operator.

Derivation of the desired equation was began by Monin with interrelation between concentration fields at the observation time $q_\omega(t) \equiv q_\omega(\mathbf{x}, t)$, $\mathbf{x} = \{x_1, x_2\}$, and at the initial time $q_\omega(0)$ for a fixed realization ω of statistical turbulent ensemble. If the molecular diffusion is neglected (as Monin supposed), $q_\omega(t)$ is uniquely determined by $q_\omega(0)$,

$$q_\omega(t) = A_\omega^t q_\omega(0).$$

Averaging this equation over ensemble of realizations,

$$\langle q_\omega(t) \rangle = \langle A_\omega^t q_\omega(0) \rangle,$$

Monin *defines* a new operator A^t via relation

$$\langle q_\omega(t) \rangle = A^t \langle q_\omega(0) \rangle. \tag{2.16}$$

This operator is assumed to be linear. It is the first Monin's hypothesis. The author doesn't discuss its validity in his paper, but, as usual, this means the absence of influence of admixture particles in each other both by direct interactions and through the liquid.

Further Monin notes that stationary homogeneous and locally isotropic turbulence A^t is a space-time-translationally and rotationally invariant operator depending on a unique material parameter ϵ, so the averaged concentration will be axially symmetrical,

$$f(r, t) = \langle q_\omega(t) \rangle, \quad r = |\mathbf{x}| = \sqrt{x_1^2 + x_2^2},$$

so Eq. (2.16) becomes

$$f(r, t) = A^t f(r, 0). \tag{2.17}$$

Under these conditions, the Fourier transform

$$f(r, t) \mapsto \widetilde{f}(k, t) = \int_{\mathbb{R}^2} e^{i\mathbf{k} \cdot \mathbf{x}} f(r, t) d\mathbf{x},$$

converts operator equation (2.16) into algebraic one,

$$\widetilde{f}(k,t) = a(k,t,\epsilon)\widetilde{f}(k,0).$$

Dimensionality consideration leads us to conclude that

$$a(k,t,\epsilon) = a(\theta), \tag{2.18}$$

where

$$\theta = \epsilon^{1/3}k^{2/3}t.$$

The solution of this equation is of the form,

$$\widetilde{f}(k,t) = a(\epsilon^{1/3}k^{2/3}t)\widetilde{f}(k,0). \tag{2.19}$$

In addition to linearity of A, Monin set up one more hypothesis:

$$A^{t_1+t_2} = A^{t_1}A^{t_2}. \tag{2.20}$$

It would be unconditionally valid if the turbulent diffusion was a process with independent increments, say, like Brownian motion, but in reality motions of a tracer during neighboring intervals t_1 and t_2 are statistically linked through the evolution of the turbulent medium. For this reasoning, Eq. (2.20) seems to be rather wobbly statement. Nevertheless, the author put it in the basement of his theory and prescribed to function (2.18) the property

$$a(\theta_1 + \theta_2) = a(\theta_1)a(\theta_2). \tag{2.21}$$

Replacing $t \mapsto t + dt$ in Eq. (2.19), inserting there (2.21) with $t_1 = t$ and $t_2 = dt$, and passing to $(dt \to 0)$-limit in a usual manner yields the ordinary differential equation

$$\frac{\partial \widetilde{f}}{\partial t} = -\epsilon_1^{1/3}k^{2/3}\widetilde{f}(k,t), \tag{2.22}$$

where $\epsilon_1 = c\epsilon$ and c is a constant value of order 1. Its solution under a given initial condition $\widetilde{f}(k,0)$ is represented by the exponential function

$$\widetilde{f}(k,t) = \exp\left(-\epsilon_1^{1/3}k^{2/3}t\right)\widetilde{f}(k,0), \quad t \geq 0. \tag{2.23}$$

After inversion of Fourier images (2.22), (2.23), Monin gets

$$\frac{\partial f(r,t)}{\partial t} = -\epsilon_1^{1/3}\int_0^\infty\int_0^\infty k^{2/3}qJ_0(kr)J_0(kq)f(q,t)dkdq,$$

and

$$f(r,t) = \int_0^\infty\int_0^\infty kqJ_0(kr)J_0(kq)e^{-\epsilon_1^{1/3}k^{2/3}t}f(q,0)dkdq,$$

where J_0 is the Bessel function.

In order to have a more handy result, Monin returns to Eq. (2.22) and twice differentiate it with respect to time:

$$\frac{\partial^3 \widetilde{f}}{\partial t^3} = -\epsilon_1 k^2 \widetilde{f}(k, t).$$

Because $-k^2 = -(k_1^2 + k_2^2)$ corresponds to two-dimensional Laplace operator $\Delta_2 = \partial^2/\partial x_1^2 + \partial^2/\partial x_2^2$, the equation itself is the Fourier transform of the equation in spatial variables

$$\frac{\partial^3 f}{\partial t^3} = \epsilon_1 \Delta_2 f(r, t). \qquad (2.24)$$

Evidently, this equation contains among others the solution of turbulent diffusion equation (2.22). Monin hound the solution for two-dimensional turbulent diffusion from instantaneous point source placed at the origin which has the self-similar form

$$f_2(r, t) = \frac{1}{\pi \epsilon_1 t^3} \phi_2(z), \quad z = \frac{r^2}{\epsilon_1 t^3}, \qquad (2.25)$$

with scaling function $\phi_2(z)$ expressed through the Whittaker function. At large distances $\phi(z) \propto z^{-8/3}$, this gives a ground for talking on enhanced type of the turbulent diffusion. This is a result of accelerating action of eddies sizes of which grow with growing size the diffusion cloud itself.

The further development of this idea we observe in the Monin-Yaglom book [Monin and Yaglom (1975)]. The authors give three-dimensional solution of this problem in the form

$$f(r, t) = \frac{A}{(\epsilon_1 t^3)^{3/2}} \left(\frac{r^2}{\epsilon_1 t^3}\right)^{-3/2} \exp\left(\frac{2\epsilon_1^3 t^3}{27 r^2}\right) W_{-3/2, 1/6}\left(\frac{4\epsilon_1 t^3}{27 r^2}\right),$$

where $W_{\mu, \lambda}$ is the Whittaker function, and A is the normalizing constant. But another fact is more notable in this publication. The inverted Eq. (2.22) was represented in the form

$$\frac{\partial f}{\partial t} = \mathsf{B} f(r, t),$$

where B was claimed as "*a linear operator proportional to $1/3$-degree of the Laplace operator*". This is the first mention of fractional Laplacian in the turbulence theory which allows us to represent this equation in the form

$$\frac{\partial f}{\partial t} = -K_\alpha (-\Delta)^{\alpha/2} f(r, t), \qquad (2.26)$$

with $\alpha = 1/3$ and $C(\alpha)$ is some normalized constant. When $\alpha = 2$, Eq. (2.26) becomes a normal diffusion equation with the diffusion coefficient $D = C(2)$. The main use of α: shifting it from 2 to smaller values introduces long-range correlations in particle displacements inherent to the turbulent motion.

2.6 The Heisenberg-Kolmogorov-Weizsäcker formalism

Two Monin's hypotheses (2.16) and (2.20) seem to be *ad hoc* statements without any direct link to hydrodynamical laws. This gap has been filled by Chan-Mou Tchen in [Tchen (1954, 1959)]. The author referred to works of Monin and Heisenberg, the latter in its turn referred to Kolmogorov and von Weizsäcker [Heisenberg (1948, 1985)].

Let us follow their common logic of reasoning. The existence of the stationary state of a turbulent medium assumes stationary support of energy balance: the energy coming into this medium transfers from big patterns to smaller and smaller ones and finally disappears due to dissipation. This process is controlled by *transfer function* $W(k)$ describing the rate of the energy transfer from wave numbers lower than k to wave numbers higher than k. This function is connected with another important function called *spectral function* $F(k)$ such that $F(k)dk$ is the kinetic energy associated with wave numbers interval $(k, \ k + dk)$ (here and below we neglect molecular motion). This link is expressed via the balance equation (see, for instance, [Hinze (1959)])

$$\frac{\partial F(k)}{\partial t} + \frac{\partial W(k)}{\partial k} = Q(k) - 2\nu k^2 F(k). \qquad (2.27)$$

Here $Q(k)$ is the production of turbulence at wave number k, which vanishes for isotropic turbulence dominating the smaller scales, such that turbulence production is dominant on the larger scales. The last term in (2.27) is the spectral formulation of the turbulent dissipation rate. Under these conditions, $Q(k) = 0$, $W(0) = W(\infty) = 0$, and integrating (2.27) over the whole spectrum yields

$$\int_0^\infty \left(\frac{\partial F(k)}{\partial t} \right) dk = \frac{\partial}{\partial t} \left(\int_0^\infty F(k)dk \right) = \epsilon. \qquad (2.28)$$

From Eq. (2.27) we can find that ϵ may also be written in the form

$$\epsilon = 2\nu \int_0^\infty F(k)k^2 dk. \qquad (2.29)$$

Reducing the integration domain to interval $(0, k)$, Heisenberg expressed the transfer function in the form

$$W(k) = 2\nu(k) \int_0^k F(q)q^2 dq$$

with $\nu(k)$ defined by

$$\nu(k) = \kappa \int\limits_{k}^{\infty} \sqrt{F(q)/q^3}\,dq, \tag{2.30}$$

where κ is a numerical constant. Notice that the transfer is expressed as a turbulent dissipation which, similarly to the viscous dissipation, has the form of the product of the turbulent viscosity by the square of the vorticity. The turbulent diffusion coefficient is determined by eddies up to a certain size. An important role of small eddies in the turbulent diffusion is confirmed by deeper kinetic analysis and phenomenological reasoning.

Now, in order to continue our consideration, we are in need of an explicit expression for $F(q)$. This problem has been discussed in many works (see, for instance, [Obukhov (1941); von Weizsäcker (1948); Tchen (1954); Monin and Yaglom (1975)]. We choose here the way of von Weizsäcker which seems to be more instructive and closely connected to the main paradigm of our book: concept of similarity.

According to the Kolmogorov hypotheses, the short wave number subrange of the equilibrium range (the inertial subrange) is not affected by viscosity and described by one-parameter function,

$$F = F(k, \epsilon), \quad \Lambda^{-1} \ll k \ll \lambda^{-1}.$$

Taking into account dimension of the quantities involved into the equality,

$$[F] = L^3 T^{-2},$$

$$[\epsilon] = L^2 T^{-3},$$

$$[k] = L^{-1},$$

and representing F in the form

$$F(k) = K_0 \epsilon^a k^b \psi(\lambda k), \tag{2.31}$$

where ψ is a dimensionless *window function*, we obtain from Eq. (2.18)

$$F(k) = K_0 \epsilon^{2/3} k^{-5/3} \psi(\lambda k) \tag{2.32}$$

with K_0 being a positive constant, ranging between 1.4 and 1.8 (in dependence of the Reynolds number).

There exists a few analytical approximations for the window function (see [Kovasznay (1948)], [Heisenberg (1948)], [Corrsin (1964)], [Pao (1965)]). In the roughest approximation,

$$\psi(z) = \begin{cases} 0, & z < \zeta \equiv \lambda/\Lambda; \\ 1, & z > \zeta; \end{cases} \tag{2.33}$$

and Eq. (2.32) becomes

$$F(k) = K_0 \epsilon^{2/3} k^{-5/3}, \quad \Lambda^{-1} < k < \lambda^{-1}. \tag{2.34}$$

This expression is often referred to as the *Kolmogorov-Obukhov* $k^{-5/3}$ *law.*

Let us draw the reader's attention to the Weizsäcker paper published a few years later (in 1948) than Kolmogorov and Obukhov's papers. He obtained the same spectral function $F(k)$ (up to the window function ψ) using the self-similarity idea. Namely, the author divided an arbitrarily chosen cubic volume of size L_0 into equal parts of the level of size L_1 and successively repeated this procedure to the objects of next levels, $L_2, L_3, \ldots, L_n, \ldots$ so that

$$\frac{L_{n+1}}{L_n} = \delta = \text{const.}$$

Each element was interpreted as an eddy and the turbulent system was represented as a set of eddies of all sizes from the spatial dimensions of the entire system down to a limit set by the molecular viscosity. These eddies enclosed in each other form what is called the 'hierarchy of eddies'. The turbulent motion inside each eddy is characterized by the average relative velocity v_n of two points in the fluid at a distance L_n. The average kinetic-energy density of the eddies of size L_n is $(1/2)\rho v_n^2$, so its specific (per unit mass) value is linked to spectral function $F(k)$ via relation

$$\frac{1}{2} v_n^2 = \int_{k_n}^{\infty} F(k) dk. \tag{2.35}$$

The rate of specific energy transferred through the eddies of size L_n is

$$W_n = \nu |\text{rot} \mathbf{v}_n|^2 = L_n v_n |\text{rot} \mathbf{v}_n|^2 = L_n v_n \left(\frac{v_n}{L_n}\right)^2 = \frac{v_n^3}{L_n}.$$

In the stationary case, the energy entering the system with a constant rate penetrates the eddies of all scales with the same rate up to dissipation. This means that W_n must be independent of n, i.e.,

$$v_n \propto L_n^{1/3} = k_n^{-1/3}.$$

Under this condition Eq. (2.35) takes the form

$$\frac{1}{2} k_n^{-2/3} \propto \int_{k_n}^{\infty} F(k) dk, \tag{2.36}$$

which after differentiation with respect to k_n becomes an analog of (2.34). Weizsäcker wrote: "This law, which in a different mathematical formulation, was first given by Kolmogoroff, can be used to estimate all other statistical properties of the eddies by purely dimensional arguments."

2.7   Tchen's interpretation

In order to continue our way to fractional diffusion equation, we resort to the basic equations of incompressible fluid dynamics in the absence of the mean motion and neglecting molecular diffusion. Denoting the turbulent velocity field by $\mathbf{u}$ and splitting the total concentration into component N averaged over some statistical ensemble and the pulsing component n considering as fluctuating part of the random field, one can bring the system of equations to the form

$$\frac{\partial N}{\partial t} + \left\langle u_j \frac{\partial n}{\partial x_j} \right\rangle = 0,$$

$$\frac{\partial n}{\partial t} + u_j \frac{\partial n}{\partial x_j} - \left\langle u_j \frac{\partial n}{\partial x_j} \right\rangle + u_j \frac{\partial N}{\partial x_j} = 0,$$

$$\frac{\partial u_i}{\partial t} + u_j \frac{\partial u_i}{\partial x_j} - \left\langle u_j \frac{\partial u_i}{\partial x_j} \right\rangle = -\frac{1}{\rho} \frac{\partial p}{\partial x_i} + \nu \frac{\partial^2 u_i}{\partial x_j^2}.$$

Passing to the Fourier representation with the use of incompressibility condition

$$ik_j \widetilde{u}_j(\mathbf{k}) = 0$$

yields

$$\frac{\partial^2 \widetilde{N}(\mathbf{k})}{\partial t^2} = ik_j \int d\mathbf{q} \frac{\partial \langle \widetilde{u}_j(\mathbf{k}-\mathbf{q})\widetilde{n}(\mathbf{q})\rangle}{\partial t}, \tag{2.37}$$

$$\frac{\partial \widetilde{n}(\mathbf{k})}{\partial t} = -i \int d\mathbf{q} q_j \left[\widetilde{u}_j(\mathbf{k}-\mathbf{q})\widetilde{n}(\mathbf{q}) - \langle \widetilde{u}_j(\mathbf{k}-\mathbf{q})\widetilde{n}(\mathbf{q})\rangle\right]$$

$$-i \int d\mathbf{q} q_j \widetilde{u}_j(\mathbf{k}-\mathbf{q})\widetilde{N}(\mathbf{q}), \tag{2.38}$$

and

$$\frac{\partial \widetilde{u}_i(\mathbf{k})}{\partial t} = -i \int d\mathbf{q} q_j \left[\widetilde{u}_j(\mathbf{k}-\mathbf{q})\widetilde{u}_l(\mathbf{q}) - \langle \widetilde{u}_j(\mathbf{k}-\mathbf{q})\widetilde{u}_l(\mathbf{q})\rangle\right]\left(\delta_{il} - \frac{k_i k_l}{k^2}\right).$$
$$\tag{2.39}$$

Up to small terms, two last equations lead to expression of a more compact form

$$\frac{\partial \widetilde{u}_j(\mathbf{k}-\mathbf{q})\widetilde{n}(\mathbf{q})}{\partial t} = iq_l \widetilde{u}_j(\mathbf{k}-\mathbf{q}) \int d\mathbf{q}' \widetilde{u}_l(\mathbf{q}-\mathbf{q}')\widetilde{N}(\mathbf{q}'). \tag{2.40}$$

By using the *exchange phase formula* (see [Tchen (1954, 1959)])

$$\int dq \widetilde{u}_l(\mathbf{k} - \mathbf{q}) \widetilde{N}(\mathbf{q}) = -ik_l \nu(k) \widetilde{N}(\mathbf{k}) \qquad (2.41)$$

one gets from (2.40)

$$\frac{\partial \widetilde{u}_j(\mathbf{k} - \mathbf{q}) \widetilde{n}(\mathbf{q})}{\partial t} = k^2 \nu(k) \widetilde{N}(\mathbf{k}) \widetilde{u}_j(\mathbf{k} - \mathbf{q}). \qquad (2.42)$$

Now, averaging both sides of the equation over fluctuations, replacing the first-order time derivative in Eq. (2.37) by right-hand side of averaged Eq. (2.42) and secondarily using (2.41), we arrive at the equation

$$\frac{\partial^2 \widetilde{N}(\mathbf{k})}{\partial t^2} = [k^2 \nu(k)]^2 \widetilde{N}(\mathbf{k}).$$

The family of its solutions also contains solutions of the standard relaxation equation

$$\frac{\partial \widetilde{N}(\mathbf{k})}{\partial t} = -R(k) \widetilde{N}(\mathbf{k}) \qquad (2.43)$$

with inverse relaxation time

$$R(k) = k^2 \nu(k) = \kappa k^2 \int\limits_k^\infty \sqrt{F(q)/q^3} dq. \qquad (2.44)$$

Finally, using anzatz (2.35), we obtain for $R(k)$ expression

$$R(k) = (3\kappa/4) \sqrt{K_0} \left(\epsilon k^2\right)^{1/3}, \quad \Lambda^{-1} \ll k \ll \lambda^{-1}.$$

If not for the last limitation, factor $(k^2)^{1/3}$ would correspond to one third power of the Laplace operator, $\Delta^{1/3}$.

Table 2.1. Turbulent regimes and indexes.

Turbulence regime	Spectral index b	Lévy-index α
Isotropic turbulence	$-5/3$	$2/3$
Turbulence in flow with a velocity gradient	-1	1
Molecular diffusion	1	2

One should also notice that the law (2.32) relates only to the isotropic turbulence, whereas a more general form (2.31) can be applied to other diffusion regimes depending on b (Table 2.1). Solution to (2.30) is of the form

$$\widetilde{N}(\mathbf{k}, t) = e^{-t/R(k)} \widetilde{N}(\mathbf{k}, 0), \qquad (2.45)$$

with

$$R(k) = R(1) \cdot k^{\alpha}, \quad \alpha = (3+b)/2 \tag{2.46}$$

(if only ψ to be taken equal to 1). Tchen [Tchen (1959)] has inverted the Fourier image (2.45) with ansatz (2.46) under initial condition $\widetilde{N}(\mathbf{k}, 0) = 1$ corresponding to a point instantaneous source for one-, two-, and three-dimensional diffusion processes. In our notation it looks as follows:

$$\frac{\partial \bar{G}(\mathbf{x}, t)}{\partial t} = -R(1)(-\Delta)^{\alpha/2} \bar{G}(\mathbf{x}, t). \tag{2.47}$$

Recall that its solution takes the self-similar form with the scaling variable $\xi = rt^{-1/\alpha}$ with the Lévy-index α ($\alpha \in (0, 2]$).

Chapter 3

Interstellar medium

This chapter begins with listing main interstellar medium components (Section 3.1), almost common property of which is a self-similar heterogeneity called fractality. Section 3.2 introduces the magneto-hydrodynamic (MHD) system of equations and concerns the phenomena of the freeze-in of magnetic field lines into the plasma. Section 3.3 discusses the interstellar turbulence spectra including magnetic line diffusion and the Corrsin independence hypothesis. In Section 3.4, we deal with two fractional models in plasma expressed by the Balescu hybrid equation for the spatial distribution and by the Lévy-Feldheim representation of equilibrium velocity distribution. Sections 3.5 and 3.6 are devoted to fractional description of diffusion in molecular clouds and diffusion blurring of the image of astrophysical objects in ISM respectively.

3.1 Interstellar medium

Components

The interstellar medium (ISM) is a complex dynamic environment. It consists of the following components interacting with each other.

(1) **The neutral gas.** This is a set of hierarchically clumped irregular structures including low-densities gas between shells, the gas inside the shell periphery, the gas inside spiral arms, and everywhere else. Many astrophysicists believe that the gas may be treated as fractal far from its boundaries [Scalo (1990); Zimmermann and Stutzki (1992); Elmegreen (1998)] and gave for the projected fractal dimension the value 1.3. The origin of fractality has generally been thought to lie in turbulence [Larson (1981); Falgarone *et al.* (1991)]. The energy of turbulence could come from galactic rotation at large scale, then cascade down to be dissipated on small scales by viscosity; it has been suggested that such

turbulence helps to prevent massive molecular clouds from collapsing in response to their own gravity [De Vega *et al.* (1998)].

Fig. 3.1   The Hubble Space Telescope image of the Carina Nebula. Violent stellar winds and powerful radiation from massive stars are sculpting the surrounding nebula. Inside the dense structures, new stars may be born. Credit: NASA/ESA/Hubble Heritage Project (STScI/AURA), M. Livio (STScI), N. Smith (University of California, Berkeley, USA).

(2) **Dust grains**. They are small solid particles with sizes from the micron down to fractions of nanometers. Most theoretical studies of processes with interstellar dust grains use the Mie theory, which assumes a spherical or cylindrical form of dust grain. But interstellar dust grains are very unlikely to be spherical, and the most likely situation when they change their size and form due to colliding and sticking or shattering. Nevertheless, one can assume that dust grains concentration approximately repeat those for gas molecules.

(3) **Plasmas** consisting of partially or totally ionized molecules of gases, play very important role in the astrophysics, and particularly in the case of interstellar turbulence. Similarly other components of interstellar medium, plasmas reveal a hierarchy of density structures, viewed as dense cores nested in less dense regions, which are in turn embedded in low-density regions and so on. Such a hierarchical structure was described by [von Weizsäcker (1951)].

Figure 3.1 demonstrates large panoramic view of Carina nebula, a large cloud of gas and dust, with a mix of hot, young stars, dying stars and regions of starbirth.

(4) **Magnetic field**. Magnetic fields permeating almost every place in the Universe, but its evolution, structure and origin are still open problems in fundamental physics and astrophysics. The average energy density of the magnetic field is comparable to that of the kinetic energy of the turbulent gas motions and is about one order of magnitude larger than the thermal energy. The average strength in the galactic plane is of the order of $1G$, while the amplitude of the fluctuations is of the order of $5G$. The fields also form irregular structures but of another kind: they are vectorial, reveal long-range correlations and are directed mostly parallel to the galactic plane. For this reason, it is very convenient to describe interstellar magnetic fields in terms of *magnetic force lines* and to interpret their turning and other irregularity in terms of *random walks*.

The randomization is driven by turbulence playing an important role in the astrophysics of the ISM. Turbulence-associated magnetic field fluctuations govern the diffusion and propagation of charged particles in the galaxy. Similarly to the case in the solar wind and corona, interstellar turbulence is almost certainly responsible for determining fundamental transport coefficients such as viscosity, resistivity, and thermal conductivity. Knowledge of the values of these transport coefficients is crucial for developing credible mathematical models of the ISM.

All charged particles are constrained in their movement by the local magnetic field. First of all, it relates to cosmic rays.

(5) **Cosmic rays** including neutral (photons, neutrinos, etc.) and charged (electrons, protons, other nuclei) particles mainly of high energy. Galactic CRs are born in supernova remnants (Fig. 3.2) and then propagate throughout the Galaxy. The gyration radii for CRs with $E < 10^7$ GeV are much smaller compared with the coherence lengths of ISM magnetic fields, they move approximately along magnetic field lines. They may occasionally move onto other magnetic field lines by scattering from MHD waves or from ISM nuclei. Because of the turbulent character of ISM magnetic field, the motion of each of charged particle is a very complicated process described usually in terms of one or the other random process. The very small anisotropy of CR flux justifies (as a rough approximation) the diffusion model, but calculations of mass- and time-characteristics need more fine-tuning of the stochastic model.

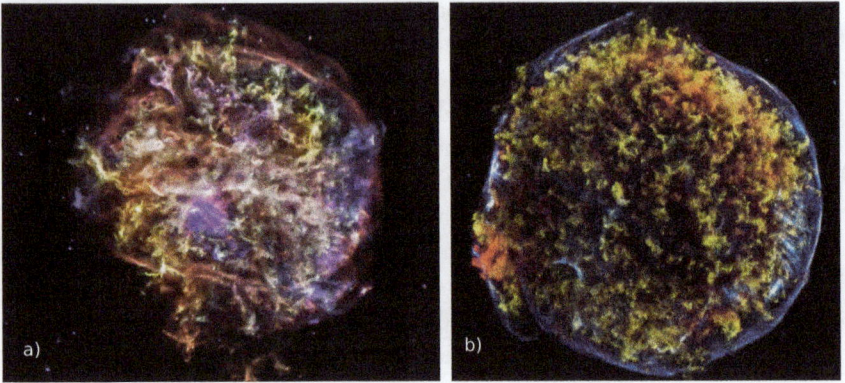

Fig. 3.2   a) Supernova remnant G292.0+1.8, one of only three remnants (in the Milky Way) known to contain large amounts of oxygen. These supernovas are of great interest because they are one of the primary sources of the heavy elements. Image credit: NASA/CXC/SAO. b) The Tycho supernova remnant. There is evidence from the Chandra data that shock waves observed in this image are responsible for acceleration of the cosmic rays. Image credit: NASA/CXC/SAO.

One of the most intriguing questions concerns the origin and acceleration of cosmic rays. Nowadays, it is assumed that the acceleration is caused by multiple scattering by the irregularities of the interstellar magnetic fields, either in supernova shock waves, or in the fluctuating magnetic field of the magnetohydrodynamic turbulence of the ISM. [Blasi (2013)] gives a review of the theory of the origins of cosmic rays. There is abundant observational evidence that stellar winds can produce cavities in the interstellar medium. On larger scales, [Everett *et al.* (2008)] have shown that steady-state galactic winds can be driven from the inner Milky Way by a combination of thermal and cosmic ray pressures. In this case, the wind should draw up the interstellar magnetic field lines, creating apparent holes in the horizontal component of the galactic magnetic field. Such holes could provide (a) a possible escape route for the cosmic rays, and (b) an impact on galactic dynamos. Whether this occurs in our Galaxy or not seems uncertain, and depends on the ability of stellar winds and supernovae to keep a "galactic fountain" going.

(6) **The radiation field**. This radiation field covers the full electromagnetic spectrum, and different wavelength ranges are dominated by different types of sources. The polarized radio emission is one of the more

fruitful diagnostics of cosmic magnetic fields. This emission comes from extragalactic (radio galaxies and quasars) and Galactic sources (pulsars and diffuse synchrotron emission from cosmic ray electrons). The emission is subject to Faraday rotation and is partially depolarized during propagation through the Galaxy. This rotation and depolarization yield information on the strength and structure of the Galactic field. In the domain of microwave radiations, the interstellar radiation field is dominated by the cosmic microwave background (CMB) emitted at the recombination era, and redshifted to the microwave domain by the expansion of the Universe. The visible and UV domains are dominated by the light of stars, with more massive stars contributing to lower wavelengths. There are many types of radio and optical observations which help us diagnose interstellar turbulence, such as radio scintillations and the analysis of spectral data cubes of 21 cm HI observations [Lazarian (2008)].

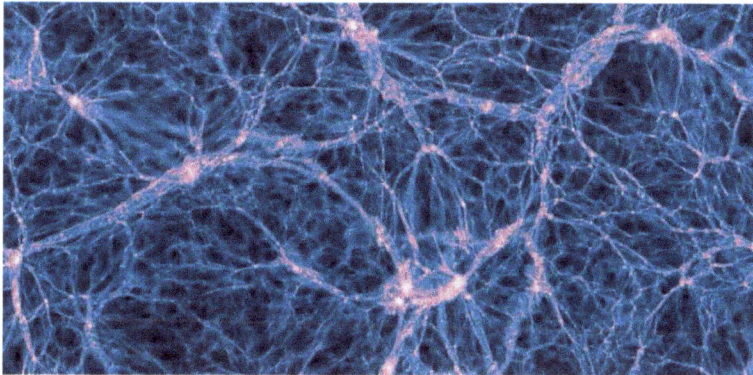

Fig. 3.3 Visualization of simulation [Vogelsberger *et al.* (2014); Genel *et al.* (2014)] of dark matter clusterization. Credit: Illustris Collaboration/Illustris Simulation.

(7) **Dark matter**. In 1925, Knut Lundmark tried to estimate the Milky Way mass using information on stellar velocities and found that 10^{12} solar masses would explain the visible motions, whereas the number of visible stars was a hundred times less. Lundmark concluded that dark stars and *dark matter* exist. Now, the inflation model adopted by most cosmologists insists that the dark matter makes the most part of the total mass of the universe, and the theory of primordial nucleosynthesis demands that ordinary matter, such as atoms and molecules, can form only a few percent of the total mass [Baryshev and Teerikorpi (2002)].

According to the Wilkinson Microwave Anisotropy Probe data, most (about 96%) of the universe's matter and energy are invisible. The components of the visible matter (stars, diffuse gas, accreting black holes, etc.) are organized in a web consisting of filaments, and voids. Testing different scenarios of the evolution of dark matter and dark energy requires accurate description of related formation of structure of the visible matter [Vogelsberger *et al.* (2014); Genel *et al.* (2014)]. Result of a sample simulation of the dark matter density distribution is shown in Fig. 3.3.

3.2   Physics of anomalous diffusion

MHD-equations

Distribution of high-energy charge particles in space is linked with part of interstellar medium described in terms of magnetodynamic theory. The MHD fluid is perturbed by shock waves produced by supernova explosions. The waves make the ISM turbulent.

Recall that an ideal magnetized fluid is governed by the MHD-equation system,

$$\frac{d\rho}{dt} + \rho \text{div } \mathbf{v} = 0, \tag{3.1}$$

$$\rho \frac{d\mathbf{v}}{dt} = -\text{grad } P + \frac{1}{4\pi} \left[\text{rot } \mathbf{H}, \mathbf{H} \right], \tag{3.2}$$

$$\frac{\partial \mathbf{H}}{\partial t} = \text{rot } \left[\mathbf{v}, \mathbf{H} \right] + \frac{c^2}{4\pi\sigma} \Delta \mathbf{H}, \tag{3.3}$$

$$\text{div } \mathbf{H} = 0. \tag{3.4}$$

Interstellar magnetic fields possess a complicated structure, consisting of set of random curve lines filling the space, moving, drawing closer to one another, taking away or coming back and reconnecting with each other. From time to time the shock waves generated by supernova explosions in different places, cross the space accelerating charged particles. It is impossible to describe this process in a single section or even in a chapter, and we only briefly point out here on the important aspects for future discussion.

In case, when a perfectly conducting fluid (plasma) is at rest ($\mathbf{v} = 0$), Eq. (3.3) takes the form

$$\frac{\partial \mathbf{H}}{\partial t} = \frac{c^2}{4\pi\sigma} \Delta \mathbf{H}.$$

The magnetic field concentrated initially in a certain region spreads out in space and decreases in magnitude. One can estimate time τ, for which the magnetic field can go beyond the area with a characteristic size of L, with the following stratagem, often used in such assessments.

If some function changes monotonically from the initial value to a relatively small, almost zero, one, then the derivative with respect to this argument can be evaluated (roughly in order of magnitude) by dividing this initial value to the characteristic value of argument. In our case, it gives (we take the absolute values)

$$\frac{\partial H}{\partial t} \simeq \frac{H}{\tau}, \quad \Delta H \simeq \frac{H}{L^2}.$$

Substituting the latter result in

$$\frac{H}{\tau} \simeq \frac{c^2}{4\pi\sigma}\frac{H}{L^2}$$

gives the assessed value

$$\tau \simeq (4\pi/c^2)\sigma L^2.$$

This result is important, particularly to understand the process of formation of stars from gas clouds permeated by a magnetic field.

Solar wind turbulence

Solar wind (see for review [Goldstein *et al.* (1995)]) flows nearly radially away from the Sun, at up to about 700 km/s. This is much faster than both spacecraft motions and the Alfvéen speed. Therefore, the turbulence is "frozen" and the fluctuations at frequency f are directly related to fluctuations at the scale k in the direction of the wind, as $k = 2\pi f/v$, where v is the solar wind velocity [Horbury (1999)].

Usually two types of solar wind are distinguished, one being the fast wind which originates in coronal holes, and the slower bursty wind. Both of them show, however, $f^{-5/3}$ scaling on small scales. The turbulence is strongly anisotropic with the ratio of power in motions perpendicular to the magnetic field to those parallel to the magnetic field being around 30. The intermittency of the solar wind turbulence is very similar to the intermittency observed in hydrodynamic flows [Horbury and Balogh (1997)].

Interstellar turbulence

Interstellar medium (ISM) is filled with magnetic field, plasma, gas, dust and cosmic rays. Magnetic field statistics are the most poorly constrained aspect of ISM turbulence. The polarization of starlight and of the Far-Infrared Radiation (FIR) from aligned dust grains is affected by the ambient magnetic fields. Assuming that dust grains are always aligned with their longer axes perpendicular to magnetic field, one gets the 2D distribution of the magnetic field directions in the sky. Note that the alignment is a highly nonlinear process in terms of the magnetic field and therefore the magnetic field strength is not available.

The statistics of starlight polarization is rather rich for the Galactic plane and it allows to establish the spectrum $E(\mathbf{K}) \sim K^{-1.5}$, where $\mathbf{K}$ is a two-dimensional wave vector describing the fluctuations over sky patch, $K = |\mathbf{K}|$.

For uniformly sampled turbulence $E(\mathbf{K}) \sim K^{\alpha}$ for $K < K_0$ and K^{-1} for $K > K_0$, where K_0^{-1} is the critical angular size of fluctuations which is proportional to the ratio of the injection energy scale to the size of the turbulent system along the line of sight. For Kolmogorov turbulence $\alpha = -11/3$ (see for more details [Cho *et al.* (2002)].

Truncated Kolmogorov-Obukhov energy spectrum and tempered fractional equation

Observe that we ignored restrictions (Fig. 3.4) imposed on the power spectrum by windows function, and this is why we obtained a pure fractional Laplacian. Now, we take into account this effect at least in the simplest form of a one-sided truncation (2.33). In this case we arrive at the integral

$$R(k) = \kappa\sqrt{K_0\epsilon^{2/3}}k^2 \int_{k}^{\infty} q^{-7/3}\sqrt{\psi(\lambda q)}dq = \kappa\sqrt{K_0\epsilon^{2/3}}\lambda^{4/3}k^2 \int_{\lambda k}^{\infty} z^{-7/3}\sqrt{\psi(z)}dz.$$

Recall that $\psi(z) = 0$ if $z < \zeta$, thus for $\lambda k < \zeta$ (large spacial scales), the lower limit of the integral should be replaced by ζ and we obtain

$$R(k) = \left[\kappa\sqrt{K_0\epsilon^{2/3}}\lambda^{4/3}\int_{\zeta}^{\infty} z^{-7/3}dz\right]k^2 = R(1)k^2.$$

Otherwise, if $\lambda k > \zeta$ (relatively small spatial scales), the lower limit is λk and we have

$$R(k) = \left[\kappa\sqrt{K_0\epsilon^{2/3}}\lambda^{4/3}\int_{\lambda k}^{\infty} z^{-7/3}dz\right]k^2 = R(1)k^{2/3}.$$

These results give us reason to write the equation of turbulent diffusion propagator in the form

$$\frac{\partial f}{\partial t} = \mathsf{R}_L^\alpha f(\mathbf{x}, t) + \delta(\mathbf{x})\delta(t),$$

where R_L^α is a nonlocal operator dependent on spatial parameter L and satisfying the asymptotical conditions

$$\mathsf{R}_L^\alpha \sim \begin{cases} \Delta, & \text{on scales much larger than } L; \\ -(-\Delta)^{\alpha/2}, & \text{on scales much smaller than } L. \end{cases}$$

We call it the *tempered fractional Laplacian*.

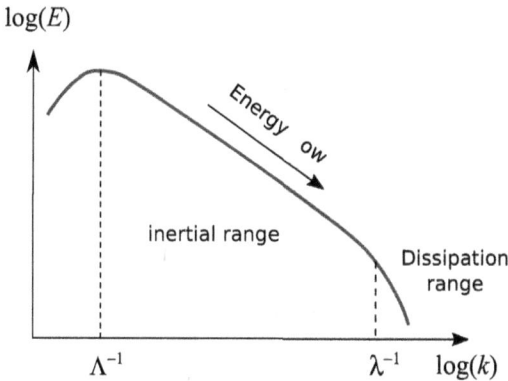

Fig. 3.4 A typical plot of the Kolmogorov-Obukhov energy spectrum leading to a tempered fractional equation.

The equation with the tempered fractional Laplacian describes asymptotical regime of flights with smoothly truncated power law distribution of free path lengths. Truncated Lévy flights showing a slow convergence to a Gaussian was introduced by Mantegna and Stanley in 1994 [Mantegna and Stanley (1994)]. Smoothly (exponentially) truncated Lévy flights, introduced by Koponen in 1995 [Koponen (1995)], constructed on Mantegna and Stanleys ideas, allowed to get a convenient analytic representation of results.

[Sabzikar *et al.* (2015)] note: "Unlike the truncated model, tempered Lévy flights offer a complete set of statistical physics and numerical analysis tools. Random walks with exponentially tempered power law jumps converge to a tempered stable motion [Chakrabarty and Meerschaert (2011)]. Probability densities of the tempered stable motion solve a tempered fractional diffusion equation that describes the particle plume shape [Baeumer and Meerschaert (2010)], just like the original Einstein model for traditional diffusion. Tempered fractional derivatives are approximated by tempered fractional difference quotients, and this facilitates finite difference schemes for solving tempered fractional diffusion equations".

3.3   Walking magnetic field lines

The pressing need for taking into account the turbulent nature of magnetic fields in cosmic-ray propagation calculations gave rise to an alternative method of modeling, which does not require a full spectral analysis of fluctuations of the field: this is magnetic field lines method.

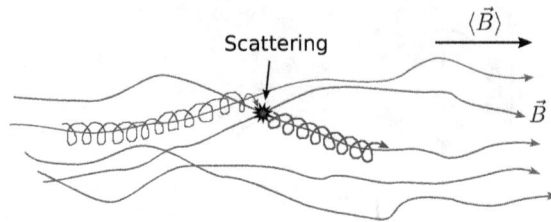

Fig. 3.5   Transition between magnetic field lines due to scattering.

By considering the motion of cosmic-ray particles in regions of limited sizes where fluctuations of the direction and strength of the interstellar magnetic field are relatively small, we imagine magnetic field lines on which the trajectories of charged particles are wound. The leading centers of particles move along these lines, slowing down their motion in front of 'plugs', reflect from them, and return. In this case, the field lines do not remain motionless: they are displaced and bent, carrying the nearby charged particles with them, and the distances between line condensations and switchings change, complicating the motion of particles along *field lines randomly walking in space and time* (Fig. 3.5) [Jokipii (1967, 1973a)]. Following a set of such lines passing through the vicinity of point O in Fig. 3.6, we see how they begin to diverge from each other, demonstrating a statistical ensemble, which has been described in many papers.

A purely phenomenological model was proposed in [Getmantsev (1963)], with the ensemble of magnetic field lines represented as a family of independent three-dimensional trajectories consisting of successive independent segments with random lengths and random directions along which particles perform one-dimensional diffusive random walks. Such a compound diffusion causes the slowing down of diffusion in transverse directions:

$$\langle R^2(t) \rangle \propto t^{1/2}.$$

However, this slowing down is caused by the assumption about the diffusion motion of particles along the field lines: if we assume that particles move freely along these lines, we return to the asymptotically normal diffusion (if the root-mean-square length of segments is finite).

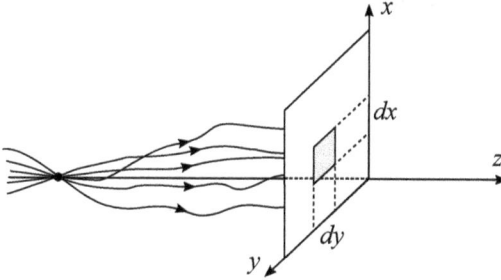

Fig. 3.6 Ensemble of magnetic field line realizations.

Later, the compound diffusion model was used for solving the problem of the motion of particles in a weakly inhomogeneous magnetic field (see, e.g., [Ginzburg and Syrovatskii (1964); Hayakawa (1969); Rechester and Rosenbluth (1978); Webb *et al.* (2006)]). In the simplest formulation, a region was considered in a homogeneous stationary random field with the mean value

$$\langle \mathbf{H} \rangle \equiv \mathbf{H}_0 = H_0 \mathbf{e}_z,$$

and the autocorrelation function

$$\langle H_i(z_2) H_j(z_1) \rangle = \langle \delta H_i(z_2) \delta H_j(z_1) \rangle = H_0^2 C_{ij}(\zeta), \quad \zeta = |z_2 - z_1|. \quad (3.5)$$

The only random process completely determined by these two characteristics (the mean value and the correlation function) is a Gaussian process, whose particular case is the Brownian motion characterized by independent increments and the distribution density satisfying the standard diffusion equation

$$\frac{\partial M}{\partial z} = D_L \Delta_\perp M(x, y, z)$$

with the diffusion coefficient $D_L > 0$. A particle diffuses (in the modified compound model [Hayakawa (1969); Webb *et al.* (2006)]) with the longitudinal diffusion coefficient $D_\parallel$ along one of the realizations of this ensemble symmetrically continued to the region $z < 0$ (in the statistical sense). In this case, again,

$$\langle R_\perp^2(t) \rangle \simeq D_L (D_\| t)^{1/2}.$$

We note, however, that an ensemble of Brownian trajectories, which are nowhere differentiable fractal curves [Zel'dovich and Sokolov (1985)] with independent increments, is difficult to reconcile with the concept of magnetic field lines. Returning to piecewise smooth lines in [Getmantsev (1963)], but assuming that particles move in one direction at a constant velocity along these lines, we obtain the normal diffusion of transverse displacements of particles in the long-time asymptotic regime. An analysis based on the kinetic equation performed in [Zybin and Istomin (1985)] yields the same result.

To approach the real properties of the magnetic field of the interstellar medium, the authors of [Shalchi and Kourakis (2007a,c); Shalchi *et al.* (2009)] represented autocorrelation function (3.5) in terms of the Fourier component of magnetic field fluctuations:

$$C_{xx}(z) = \frac{1}{(2\pi)^6} \int d\mathbf{k} \int d\mathbf{k}' \left\langle \widetilde{\delta H}_x(\mathbf{k}) \widetilde{\delta H}_x^*(\mathbf{k}') e^{-i[\mathbf{k}\cdot\mathbf{x}(z) - \mathbf{k}'\cdot\mathbf{x}(0)]} \right\rangle.$$

Using the Corrsin independence hypothesis [Corrsin (1959)] and neglecting the correlations of field components with different wave vectors $\mathbf{k}$ and $\mathbf{k}'$, the authors obtained

$$C_{xx}(z) = \int \left\langle e^{-i\mathbf{k}\cdot\Delta\mathbf{x}(z)} \right\rangle P_{xx}(\mathbf{k}) d\mathbf{k}, \quad P_{xx}(\mathbf{k}) = (2\pi)^{-6} \left\langle \left| \widetilde{\delta H}_x(\mathbf{k}) \right|^2 \right\rangle,$$

$$(3.6)$$

where $\Delta\mathbf{x}(z) = \mathbf{x}(z) - \mathbf{x}(0)$ is a random transverse displacement vector over which the exponential is averaged. As a result, they obtained the expression

$$\left\langle [\Delta X(z)]^2 \right\rangle \equiv \frac{2}{H_0^2} \int_0^z (z - \zeta) C_{xx}(\zeta) d\zeta$$

$$= \frac{2}{H_0^2} \int d\mathbf{k} P_{xx}(\mathbf{k}) \int_0^z (z - \zeta) \cos(k_\| \zeta) e^{-(1/2)\langle[\Delta X(\zeta)]^2\rangle k_\perp^2} d\zeta,$$

providing a basis for a more natural model of randomly walking field lines. They continued these calculations using the hybrid approach, where a turbulent axially symmetric field is described by a combination of the one-dimensional planar ($\mathbf{k} \parallel \mathbf{H}_0$) and two-dimensional ($\mathbf{k} \perp \mathbf{H}_0$) components:

$$P_{xx}(\mathbf{k}) = g'(k_\|) \frac{\delta(k_\perp)}{k_\perp} + g''(k_\perp) \frac{\delta(k_\|)}{k_\perp} \left(1 - \frac{k_x^2}{k^2} \right),$$

where

$$g'(k_\parallel) = \frac{c(\nu)}{2\pi} \frac{l' \delta H_{1D}^2}{(1 + k_\parallel^2 l_\parallel^2)^\nu}, \quad g''(k_\perp) = \frac{2c(\nu)}{\pi} \frac{l'' \delta H_{2D}^2}{(1 + k_\perp^2 l_\perp^2)^\nu},$$

$$c(\nu) = \frac{1}{2\sqrt{\pi}} \frac{\Gamma(\nu)}{\Gamma(1 - \nu/2)}$$

with 2ν being the spectral index in the inertial interval. The combination of these components, with an asymptotically small term neglected (as $z \to \infty$), gave the equation

$$\frac{d^2}{dz^2} \left\langle [\Delta X(z)]^2 \right\rangle = \frac{2\pi g''(0)}{H_0^2} \int_0^\infty e^{-(1/2)\langle [\Delta X(z)]^2 \rangle k_\perp^2} dk_\perp = \frac{2\pi}{H_0^2} \sqrt{\frac{\pi}{2}} \frac{g''(0)}{\sqrt{\langle [\Delta X(z)]^2 \rangle}}.$$

The solution of this equation

$$\langle [\Delta X(z))]^2 \rangle \sim \left[\sqrt{\frac{\pi}{2}} \frac{9\pi g''(0)}{2H_0^2} \right]^{2/3} z^{4/3}, \quad z \to \infty, \tag{3.7}$$

shows that taking the autocorrelations of magnetic field lines into account in the framework of the Corrsin hypothesis [Corrsin (1959)] leads to a *superdiffusion* behavior of transverse displacements. We note that the variable z is not a time but the longitudinal coordinate, and therefore the emerging "superdiffusion" relates not to the development of the diffusion process in time but to the rapid divergence of a bunch of magnetic field lines in space. As a result, the decomposition of the displacement of a particle into independent longitudinal and transverse components becomes incompatible with the asymptotic regime of the sought solution. In this case, it is convenient to pass from the linear coordinate z to the curvilinear coordinate s measured along a field line, and finally to the model of isotropically randomly walking lines. The simulation of these lines by continuous broken lines with linear segments of random lengths ("ranges") allows representing the large-scale correlations of magnetic fields by using nonexponential range distributions. Asymptotic power-law distributions, which seem to be a natural continuation of the regularities of large-scale turbulence, are convenient for this purpose.

3.4 Diffusion in molecular clouds

The interstellar gas cannot remain stable at the average local density of ~ 1 cm^{-3}. Moving in a turbulent way, it produces a variety of structures and its density becomes in the range of 10 to 100 cm^{-3} for the cool phase, and much less, such as 0.1 cm^{-3}, in the warm neutral phase, given the total interstellar pressure. Any random motion inevitably partitions the gas into clouds and an intercloud medium. Figure 3.7 demonstrates the giant Orion Molecular Cloud. What is new about the fractal interpretation is the recognition that such turbulence partitioning can lead to fractal structure with well determined and universal properties.

Fig. 3.7 Three nebulae (the Flame nebula, the Horsehead nebula and NGC 2023) as parts of the giant Orion Molecular Cloud. This mosaic image is taken by NASA's Wide-field Infrared Survey Explorer. Image credit: NASA/JPL-Caltech/UCLA.

In conclusion to his article, Elmegreen [Elmegreen (1998)] writes: "The origin of the fractal structure is probably supersonic, sub-Alfvenic turbulence driven by a variety of pressure sources and possibly Galactic rotation. The fractal dimension of interstellar structure is about the same as the fractal dimension of laboratory turbulence." Summing far-reaching implications of various observations, the author notices among others:

- The interstellar gas is highly clumped, with ∼ 3 fractal cloud complexes per kpc locally. Each complex has a hierarchical or fractal distribution of clumps within clumps.
- The extremely small-scale structure that is observed in diffuse clouds and molecular clouds is probably the result of turbulence.

- The photon mean free path in the intercloud medium is ~300 pc, allowing disk UV photons to reach the Galactic halo.

The molecular clouds dynamics underlies understanding processes of forming planets, stars and galaxies. In order not to lose touch with the logic of the concept of a turbulent turn to another work of Weizsäcker devoted to evolution of galaxies and stars [von Weizsäcker (1951)]. The author considered that the motion of the cosmic gas obeys the equations of hydrodynamics and that in most cases it is turbulent and compressible. The latter property served as a reason for making result to be more precise. Instead of assumption $\rho = $ const, the n-dependent cloud density was involved by the equation

$$\frac{\rho_{n+1}}{\rho_n} = \left(\frac{L_{n+1}}{L_n} \right)^{3c_n},$$

where the exponent c_n called the degree of compression at the level n must be smaller than unity. Referring to observation data, the author estimated a limit value $c_\infty = c$ as lying between 0.09 and 0.23. Numerical simulations of compressible turbulence performed later [Kritsuk *et al.* (2007); Kowal and Lazarian (2007)] have confirmed the scaling with $c \cong 0.15$ close to value 1/6, for which the density power spectrum becomes flat.

The spectral function then reads

$$F(k) \propto k^{-5/3-2c},$$

and the diffusion equation in the Fourier space has the form of (2.43) with the relaxation factor

$$R(k) \propto k^{2/3-c}, \quad \Lambda^{-1} \ll k \ll \lambda^{-1}.$$

If the limitation in k can be neglected, this fact leads to fractional diffusion equation

$$\frac{\partial n}{\partial t} = -C(-\Delta)^{1/3-c/2} n(\mathbf{x}, t).$$

[Martinell *et al.* (2006)] used this kind of a nonlocal transport model to study the chemical structure of a molecular cloud. Recognizing it as rather idealized model, the authors find analytic results for both the 'classical' turbulent diffusion model as well as for the nonlocal transport model. The latter is based on the one-dimensional steady-state fractional equation

$$\frac{d^\alpha n(x)}{dx^\alpha} \equiv \frac{1}{\Gamma(2-\alpha)} \int_0^x \frac{d^2 n(x')}{dx'^2} \frac{dx'}{(x-x')^{\alpha-1}} = \gamma n(x), \quad 1 < \alpha < 2, \quad (3.8)$$

solution of which is expressed via two-parameter Mittag-Leffler's functions:

$$n(x) = n(0) \left[E_{\alpha,1} \left(\gamma x^\alpha \right) - \gamma^{1/\alpha} x E_{\alpha,2} \left(\gamma x^\alpha \right) \right].$$

Asymptotical behavior of this solution

$$n(x) \sim \frac{\gamma^{(1-\alpha)/\alpha}}{\Gamma(2-\alpha)x^{\alpha-1}}, \quad x \to \infty,$$

describes the slow (algebraic) decay instead to fast (exponential) one predicted by the normal diffusion model. As the authors remark in conclusions, in the case $\alpha = 0.25$ they obtained desirable interrelation $l \approx L$ between the penetration length l and the characteristic size of the cloud L.

It should be noted that the identity

$$\frac{d^2 f}{dx^2} = \frac{d^2 f}{d|x|^2}$$

does not automatically extend to fractional-order derivatives,

$$\frac{d^\alpha f}{dx^\alpha} \neq \frac{d^\alpha f}{d|x|^\alpha}, \quad \alpha \neq 2,$$

and the choice of the fractional operator in Eq. (3.8) needs additional verification.

The authors compensate the lack of physical ground of the model by referring to paper [del Castillo-Negrete *et al.* (2003)] where nonlinear counterpart of fractional model has been developed for consideration of reaction-diffusion problems at shock fronts. Here is a quotation from this paper.

"The origin of non-Gaussian diffusion can be traced back to the existence of long-range correlations in the dynamics, or the presence of anomalously large particle displacements described by broad probability distributions. Here we are interested in the second possibility. In particular, we focus on systems that exhibit anomalous diffusion caused by Lévy flights, for which the probability distribution of particle displacements, $p(l)$, is broad in the sense that $\langle l^2 \rangle = \infty$. As is well known, for these kinds of systems the central limit theorem cannot be applied; and as $N \to \infty$, the probability distribution function of $x = \sum_{n=1}^{N} l_n$, rather than being Gaussian, is an α-Lévy distribution."

Pointing out an important role of asymmetric walks in astrophysical problems such as beam propagation observations and so forth, the authors focus on anomalous diffusion processes that exhibit Lévy flights in one (positive) direction, but Gaussian behavior in the opposite. Thus, instead of classical reaction-diffusion equation

$$\frac{\partial n}{\partial t} = D \frac{\partial^2 n}{\partial x^2} + F(n)$$

for concentration $n = n(x, t)$ its space-fractional counterpart is used:

$$\frac{\partial n}{\partial t} = D \,_a D_x^\alpha n + F(n).$$

Here

$$_a D_x^\alpha n = \frac{\gamma}{\Gamma(2-\alpha)} \frac{\partial^2}{\partial x^2} \int_a^x \frac{n(\xi, t)d\xi}{(x-\xi)^{\alpha-1}}, \quad 1 \le \alpha < 2,$$

and

$$F(n) = \gamma(1-n)n.$$

[Marschalkó *et al.* (2007)] considered a 3-dimensional symmetric version of anomalous diffusion in the same problem. Transforming the initial partial differential equation

$$\frac{\partial n}{\partial t} + \frac{n}{\tau_c} = D\Delta n(r, t) + \frac{n_0(r)}{\tau_e} \tag{3.9}$$

into the Fourier space

$$\frac{\partial \tilde{n}}{\partial t} + \frac{\tilde{n}}{\tau_c} = -Dk^2\tilde{n}(k, t) + \frac{n_0(r)}{\tau_e}$$

where τ_e stands for a relaxation time, during which the solution converges to $n_0(r)$ without diffusion. For the steady-state regime, the authors obtained the solution

$$\tilde{n}(k) = \frac{\tilde{n}_0(k)}{1 + \tau_e Dk^2}. \tag{3.10}$$

Backward Fourier-Hankel inversion of Eq. (3.10)

$$n(r) = \int_0^\infty \tilde{n}(k)J_{1/2}(kr)k^{3/2}dk$$

led to the classical result for molecular diffusion in a homogeneous medium. The authors give the numerical results obtained with the choice of the 'fiducial' values $D = 10^{23}$ cm^2/sec (on assumption that $v \approx 1$ km/s and $l \approx 0.1$ pc) and $\tau_e = 10^6$ yrs. Further, the authors write:

"In a turbulent cascade regime where turbulence cannot be characterized by a single dominant scale anymore, equation (3.9) does not hold. However, its Fourier transform, with the appropriate scale-dependent diffusivity value $D(k)$ will still hold for each Fourier component of the passive scalar field. For the wave number dependence of the diffusivity we use the Kolmogorov form $D(k) = D_0(k/k_0)^{-4/3}$ throughout this paper, D_0 being the value of the diffusivity at the wave number k_0, corresponding to the fiducial scale l." Thus, Marschalkó *et al.* came to the fractional superdiffusion equation with the solution having the Fourier image

$$\widetilde{n}(k) = \frac{\widetilde{n}_0(k)}{1 + \tau_e D' k^{2/3}}, \quad D' = D k_0^{4/3}. \tag{3.11}$$

On the base of numerical calculations the authors arrive at the conclusion that there is a significant difference between the diffusive and superdiffusive case. In the superdiffusive case the concentration of the tracer in the cloud core is much more pronounced, as a consequence of the reduced diffusivity at small scales. Interestingly, an overall increase of the diffusivity leads to an increase in the central number density of the tracer in the diffusive case, while it leads to a decrease in the superdiffusive case.

At the end of their article, the authors noticed that to take into account the non-uniformity more consistently, it would be necessary to replace $D\Delta n(r,t)$ by $\nabla(D(\mathbf{x})\nabla n(\mathbf{x},t))$ but in this case the problem becomes mathematically much more complicated.

3.5   Fractional diffusion in plasma

The Balescu hybrid equation

One can think that the plasmas physics is the second physical direction in which the fractional calculus has achieved significant results (the first place is occupied by the rheology, we suppose). However, there is no any possibility to list here not only all these works, but even the most significant of them. Our aim is much modest: to give the most clear physical basis for fractional calculus in the plasmas physics and to comment a few of them directly connected with the interstellar medium.

For physical introduction of fractional operators into the plasma dynamics, we take the Balescu work [Balescu (2000)]. Considering the collisionless motion in plane (x,y), perpendicular to magnetic field, he write the continuity equation

$$\frac{\partial N(\mathbf{x},t)}{\partial t} + \mathbf{v}(\mathbf{x},t) \cdot \nabla N(\mathbf{x},t) = 0,$$

for ion concentration $N(\mathbf{x},t)$. The ions move according to the random velocity field $\mathbf{v}(\mathbf{x},t)$, given by means of potential $\phi(\mathbf{x},t)$:

$$\mathbf{v}(\mathbf{x},t) \propto [\nabla\phi(\mathbf{x},t), \mathbf{e}_z].$$

This potential is determined as a Gaussian stationary homogeneous and isotropic process with zero mean and two-point correlation function

$$\langle \phi(\mathbf{x}_1, t_1)\phi(\mathbf{x}_1 + \mathbf{x}, t_1 + t)\rangle = \epsilon^2 E(\mathbf{x}/\lambda_c)h(t/\tau_c),$$

where E and h are dimensionless functions of space and time falling to zero with raising arguments, ϵ denotes the measure of intensity of the potential fluctuations, and λ_c and τ_c stand for space- and time-scales of correlations respectively. The plasma is assumed to be incompressible and its turbulence to be stochastically homogeneous and isotropic. Separating the concentration N into average component n and fluctuations n', Balescu writes the master equation in the form

$$\frac{\partial n(\mathbf{x}, t)}{\partial t} = \int_0^t d\tau \Lambda(\tau) \Delta n(\mathbf{x}, t - \tau) + S(\mathbf{x}, t),$$

where

$$\Lambda(\tau) = \langle \mathbf{v}(\mathbf{x}, \tau)\mathbf{v}(\mathbf{X}(\tau|0), 0)\rangle,$$

$$S(\mathbf{x}, t) = -\nabla \langle \mathbf{v}(\mathbf{x}, t)n'(\mathbf{X}(t|0), 0)\rangle,$$

and

$$\mathbf{X}(t|t') = \mathbf{x} - \int_{t'}^t d\tau \mathbf{v}(\mathbf{X}(t|\tau), \tau))$$

is the trajectory inversed in time (in other words, the position at time t' of that particle which is observed at point $\mathbf{x}$ at time t). Evidently, this equation is not Markovian, because the change in concentration at time t depends on its prehistory. Such effect may be explained by trapping phenomena: some ions turn out in traps (domains of field) where they perform a confined motion along closed curves for a long time. This phenomenon was analytically discovered in [Bernstein *et al.* (1957)] and later was confirmed by numerical simulations [Berk and Roberts (1967); Morse and Nielson (1969)]. A significant role is attributed to the interstellar medium traps in the cosmic electrodynamics and particularly in cosmic rays physics [Zelenyi and Milovanov (2004); Dorman (1975)].

The degree of influence of the non-Markovian property on the turbulent diffusion process is determined by the Kubo number

$$\mathsf{Ku} = \frac{\epsilon \tau_c}{\lambda_c^2}.$$

In case of a weak turbulence (Ku $\ll$ 1) retardation in the master equation may be neglected ($n(\mathbf{x}, t - \tau) \approx n(\mathbf{r}, t)$) and integration with respect of τ can be extended to infinity. The process becomes Markovian which is described by the ordinary diffusion equation:

$$\frac{\partial n(\mathbf{x}, t)}{\partial t} = K\Delta n(\mathbf{x}, t) + S(\mathbf{x}, t).$$

While analyzing strong turbulence (Ku $\gg$ 1) one need to account the retardation function $\Lambda(\tau)$ form. It is convenient to be done if passing to Laplace images:

$$\lambda \widehat{n}(\mathbf{x}, \lambda) = \widehat{\Lambda}(\lambda)\Delta_2 \widehat{n}(\mathbf{x}, \lambda) + \widehat{S}(\mathbf{x}, \lambda) + \delta(\mathbf{x})$$

(a particle is supposed to take the origin at the initial instant).

Following the self-similarity principle, confirmable by numerical modeling, choose the kernel transform in the power function form

$$\Lambda(\lambda) = K\lambda^{1-\beta}, \quad 0 < \beta \leqslant 1,$$

and multiply both parts of the equation by $\lambda^{\beta-1}$. Inverse Laplace transform leads to a subdiffusion equation

$$_0\mathsf{D}_t^\beta n(\mathbf{x}, t) = K\Delta_2 n(\mathbf{x}, t) + Q(\mathbf{x}, t).$$

Its solution is expressed via isotropic fractionally stable density $\psi^{(2,\beta)}$ and diffusive packet length increases proportionally to $t^{\beta/2}$.

Thus, memory influence slows down a diffusion process. Within Balescu's theory the limit case $\beta \to \infty$ corresponds to normal diffusion. This is because we didn't introduce the turbulent character of the interstellar plasma tightly connected with other turbulent components in the interstellar medium. Doing this as in the preceding section, we generalize the master equation to the bifractional form

$$_0\mathsf{D}_t^\beta n(\mathbf{x}, t) = -K(-\Delta_2)^{\alpha/2} n(\mathbf{x}, t) + Q(\mathbf{x}, t).$$

Fractional equation for velocity distribution in plasma

Chechkin with coauthors [Chechkin and Gonchar (2000); Chechkin *et al.* (2002)] used fractional Fokker-Planck equation to describe 3d motion of a charge in magnetic field $\mathbf{H} = H\mathbf{e}_z$ taking into account a friction force $-\eta m\mathbf{v}$ and a random electric field $\mathbf{E}$. The latter was represented by homogeneous, isotropic, stationary Lévy white noise with intensity K and exponent α. Lévy noise is the sequence of independent stationary increments of Lévy motion in like manner as white Gaussian noise in the sequence of Brownian motion increments. Increment characteristic function is

$$\widetilde{p}(\mathbf{k}, \Delta t) = e^{-K|\mathbf{k}|^{\alpha} \Delta t}, \quad 0 < \alpha \le 2.$$

When $\alpha \to 2$ Lévy noise becomes Gaussian one.

The corresponding equation for velocity distribution density $f(\mathbf{v}, t)$ of a charged particle has the form:

$$\frac{\partial f}{\partial t} + \mathbf{\Omega}[\mathbf{v}, \mathbf{e}_z] \nabla_{\mathbf{v}} f = \eta \nabla_{\mathbf{v}}(\mathbf{v} f) - K(-\Delta_{\mathbf{v}})^{\alpha/2} f.$$

Using the Fourier transform

$$\widetilde{f}(\mathbf{k}, t) = \int d\mathbf{v} e^{i\mathbf{k}\mathbf{v}} f(\mathbf{v}, t)$$

it can be reduced to the form

$$\frac{\partial \widetilde{f}}{\partial t} + (\mathbf{\Omega}[\mathbf{k}, \mathbf{H}] + \eta \mathbf{k}) \nabla_{\mathbf{k}} \widetilde{f} = -K|\mathbf{k}|^{\alpha} \widetilde{f}.$$

The solution of the latter equation (with initial condition $\mathbf{v}(0) = 0$) is given by the formula

$$\widetilde{f}(\mathbf{k}, t) = \exp\left\{-(K/\alpha\eta)\left(1 - e^{-\alpha\eta t}\right)|\mathbf{k}|^{\alpha}\right\},$$

so that the velocity distribution is expressed via isotropic stable density of Lévy-Feldheim:

$$f(\mathbf{v}, t) = \left[(K/\alpha\eta)\left(1 - e^{-\alpha\eta t}\right)\right]^{-3/\alpha} g_3\left(\left[(K/\alpha\eta)\left(1 - e^{-\alpha\eta t}\right)\right]^{-1/\alpha} \mathbf{v}; \alpha\right).$$

Within small times

$$f(\mathbf{v}, t) \sim (Kt)^{-3/\alpha} g_3((Kt)^{-1/\alpha} \mathbf{v}; \alpha),$$

and we deal with the Lévy motion, in large times limit the time dependence disappears and we come to stationary distribution over velocities

$$f(\mathbf{v}, \infty) = (K/\alpha\eta)^{-3/\alpha} g_3((K/\alpha\eta)^{-1/\alpha} \mathbf{v}; \alpha).$$

When $\alpha = 2$ it coincides with equilibrium Maxwell distribution, in other cases it essentially differs from equilibrium probability redistribution from intermediate area of velocities, producing power type asymptotic tails at small and large values:

$$f(\mathbf{v}, \infty) \propto |\mathbf{v}|^{-\alpha - d}, \quad |\mathbf{v}| \to \infty.$$

3.6   Radio-signals from pulsars

Known radio pulsars appear to emit short pulses of radio radiation with
pulse periods between 1.4 ms and 8.5 seconds. The word "pulsar" is a combination of "pulse" and "star", however actually pulsars are not pulsating
stars. Their radio emission is actually continuous but concentrated in a
narrow beam rotating together with the source (a spinning neutron star)
like a beacon beam scanning night space so a fixed observer sees only short
repeated flashes (Fig. 3.8). The intensity $I(t)$ would coincide in its shape
with the signal emitted by the pulsar in absence of the interstellar medium.
In fact, the shape of the observed pulses is changed due to broadening by
fluctuating electron density of the ISM. The deformation of this signal de-
livers information about the properties of the medium whose extraction is
one of the tasks of pulsar astronomy.

Fig. 3.8   Artwork of a pulsar. Radio and optical beams of radiation, emitted from the
pulsar's magnetic poles, flash across our line of sight and the star appears to blink on
and off as it spins. Credit: Mark Garlick / Science Photo Library.

Pulsars emit not only radio-wave but also optical, X-ray and gamma
radiation. Propagation of radiation through a turbulent medium can often
be described in terms of particles (photons) scattering theory and in terms
of wave scattering theory. The differential scattering cross-section $w(\Theta)$
per unit length and unit solid angle is connected to the spectral density
(Fourier transform) of the dielectric correlations function

$$\Phi_\varepsilon(\mathbf{q}) = \frac{1}{8\pi^3} \int_{\mathbb{R}^3} e^{-i\mathbf{q}\mathbf{x}} \langle [\varepsilon(\mathbf{x}_1) - \overline{\varepsilon}][\varepsilon(\mathbf{x}_1 + \mathbf{x}) - \overline{\varepsilon}] \rangle d\mathbf{x}.$$

In a statistically homogeneous and isotropic turbulent medium, the spectral
density depends only on the absolute value of the vector argument, $q = |\mathbf{q}|$,

and this link is expressed as

$$w(\Theta) = (1/2)\pi k_0^4 \Phi_\varepsilon(q),$$

where $q = 2k_0 \sin(\Theta/2)$, and k_0 is a wave number in the homogeneous medium with the dielectric constant $\bar{\varepsilon}$. According to Kolmogorov's two-thirds law,

$$\Phi_\varepsilon(q) = Cq^{-11/3}, \quad C = \text{const},$$

in the inertial interval of wave numbers (see Eq. (26.31) in the book [Rytov *et al.* (1978)]. This validates the small angle approximation with

$$w(\Theta) \approx A\Theta^{-\alpha-2}, \quad A = \text{const}, \quad \alpha = 5/3. \tag{3.12}$$

The more detailed description of the interstellar turbulence can be found in many works (see, e.g., [Xu and Zhang (2017)]). For illustrative aims, we restrict ourselves by using the simplest representation (3.12).

As was repeatedly noted in the first chapter, an undoubted sign of the applicability of a fractional calculus to description of transport processes is appearance of distributions with power tails (such as the Lévy distribution). In the problem of pulsars, such a step was taken in the work [Boldyrev and Gwinn (2003)]. But before discussing this point, we consider the multiple small-angle scattering from stationary source.

Let us assume that a one-directional point source is placed at origin and emits a photon along z-axis. In the small-angle approximation, z-coordinate of the photon is equated to its travel. For characterizing the deviation of the photon from the initial direction, the two-dimensional vector $\mathbf{u}$ is used. The kinetic equation for the angular distribution density $f(\mathbf{u}, z)$ reads

$$\frac{\partial f(\mathbf{u}, z)}{\partial z} = \int_{\mathbb{R}^2} [f(\mathbf{u} - \mathbf{u}', z) - f(\mathbf{u}, z)]w(|\mathbf{u}'|)d\mathbf{u}', \quad f(\mathbf{u}, 0) = \delta(\mathbf{u}). \tag{3.13}$$

Developing as series in $\mathbf{u}'$ and making integration (i.e., averaging over the scattering angle), one arrives at the diffusion equation for the main asymptotical part of the solution $f^{\text{as}}(\mathbf{u}, z)$:

$$\frac{\partial f^{\text{as}}(\mathbf{u}, z)}{\partial z} = \frac{\langle u^2 \rangle}{2} \Delta_{\mathbf{u}} f^{\text{as}}(\mathbf{u}, z),$$

where $\Delta_{\mathbf{u}} = \partial^2/\partial u_x^2 + \partial^2/\partial u_y^2$ and

$$\langle u^2 \rangle = \int_{\mathbb{R}^2} w(|\mathbf{u}|)u^2 d\mathbf{u} \tag{3.14}$$

is the mean square scattering angle per unit length.

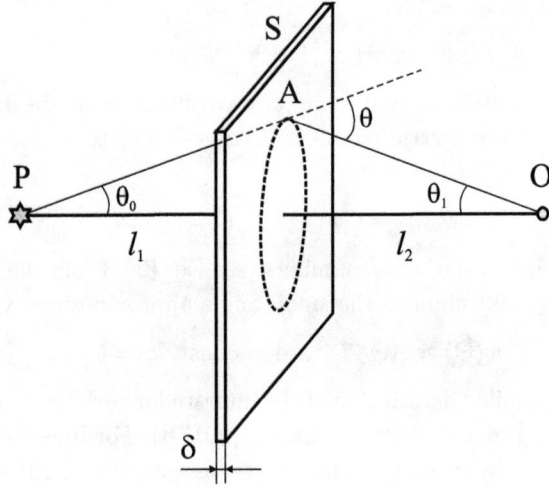

Fig. 3.9   A single-screen approximation.

In case $\alpha \in (0,1)$ integral (3.13) diverges and instead of expanding into series, we insert ansatz (3.12) directly into integral term in the right-hand side of Eq. (3.13):

$$\frac{\partial f^{\mathrm{as}}(\mathbf{u},z)}{\partial z} = AJ(\mathbf{u}), \quad J(\mathbf{u}) = \int_{\mathbb{R}^2} [f^{\mathrm{as}}(\mathbf{u}-\mathbf{u}',z) - f^{\mathrm{as}}(\mathbf{u},z)]|\mathbf{u}'|^{-\alpha-2}d\mathbf{u}'.$$

When $\alpha \geq 1$, the collision integral $J(\mathbf{u})$ diverges again and we have to use the Hadamard regularization procedure, replacing the divergent integral $J(\mathbf{u})$ by its finite (in Hadamard's sense) part:

$$J(\mathbf{u}) \to \mathrm{p.f.}\ J(\mathbf{u}) \equiv \int_{\mathbb{R}^2} [f^{\mathrm{as}}(\mathbf{u}-2\mathbf{u}',z)-2f^{\mathrm{as}}(\mathbf{u}-\mathbf{u}',z)+f^{\mathrm{as}}(\mathbf{u},z)]|\mathbf{u}'|^{-\alpha-2}d\mathbf{u}'$$

$$= -c(\alpha)(-\Delta_{\mathbf{u}})^{\alpha/2},$$

where

$$c(\alpha) = \frac{\pi^2(1-2^{1-\alpha})}{[\Gamma(1-\alpha/2)]^2 \sin(\alpha\pi/2)}.$$

Consequently, the asymptotical behavior of the angle distribution $f^{\mathrm{as}}(\mathbf{u},z)$ of photons traversed path z in a turbulent medium is described by the equation[1]

$$\frac{\partial f(\mathbf{u},z)}{\partial z} = -c(\alpha)A(-\Delta_2)^{\alpha/2}f(\mathbf{u},z), \quad f(\mathbf{u},0) = \delta(\mathbf{u}).$$

---

[1] We drop 'as' here and below for short.

The solution to this equation is expressed through the two-dimensional isotropic Lévy-Felgheim stable density with the characteristic exponent $\alpha = 5/3$

$$g_2(r;\alpha) = \frac{1}{2\pi} \int_0^\infty e^{-k^\alpha} J_0(kr)k\,dk \qquad (3.15)$$

by the relation

$$f(\mathbf{u}, z) = [c(\alpha)Az]^{-2/\alpha} g_2([c(\alpha)Az]^{-1/\alpha}|\mathbf{u}|; \alpha). \qquad (3.16)$$

Let us note three characteristic differences of scattering in a turbulent medium from analogous process in a medium with small-scale fluctuations of the refraction coefficient. The angle distribution width of the scattered photons grows with depth proportionally to $z^{3/5}$, but not to $z^{1/2}$, distribution tails have the power-law form but not the Gaussian one, and x- and y-projections of the vector $\mathbf{u}$ are not statistically independent anymore.

Let us come back now to the temporal broadening problem. The pulse shape $I(t)$ observed by means of the radio-telescope is a convolution of the intrinsic one with the impulse response of the scattering interstellar medium. The scattered radiation rays traverse a longer path than the rays coming directly from the pulsar to the observer [Lee and Jokipii (1975)]. The analytical form of the ISM response is determined by the distribution of scattering material between the pulsar and the observer. The case, when the material is concentrated in a region being thin in comparison to the pulsar-screen (PS) and screen-observer (SO) distances (Fig. 3.9) is normally considered in framework of *thin screen approximation* (see, for example, [Koay and Macquart (2015)]).

Ignoring perturbations of the instrumental origin reduces the problem of the intrinsic pulse $I_0(t)$ recovery from signal $I(t)$ to solving the convolution equation

$$I(t) = G * I_0(t) = \int_{-\infty}^t G(t - t')I_0(t')dt' \qquad (3.17)$$

with respect to $I_0(t)$. The kernel $G(t - t')$ (the *memory kernel*) describes the delay effect caused by the passage time increase due to the scattering.

We restrict ourselves by the model introduced in [Boldyrev and Gwinn (2003)] in framework of which (a) the continuous ISM is replaced be a thin turbulent layer and (b) the rotating stationary source is replaced by an instantaneous mono-directed point source. Its peculiarity is that 'thin' is understood only in the geometrical sense (compared with the distances) whereas the layer thickness δ is assumed to be much longer than the typical distance between successive events of scattering (Fig. 3.9). This condition looking as a quite natural one allows us to use for the angular distribution of photons scattered in the screen the asymptotical expression obtained above in the framework of the fractional approach. Let us turn to Fig. 3.9. The total time of passage of photon along PAO

$$t = \frac{1}{c}\left(\sqrt{l_0^2 + r^2} + \sqrt{l_1^2 + r^2}\right) \approx \frac{L}{c} + \frac{r^2}{2cl}, \qquad (3.18)$$

where $L = l_0 + l_1$ is the shortest distance between P and O, $l = (l_0 + l_1)/(l_0 l_1)$ is a reduced distance[2]. Designating the random passage time by T, the scattering angle by Θ and its value for the trace PAO by Θ_t, one can write the obvious equality

$$\mathsf{P}(T > t) = \mathsf{P}(\Theta > \Theta_t) = \mathsf{P}(\theta_0 + \theta_1 > \Theta_t). \qquad (3.19)$$

Inserting (3.18) into the left-hand side of Eq. (3.19) yields

$$\mathsf{P}(T - L/c > \tau) = \mathsf{P}(\Theta > r/l).$$

Observe that the last term in (3.18) presents the delay time for the trace PAO,

$$\tau = \frac{r^2}{2cl},$$

whereas

$$\mathsf{P}(\Theta > \theta) = 2\pi \int_{r/l}^{\infty} f(\theta; \delta)\theta d\theta = 2\pi \int_{\sqrt{2c\tau/l}}^{\infty} f(\theta; \delta)\theta d\theta$$

(recall that $2\pi \int_0^\infty f(\theta; \delta)\theta d\theta = 1$). As a result, pdf of the delay time is written in the form

$$G(\tau) = -2\pi \frac{d}{d\tau} \int_{\sqrt{2c\tau/l}}^{\infty} f(\theta; \delta)\theta d\theta = \frac{\pi}{\tau_0} f\left(\sqrt{\frac{\tau}{\tau_0}}; \delta\right),$$

where $\tau_0 = l/(2c)$.

In case of a homogeneous screen, passage of photons in which is described by the normal model of angular diffusion,

$$f(\theta; \delta) = \frac{1}{4\pi\sigma^2\delta} e^{-\theta^2/4\sigma^2\delta},$$

---

[2]In the small-angle approximation the time of penetration through the thin turbulent screen is also negligible compared with the total time.

and consequently

$$G(\tau) = \frac{1}{\tau_*} e^{-\tau/\tau_*}, \ \tau_* = (4\sigma\delta)\tau_0 > 0.$$

This is a known result (see formula (15) in [Cordes and Lazio (2001)].

However, in case of turbulent screen, the angular diffusion in which is described, say, by two-dimensional Cauchy motion,

$$f(\theta; \delta) = \frac{\delta}{2\pi(\delta^2 + \theta^2)^{3/2}}, \tag{3.20}$$

we obtain

$$G(\tau) = \frac{\delta}{2\tau_0(\delta^2 + \tau/\tau_0)^{3/2}}. \tag{3.21}$$

Observe that function (3.20) obeys the fractional equation with $\alpha = 1$. Taking solutions with other possible values $\alpha \in (0, 2)$, we obtain the whole family of various transfer functions G.

Once again, emphasize that assumptions (a) and (b) made within the framework of this model are inadequate to the real picture and should be revised, so that the conclusions on G should be considered only as qualitative. For better approximation of the signal at short times, one should come back to Eq. (3.17). Some of authors (for example, [McKinnon (2014)]) approximate the intrinsic pulse by Gaussian shape. This approach can effectively be used for simulation of intergalactic refractive scintillation effects of distant sources, say quasars [Pallottini *et al.* (2013)].

3.7 Astrophotography and image processing

The aim of astrophotography is to obtain as much detail as possible on a targeted cosmic image. Long exposure astrophotography required for achieving this is vulnerable to various types of atmospheric noise. Removing this noise depletes the image detail and in consequence objects appear fainter and weaker in contrast. One of the most important problems is the edge detection of an image. Ordinary methods are based on using expressions containing the first-order derivatives. The second-order derivatives applications lead to the Laplacian calculations. There exists some experience in using dipole and quadrupole moments of the image maps. Fractional Fourier transformations are used for this aim as well.

As proposed in [You *et al.* (1997)], fractional differentiation can help to scan and examine an image to detect the faint, weaker edges and details to enhance and display them. In outline, the algorithm looks as follows. Let

$f(x, y)$ be a two-dimensional map requiring such processing. The partial fractional derivatives ${}_0D_x^\nu f(x, y)$ and ${}_0D_y^\nu f(x, y)$ are numerically calculated and the fractional gradient is defined as

$$\mathbf{G}^\nu \equiv \nabla^\nu f(x, y) = \mathbf{e}_x \, {}_0D_x^\nu f(x, y) + \mathbf{e}_y \, {}_0D_y^\nu f(x, y).$$

In image processing, the magnitude of the gradient $G^\nu(x, y.c) = \sqrt{[G_x^\nu(x, y, c)]^2 + [G_y^\nu(x, y, c)]^2}$ is evaluated on the function $f(x, y)$ given by the image map $b(x, y, c)$ for each color tone c. For each color, the maximum value on the image map is defined and then the output map is determined as

$$b_G(x, y, c) = N \left(\frac{G^\nu(x, y, c)}{G_{\max}^\nu(c)} \right)^\alpha$$

where $N(= 255)$ is the total number of color tones and α is a parameter suitable to adjust the image visibility.

Authors [Sparavigna and Milligan (2009)] explain the role of fractional differentiation in the following way. "We observe a galaxy and stars in the background and in front of it. We evaluated the fractional gradient and obtained images with different values of the fractional parameter ν. The value of α is fixed at 0.5. The map, obtained for $\nu = 1$, which is working as the usual edge detector based on the ordinary gradient, almost removes the galaxy and puts in evidence the stars. The galaxy regains its visibility, when parameter ν approaches the zero value. In this case we have the original image with a resulting appearance according to α parameter."

The pictures shown in [Mathieu *et al.* (2003); Marazzato and Sparavigna (2009); Sparavigna and Milligan (2009)] look very convincing and the method itself seems handy enough.

In a deeper statement of the problem it becomes the problem of transfer of the optical image through a scattering turbulent medium and, in principle, an inverse transfer problem. This direction is developed with the use of fractional differential equations in the works of Blackledge [Blackledge (2007, 2009); Blackledge and Blackledge (2010)], where the process of image forming is considered in terms of wave scattering theory.

Applying the standard technique of transformation, one can easily arrive at the diffusion equation for the intensity of light carrying an image along z-axis in a medium with weak small-scale fluctuations of the refraction coefficient:

$$\frac{\partial I(x, y; z)}{\partial z} = D \left[\frac{\partial^2}{\partial x^2} + \frac{\partial^2}{\partial y^2} \right] I(x, y; z) \tag{3.22}$$

subjected to the initial condition $I(x, y; 0) = I_0(x, y)$. The resulting image at time t is represented by the two-dimensional convolution of

$$I(x, y; z) = \frac{1}{4\pi Dz} \int\limits_{-\infty}^{\infty} \int\limits_{-\infty}^{\infty} \exp\left[-\left(\frac{(x-x')^2 + (y-y')^2}{4Dz}\right)\right] I_0(x', y') dx' dy'$$

(see [Blackledge (2007); Blackledge and Blackledge (2010)]).

Comparing two types of equations, the diffusion equation

$$\left\{\frac{1}{D}\frac{\partial}{\partial z} - \frac{\partial^2}{\partial x^2} - \frac{\partial^2}{\partial y^2}\right\} I_1(x, y; z) = 0$$

and the wave equation

$$\left\{\frac{1}{D^2}\frac{\partial^2}{\partial z^2} - \frac{\partial^2}{\partial x^2} - \frac{\partial^2}{\partial y^2}\right\} I_2(x, y; z) = 0$$

Blackledge [Blackledge (2009)] has suggested to consider the time-fractional equation

$$\left\{\frac{1}{D^\nu}\frac{\partial^\nu}{\partial z^\nu} - \frac{\partial^2}{\partial x^2} - \frac{\partial^2}{\partial y^2}\right\} I_\nu(x, y; z) = 0 \tag{3.23}$$

as an intermediate case between diffusion and wave mechanisms of image blurring with $1 < \nu < 2$. Involving the fractional calculus is justified in terms of the generalization of a random walk model. He gives an example of application including image enhancement of star fields and other cosmic bodies imaged through interstellar dust clouds. Giving the physical validation to this approach, [Blackledge (2009)] writes:

"Fractional diffusion models apply to scattering processes that occur in a tenuous and extremely rarefied medium. In applied optics, one of the most common examples of this phenomenon occurs in astronomy and the processes associated with light scattering from cosmic dust which is composed of particles which are a few molecules to the order of 10^{-4} metres in size."

Using this approach, Blackledge has elaborated the way to solve the inverse fractional diffusion problem called *fractional de-diffusion*: finding the initial image $I(\mathbf{x}, 0)$ from the observed data $I(\mathbf{x}, t)$ [Blackledge (2007)]. He shows in [Blackledge (2009)] impressive photos demonstrating high efficiency of the method: the image of Saturn observed by a ground based telescope, and the image of a dust clouded star field in the constellation of Pegasus.

Chapter 4

Solar system scales

In this chapter, we consider applications related to the Solar system. In section 4.1, anomalous diffusion regimes in dynamics of solar flares and magnetic bright points on the solar surface are discussed. Observation of transient subdiffusion allows us to introduce diffusion equation containing tempered fractional derivatives, solutions of which are expressed through inverse tempered Lévy subordinator. In section 4.2, we consider a simple phenomenological mechanism of diffusion in a granular medium of solar surface leading to tempered subdiffusion of a tracer. Fractional kinetics of solar wind and solar cosmic rays in interplanetary medium is considered in sections 4.3 and 4.4. Recent simulation results on particle transport in turbulent interplanetary magnetic field are analyzed in section 4.5. Based on this analysis, we introduce the one-dimensional tempered Lévy walk model to describe parallel transport of charged particles in turbulent magnetic field.

4.1 Subdiffusion in photosphere

The heliospheric magnetic field reveals a complex behavior and for this reason [Leighton (1964)] suggested to describe migration of magnetic elements (magnetic bright points) over the solar surface in terms of random walk model similar to the Brownian motion. However, in spite of the similarity, this motion reveals quantitative difference; the width of the diffusion packet grows slower than in case of ordinary diffusion. Such subdiffusion regime is explained by the observation data showing that the motion is not continuous: it is interrupted on random periods (waiting times). From theoretical point of view, this model is in agreement with the assumption that surface diffusion is carried out by motions in a randomly renewing field of supergranules (Fig. 4.1) so the magnetic elements are carried passively to the boundary of a supergranule by its radial flow and remain there until the global supergranular flow field is renewed. The resulting diffusivity is determined by the size and lifetime of the supergranules. Later, observations of [Hagenaar *et al.* (1999)] supported that the process has rather subdiffusion character. The cause of distinction of this regime from the expected Brownian motion is too long time spend in traps: it is distributed according to inverse power law with pdf $q(t) \propto t^{-\beta-1}$ with $\beta \in (0,1)$. Cadavid et al. [Cadavid *et al.* (1999)] found that the transport is subdiffusive for times less than 20 minutes but normal for times larger than 25 minutes. They explained such turnover by presence of trapping time truncation. We bring below description of this process in terms of equations with fractional derivatives.

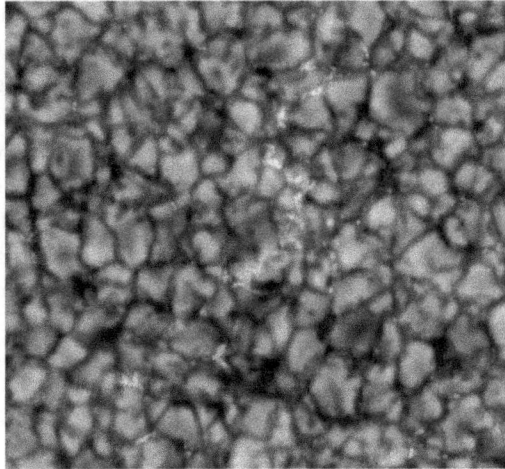

Fig. 4.1 The image from the Solar Optical Telescope showing a greatly magnified portion of the solar surface. Image credit: Hinode JAXA/NASA/PPARC.

[Abramenko *et al.* (2011)] carried out an analysis of observations of solar granulation provided by the New Solar Telescope of Big Bear Solar Observatory. They made conclusions that motion of bright points is superdiffusive. They considered random walk of BP in various regions of the solar surface: a quiet-sun area, a coronal hole, and an active region plage. Calculations of time-dependent mean-squared displacements on the base of observable data provide the following superdiffusion exponents: $\gamma \approx 1.48$ for the plage area, $\gamma \approx 1.53$ for the quiet-sun area, and $\gamma = 1.67$ for the coronal hole.

So, there are various diffusion regimes observed for motion of BPs. [Giannattasio *et al.* (2013)] note: "Several works investigated the diffusion of magnetic fields on the photosphere, and the results are often conflicting. Most studies rely on magnetic proxies like G-band bright points to track the magnetic field advection (see, e.g., [Berger *et al.* (1998); Cadavid *et al.* (1998, 1999); Lawrence *et al.* (2001); Almeida *et al.* (2010); Abramenko *et al.* (2011)]). In contrast, [Wang (1988)], [Hagenaar *et al.* (1999)], [Sainz *et al.* (2011)] used magnetograms to describe the diffusion process. ... None of those works, however, could take advantage of a data set spanning all of the spatial (from subgranular to supergranular) and temporal (from a few seconds to days) scales involved in the magnetic diffusion process."

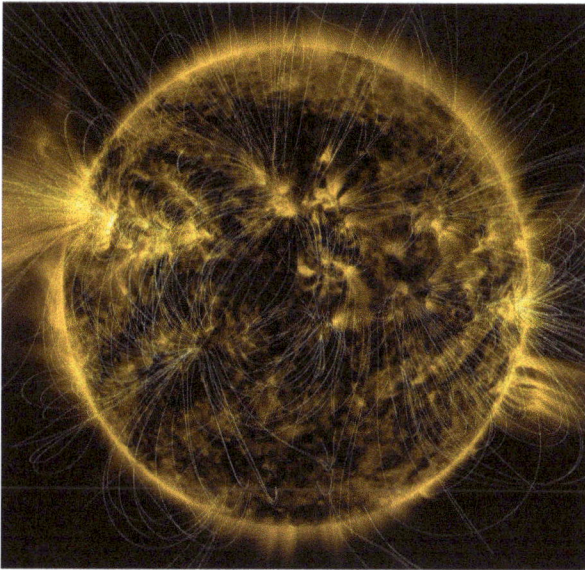

Fig. 4.2 The magnetic map of the Sun obtained using the Potential Field Source Surface model. This model based on magnetic measurements of the solar surface. Credits: NASA/SDO/AIA/LMSAL.

The subdiffusive regime dominates also in the spatio-temporal evolution of solar flares which is closely related to the structure of the magnetic field on the solar surface (Fig. 4.2). [Aschwanden (2012)] studied this evolution by fitting radial expansion function $r(t)$. He analyzed 155 events of M- and X-class flares observed with Geostationary Operational Environmental Satellite and Solar Dynamics Observatory, and found that evolution consists of two phases, an exponentially growing acceleration phase and a deceleration phase. The latter is described by the expansion function $r(t) \sim \kappa(t - t_1)^{\beta/2}$. In analyzed events, different regimes were observed. They include the logistic growth limit ($\beta = 0$), subdiffusion ($0 < \beta < 1$), classical diffusion ($\beta = 1$), superdiffusion ($1 < \beta < 2$), and the ballistic expansion ($\beta = 2$). Evolution of most flares is subdiffusive, $\beta = 0.53 \pm 0.27$ [Aschwanden (2012)]. The author explains subdiffusive behavior by "the anisotropic propagation of energy release in a magnetically dominated plasma" and the logistic growth by reaching the boundaries of energy release regions.

As was mentioned above, for random motion of BPs in the intergranular lanes at the solar photosphere, authors [Cadavid *et al.* (1999)] indicate a crossover from a subdiffusive behavior to normal diffusion at large times. In papers [Stanislavsky *et al.* (2008); Gajda and Magdziarz (2010); Sibatov and Uchaikin (2011)], the approach based on the tempered fractional calculus [Baeumer and Meerschaert (2010)] was developed to describe such crossover.

The Laplace image for the probability density function of a tempered stable variable [Rosiński (2007); Baeumer and Meerschaert (2010)] has the form

$$\tilde{g}^{(\beta,\gamma)}(s) = \exp[\gamma^\beta - (s + \gamma)^\beta].$$

The diffusion equation for the tempered subdiffusion can be derived by substitution of waiting time pdf $q(t)$ with exponentially truncated power law tails into the Montroll-Weiss formula. We use the form of a tempered fractional exponent

$$q(t) = \mu e^{-\gamma t} t^{\beta-1} E_{\beta,\beta}\left(-(\mu - \gamma^\beta)t^\beta \right) \tag{4.1}$$

having the following Laplace transform

$$\tilde{q}(s) = \frac{\mu}{\mu - \gamma^\beta + (s + \gamma)^\beta}.$$

We arrive at the tempered subdiffusion equation

$$s\tilde{\rho}(k,s) + C\mu \frac{s}{(s+\gamma)^\beta - \gamma^\beta}|\mathbf{k}|^2 \tilde{\rho}(k,s) = 1, \tag{4.2}$$

which is a particular case of equation obtained in [Gajda and Magdziarz (2010)].

The Laplace transform of the time-dependent variance is

$$\langle \mathbf{x}^2(s) \rangle = \frac{2C\mu}{s[(s+\gamma)^\beta - \gamma^\beta]}.$$

For asymptotically large times ($t \to \infty$, $s \to 0$), the law of diffusion packet spreading is normal

$$\langle \mathbf{x}^2(s) \rangle \sim \frac{2C}{\beta}\gamma^{1-\beta}s^{-2}, \quad \langle \mathbf{x}^2(t) \rangle \sim \frac{2C}{\beta}\gamma^{1-\beta}t.$$

4.2 Diffusion on a granular surface

In Fig. 4.1, the image from the Solar Optical Telescope showing granular structure of the solar surface is presented. Energy from below the surface is transported by convection and results in the convection cells, or granulation, seen in this image. The lighter areas correspond to hot gases rising from below, while the dark "intergranular boundaries" correspond to cooler gases sinking under the surface. As is known, in a granular speckle subdiffusion is often observed [Volpe *et al.* (2014)] at large fluctuations in potential energy landscape. Such behavior can be described as the Brownian motion in a nonhomogeneous (granular) medium.

Diffusion in granular systems can be approximately described by the grain-boundary diffusion. Many models of grain-boundary diffusion are based on the classical one proposed by [Fisher (1951)]. He considered the system with an isolated grain boundary (Fig. 4.3), which is a uniform and isotropic layer of a thickness δ situated in a semi-infinite medium perpendicular to the edge. The diffusion coefficient D_b along grain boundaries is usually considered to be significantly higher than the diffusion coefficient D_v in grains. Impurity atoms initially deposited on the edge of the semi-infinite medium diffuse along boundary and in grain at a constant temperature T of annealing during time t [Popov (2008)]. Impurity spreading in this system is usually described by a set of Fick diffusion equations. The accurate solutions for the Fisher model with a constant and an instant source were derived by Whipple and Suzuoka [Whipple (1954); Suzuoka (1961)]. Levin and MacCallum [Levine and MacCallum (1960)] showed that dependence of average concentration $\ln \bar{C}(z,t)$ on $z^{6/5}$ is almost linear for z large enough [Kaur and Gust (1989)].

The Fisher model can be considered in terms of the single particle tracking method. The displacement of a particle along z depends on the total time that particle spends in boundary and grain. So, knowledge about distributions of these times allows us to find pdf of z. We simplify the situation by considering only the instantaneous source in the boundary area at $z = 0$. Let $\tilde{p}_b(s_b|s)$ be the double Laplace transform of the probability density function of total sojourn time in boundary $p_b(t_b|t)$. Considering grain-boundary diffusion as a dichotomic renewal process one can write [Sibatov and Uchaikin (2010)]:

$$\tilde{p}_b(s_b|s) = \frac{1}{1 - \tilde{\psi}_b(s + s_b)\tilde{\psi}_v(s)} \frac{1 - \tilde{\psi}_b(s + s_b)}{s + s_b}$$

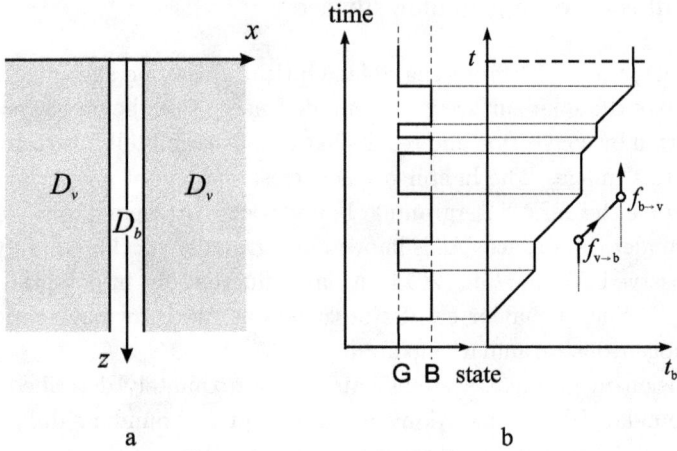

Fig. 4.3 Fisher model of grain-boundary diffusion with a single isolated boundary (left panel) and representation of the particle random walk in a granular media as a binary renewal process with two states, V (volume or grain) and B (boundary) (right panel).

$$+\frac{\tilde{\psi}_b(s+s_b)}{1-\tilde{\psi}_b(s+s_b)\tilde{\psi}_v(s)}\frac{1-\tilde{\psi}_v(s)}{s} \tag{4.3}$$

where $\psi_b(t)$ and $\psi_v(t)$ are pdf's of sojourn times in boundary and volume of grains, respectively, $\tilde{\psi}_b(s)$ and $\tilde{\psi}_v(s)$ are their Laplace transforms. We take Laplace images of waiting time pdf's in the form

$$\tilde{\psi}_b(s) \sim 1-\varepsilon\frac{\delta}{D_\perp}s, \quad \tilde{\psi}_v(s) \sim 1-\varepsilon\sqrt{\frac{2}{D_v}}\sqrt{s}, \quad \varepsilon \to 0. \tag{4.4}$$

They lead to the following expression

$$\tilde{p}_b(t_b|s) = \left(1+\frac{D_\perp\sqrt{2}}{\delta\sqrt{D_v}}s^{-1/2}\right)\exp\left\{-t_b\left(s+\frac{D_\perp\sqrt{2}}{\delta\sqrt{D_v}}s^{1/2}\right)\right\}.$$

Inverse Laplace transformation gives

$$p_b(t_b|t) = \frac{D_\perp}{\sqrt{2\pi D_v}\delta}\frac{2t-t_b}{t^{3/2}}\exp\left(-\frac{D_\perp^2 t_b^2}{2D_v\delta^2(t-t_b)}\right)$$

and for pdf of displacement along z, we have

$$p(z,t) = \int_0^t \left[p_b(z|t_b)+p_v(z|t-t_b)\right]p_b(t_b|t)dt_b.$$

For the time range

$$\bar{t}_b \ll t \ll \frac{D_b}{D_v}\bar{t}_b \quad \Rightarrow \quad \frac{2D_v\delta^2}{\pi D_\perp^2} \ll t \ll \frac{2D_b\delta^2}{\pi D_\perp^2}, \tag{4.5}$$

the solution takes the form

$$p(z,t) = \int_0^t \frac{\exp\left(-\frac{z^2}{4D_b t_b}\right)}{\sqrt{\pi D_b t_b}} \frac{2D_\perp}{\delta\sqrt{2\pi D_v t}} \exp\left(-\frac{D_\perp^2 t_b^2}{2D_v \delta^2 t}\right) dt_b \qquad (4.6)$$

which obeys the fractional equation

$$_0D_t^{1/2}\bar{p}(z,t) = \frac{D_b\delta}{D_\perp}\sqrt{\frac{D_v}{2}} \frac{\partial^2}{\partial z^2}\bar{p}(z,t) + \frac{\delta(z)}{\sqrt{\pi t}}. \qquad (4.7)$$

The variance of displacement parallel to boundary has the form

$$\langle z^2 \rangle = \frac{D_b\delta}{D_\perp}\sqrt{\frac{2D_v t}{\pi}}, \qquad \frac{2D_v\delta^2}{\pi D_\perp^2} \ll t \ll \frac{2D_b\delta^2}{\pi D_\perp^2}.$$

Solution (4.6) can be presented in the following form

$$p(\eta) = \int_0^\tau \sqrt{\frac{2}{\pi}} \frac{e^{-\xi^2/2 - \eta^2/4\xi}}{\sqrt{\pi\xi}} d\xi, \qquad \tau = \frac{D_\perp}{\delta}\sqrt{\frac{t}{D_v}}, \qquad \eta = z\sqrt{\frac{D_\perp}{D_b\delta\sqrt{D_v t}}}. \qquad (4.8)$$

For times large enough, distribution becomes universal and has an asymptotic behavior close to $\exp(-\eta^{6/5})$ as solutions obtained by Whipple [Whipple (1954)] and Suzuoka [Suzuoka (1961)].

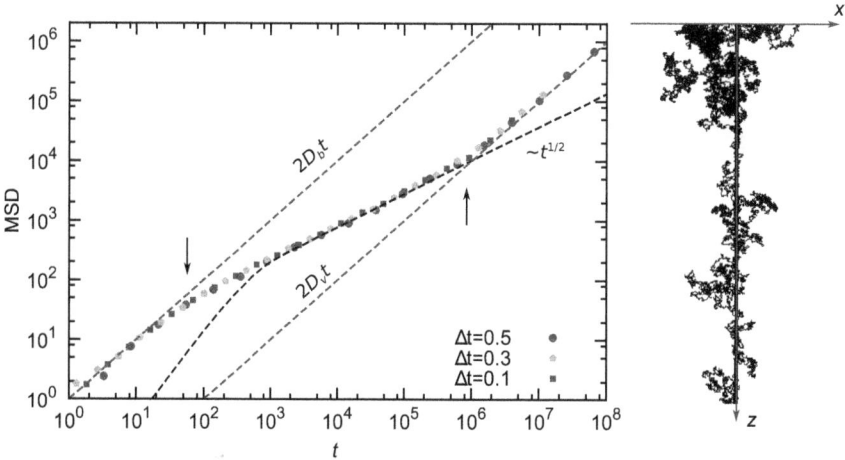

Fig. 4.4   The mean square displacement along z (left panel). Parameters: $D_v = 0.01$, $D_\perp = D_v$, $D_b = 1.0$. Dashed line corresponds to Eq. (4.6). Right panel: typical trajectory of the process.

In case of the generalized models, random sojourn times in grains have a distribution with a truncated power law tail. For example, sojourn time in the layer has the following density [Borodin and Salminen (2012)]

$$\psi_{\mathrm{v}}(t) = D_{\mathrm{v}}[ss_{(\varepsilon,d)}(D_{\mathrm{v}}t) + ss_{(d-\varepsilon,d)}(D_{\mathrm{v}}t)],$$

where $ss_{(\varepsilon,d)}(\tau)$ is a special function defined by

$$ss_{(\varepsilon,d)}(\tau) = \sum_{k=-\infty}^{\infty} \frac{d-\varepsilon+2kd}{\sqrt{2\pi\tau^3}} \exp\left\{-\frac{(d-\varepsilon+2kd)^2}{2\tau}\right\}.$$

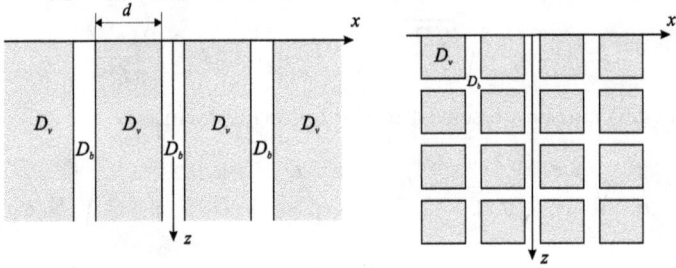

Fig. 4.5 Models of grain-boundary diffusion.

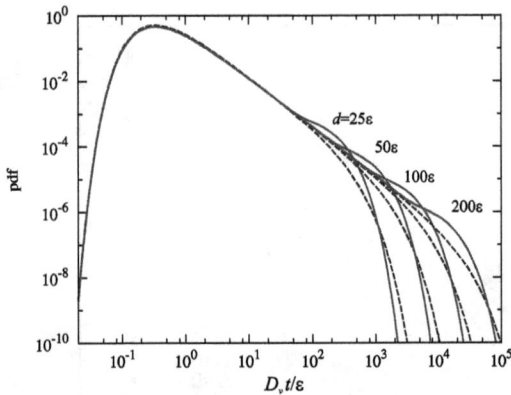

Fig. 4.6 Comparison of sojourn time pdf (red curves) for a Brownian particle in a layer of thickness d with the tempered Lévy-Smirnov stable density (blue dashed lines) (ε is an initial position of a particle in a layer).

Laplace transform of the tempered stable density has the form

$$\tilde{\psi}_{\mathrm{v}}(s) = \exp\left\{-\varepsilon\left[\sqrt{\frac{2}{D_{\mathrm{v}}}\left(s+\frac{2D_{\mathrm{v}}}{d^2}\right)} - \frac{2}{d}\right]\right\}. \tag{4.9}$$

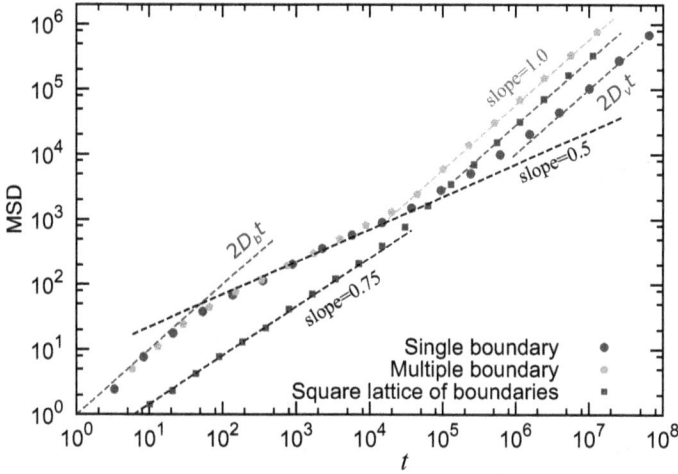

Fig. 4.7 Mean square displacement for three models of grain-boundary diffusion ($D_v = 0.01$, $D_\perp = D_v$, $D_b = 1.0$, $\Delta t = 0.3$).

The original function is a tempered Lévy-Smirnov pdf,

$$\tilde{\psi}_v(t) = \frac{\varepsilon}{\sqrt{2\pi D_v t^3}} \exp\left\{ -\frac{(2D_v t - \varepsilon d)^2}{2d^2 D_v t} \right\}.$$

For function (4.9), the following asymptotic representation can be used

$$\tilde{\psi}_v(s) \sim 1 - \varepsilon \left[\sqrt{\frac{2}{D_v}\left(s + \frac{2D_v}{d^2}\right)} - \frac{2}{d} \right], \quad \varepsilon \to 0. \tag{4.10}$$

Substituting the latter function and $\tilde{\psi}_b(s)$ from (4.4) into expression (4.3), we obtain

$$\tilde{p}_b(s_b|s) \sim \frac{\frac{\delta}{D_\perp} + \frac{1}{s}\left[\sqrt{\frac{2}{D_v}\left(s + \frac{2D_v}{d^2}\right)} - \frac{2}{d} \right]}{\frac{\delta}{D_\perp}(s_b + s) + \left[\sqrt{\frac{2}{D_v}\left(s + \frac{2D_v}{d^2}\right)} - \frac{2}{d} \right]}, \quad \varepsilon \to 0.$$

For times $t \gg D_v \delta^2 / D_\perp^2$, function $\tilde{p}_b(t_b|s)$ can be presented in the form

$$\tilde{p}_b(t_b|s) = \frac{D_\perp}{s\,\delta}\left[\sqrt{\frac{2}{D_v}\left(s + \frac{2D_v}{d^2}\right)} - \frac{2}{d} \right] \exp\left\{ -t_b \frac{D_\perp}{\delta}\left[\sqrt{\frac{2}{D_v}\left(s + \frac{2D_v}{d^2}\right)} - \frac{2}{d} \right] \right\}.$$

The inverse Laplace transformation gives us the inverse tempered stable subordinator with $\beta = 1/2$ (see [Stanislavsky *et al.* (2008); Gajda and Magdziarz (2010)]). So, for systems presented in Figs. 4.5 with condition $D_v \ll D_b$, diffusion can be described in terms of tempered subdiffusion for

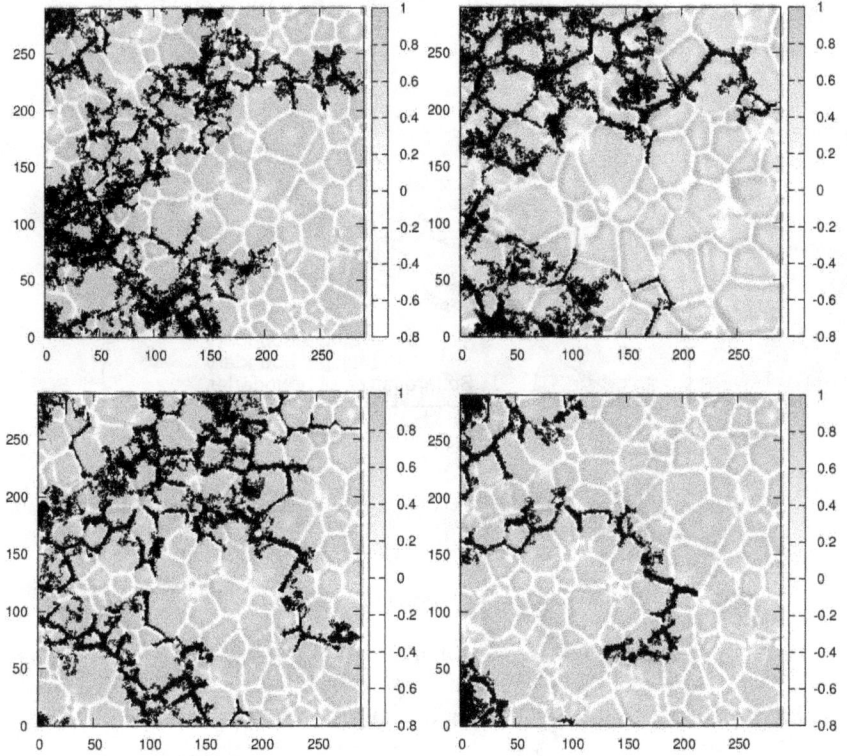

Fig. 4.8   Trajectories of surface diffusion in a granular medium.

times $t \gg D_{\mathrm{v}}\delta^2/D_{\perp}^2$. The normal diffusion is observed when $t \gg d^2/2D_{\mathrm{v}}$. Analogous reasoning can be done for sojourn times in grains of squareness or some another shape.

Mean square displacements calculated in [Sibatov and Svetukhin (2017)] for three models of grain-boundary diffusion (single boundary, set of parallel boundaries, and square lattice) are presented in Fig. 4.7. In case of multiple parallel boundaries, slope β of the transient subdiffusion law $\langle \Delta x \rangle \propto t^\beta$ equal to $1/2$, but transition to normal diffusion occurs earlier than in the case of single boundary. For square lattice of boundaries, β differs from $1/2$, and equal to 0.75, i.e. the subdiffusion exponent is defined not only by distribution of waiting times in grains, the geometry of grains and boundaries plays an important role as well. For more complex geometries, β can have arbitrary values form $(0, 1]$. Monte Carlo trajectories of surface diffusion in a complex granular medium are presented in Fig. 4.8.

4.3 Solar flares and solar wind

A solar flare is a short-lived sudden increase in the intensity of radiation emitted in the neighborhood of sunspots. Solar flares produce electromagnetic radiation across the electromagnetic spectrum at all wavelengths from long-wave radio to the shortest wavelength gamma rays. In the strong turbulized plasma of the solar atmosphere, irregular electric and magnetic fields lead to stochastic acceleration such as Fermi acceleration. Some of magnetic field lines near the acceleration region can have open configuration providing possibility to inject particles into interplanetary space. Being on a closed line, particles are trapped and their energy continuously decreases.

There are three ways of further progression of events for a particle in a magnetic trap: due to drift and scattering particle can get on an open field line and leave the area of solar flare, due to scattering in dense atmospheric layers, particle loses its energy and due to passage from one loop to another until loss of all energy or achieving an open line [Panasyuk *et al.* (2006)].

The understanding of long-term solar variability and predicting the solar activity is an actual problem for solar physics. One of the important models for simulation of these processes became non-Gaussian self-similar processes with stationary increments. The Lévy-stable processes are typical representative of this family. In particular, the periods of the Sun's irradiance influence the Earth temperature fluctuations and can be modeled with the help of the Lévy-walk statistics in the work [Burnecki *et al.* (2008)]. In this paper, the authors introduced a long-range dependent continuous FARIMA (CFARIMA) model with α-stable noise, which provided a new efficient tool for the study of the solar flare phenomenon in the fractional Langevin equation framework:

$$dX_t = b(t, X_t)dt + \sigma_t(t, X_t)dL_t^\alpha, \qquad (4.11)$$

where dL_t^α stands for the increments of Lévy α-stable motion L_t^α, $0 < \alpha \le 2$. Fractional Lévy-stable motion is described in the form of a stochastic integral. A connection with long memory or long-range dependence is also discussed and we introduce the codifference as a proper measure of long-range dependence. The obtained results are applied to determine the basic features of an empirical data series describing the energy of solar flares.

[Bian and Browning (2008)] proposed the superdiffusive model of acceleration of charged particles during solar flares. Acceleration occurs by the strong electric fields arising due to magnetic reconnection. The general point of the model is that the fluctuating electric field in turbulent

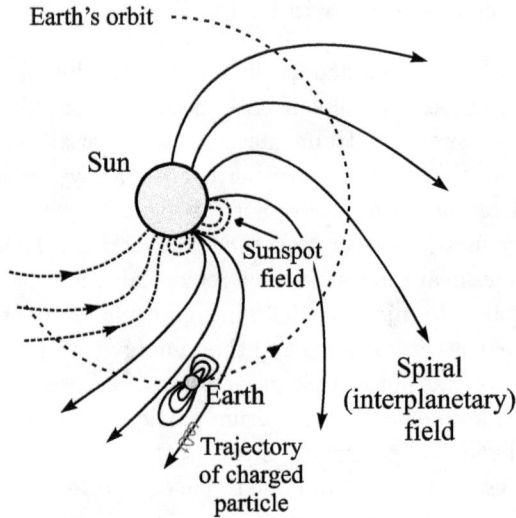

Fig. 4.9   Interplanetary magnetic field influenced by the solar rotation and solar wind (according to [Vallée (1998)]).

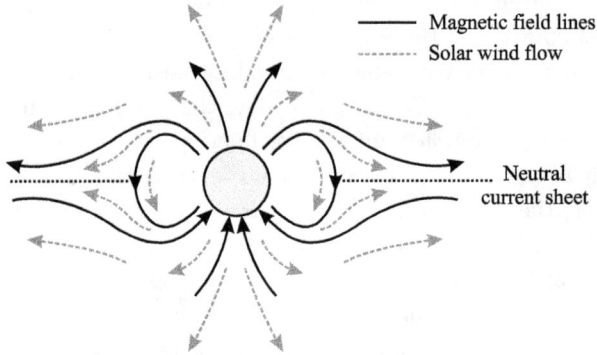

Fig. 4.10   Diagram of interplanetary magnetic field seen along the solar equatorial plane.

reconnecting plasma is strongly localized and its random magnitude has distribution of the power law type. The acceleration is described in terms of the fractional diffusion equation in velocity space. The model predicts the power law spectrum of injected particles [Bian and Browning (2008)].

Solar wind takes away matter from the upper atmosphere of the Sun (coronal matter). It consists of mostly electrons, protons and alpha particles with energies usually between 1.5 and 10 keV. Due to solar wind plasma is fully ionized, the interplanetary magnetic field (IPMF) is embedded in it. IPMF plays a major role in physical processes in solar wind, it governs the propagation of cosmic rays in the interplanetary space. The structure of

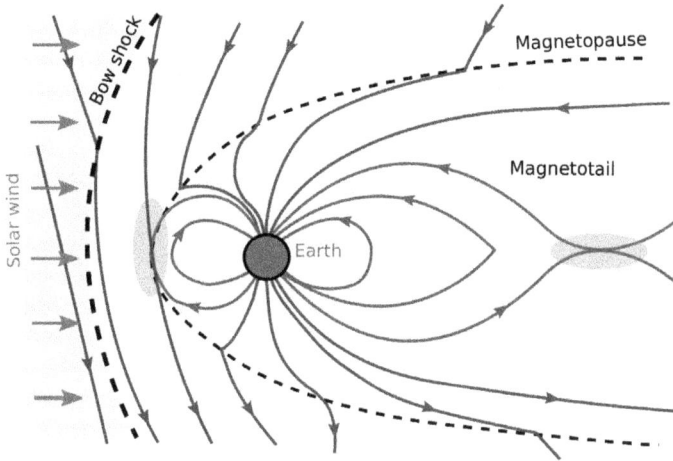

Fig. 4.11 Earth's magnetosphere.

IPMF is determined by magnetic field freezing-in solar wind with base of lines on the solar surface. In the frame of reference rotating with the Sun, trajectories of solar wind particles are congruent with IPMF lines (Figs. 4.9, 4.10 and 4.11).

For a system exhibiting self-similarity, the probability distribution function of a variable $y(t)$ can be expressed as

$$F(y,t)dy = F(y/t^\nu)dy/t^\nu = F(\xi)d\xi \quad (\xi = y/t^\nu).\tag{4.12}$$

The exponent ν is to be determined from the dependence of the first moment on t as

$$\langle |y| \rangle = \int |y| F(y,t)dy = \text{const} \cdot t^\nu,\tag{4.13}$$

with the symmetric $F(y,t)$ and the normalization condition. The crucial feature of the probability distribution function $F(y,t)$ is the existence or nonexistence of the stationary distribution $F(y)$ for $t \to \infty$. In many physical problems there is no stationary distribution and the phase space of the problem has at least one unbounded variable. The plasma density fluctuations in laboratory fusion experiments (tokamaks) have been modeled in terms of a fractional kinetic equation. Another example is the multiscale nature of the magnetosphere where the correlated data of the solar wind – magnetosphere system shows clear non-Gaussian distribution. Zaslavsky [Zaslavsky (1994)] proposed a phenomenological approach to the nonstationary case of the evolution of $F(y, t)$ in the presence of multiscale processes.

The main idea is to write down a balance equation

$$\delta_t F(y, t) = \overline{\delta_y F(y, t)}, \tag{4.14}$$

where the bar implies averaging over all admissible paths. This equation can be written as

$$\delta_t F(y, t) = \Delta_t^\beta F(y, t), \tag{4.15}$$

where Δ_t^β is a generalized difference operator for a time shift by Δt parametrized by β. Similarly

$$\delta_y F(y, t) = \sum_{\Delta y} [A(\Delta y, y) F(y, t)], \tag{4.16}$$

where Δ_y^α is a generalized difference operator for a y-variable shift by Δy along a path $A(\Delta y, y)$ in the phase space, parametrized by α, and the line in (4.14) implies averaging over all admissible paths. The main properties of Δ_t^β and Δ_y^α are

$$\Delta_t^\beta \sim (\Delta t)_0^\beta D_t^\beta, \ \Delta t \to 0; \quad \Delta_y^\alpha \sim |\Delta y|_0^\alpha D_{|y|}^\alpha, \ \Delta y \to 0 \tag{4.17}$$

where the fractional derivatives of the orders β and α, respectively, are introduced. For simplicity, the authors assumed that $F(y, t)$ is symmetric with respect to y and $A = A(\Delta y)$. Then the balance equation was written in the form of fractional equation

$$_0D_t^\beta F(y, t) = \mathcal{D}_{\alpha,\beta} \ _0D_{|y|}^\alpha F(y, t), \tag{4.18}$$

where

$$\mathcal{D}_{\alpha,\beta} = \left\langle \frac{|\Delta y|^\alpha}{(\Delta t)^\beta} A \right\rangle.$$

To the end of the article, the two equations with self-similar solutions are compared, one of which is of fractional type:

$$\frac{\partial F(y, t)}{\partial t} = \frac{\partial^\alpha F(y, t)}{\partial |y|^\alpha}$$

and

$$\frac{\partial F(y, t)}{\partial t} = \frac{\partial}{\partial y} \left[|y|^\mu \frac{\partial F(y, t)}{\partial y} \right].$$

In conclusion, the authors noticed: "In the solar wind-magneto-sphere coupling, a technique to analyze the Lévy-type processes is applied to the time series data of the solar wind electric field and the auroral electrojet index.

The probability distribution function of the flights shows similarities and differences, and provides a new insight into the origin of the multiscale behavior through the different values of the scaling indices. In a complementary approach, the fractional kinetic equations, which use fractional derivatives to represent the complexity, provide a suitable mathematical framework for the multi-scale behavior. Unlike the usual diffusion equations, the solutions of these equations yield nonconvergent moments, showing its multiscale features".

Conservation of adiabatic invariant in a decreasing regular magnetic field should lead to pitch-angles φ at 1 a.u. close to 0 for any initial values of α [Panasyuk *et al.* (2006)]. Observations often indicate almost isotropic distribution of pith-angles near the Earth. This can be explained by scattering of particles on inhomogeneities of interplanetary magnetic field. Propagation of SCRs along field lines can be interpreted as diffusion. Diffusive transport is an important aspect of the propagation of solar cosmic rays in the interplanetary medium [Parker (1958b)] (Fig. 4.12). We consider this problem in the next section.

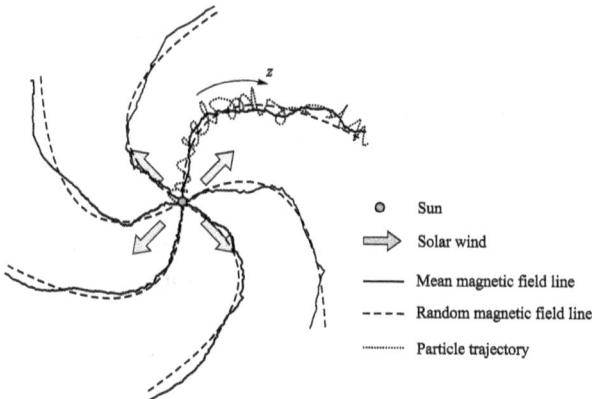

Fig. 4.12 Sketch illustrating particle motion in turbulent interplanetary magnetic field.

4.4 Solar cosmic rays

Solar cosmic rays (SCRs) are fluxes of high-energy particles ejected primarily in solar flares and coronal mass ejections. They are registered as a sudden sharp increase in the intensity of cosmic rays on the background of galactic cosmic rays (GCR). First event in SCR was registered in 1942. Now, SCRs are of particular interest and importance because fluxes of high-energy particles endanger life in outer space and normal operation of spacecrafts, primarily electronics and surface structural items.

SCRs provide a good sample of solar material. They have a composition similar to that of the Sun, they consist of protons, electrons and high-energy nuclei with energy ranging from tens of keV to GeV. SCR particles participate in a long sequence of different processes starting from nuclear, atomic, plasma and magnetohydrodynamic processes at the Sun, propagation in the interplanetary space and interaction with solar and planetary fields. Results in physics of SCRs are deeply related to other branches of space and laboratory plasma and particle physics. One of them is particle acceleration and scattering (wave-particle interaction) [Miroshnichenko (2001)].

At first, the propagation of cosmic rays was described by simple diffusion equation [Terletskii and Logunov (1952); Morrison *et al.* (1954)]

$$\frac{\partial p(\mathbf{x}, t, E)}{\partial t} = -\nabla \left(\kappa \nabla p(\mathbf{x}, t, E) \right) + S(\mathbf{x}, t, E), \qquad (4.19)$$

where κ is a diffusion coefficient and S is a source term. This equation implies that after many scattering events energy does not change. This assumption is reasonable due to the small magnitude of the electric field in a highly conducting interplanetary plasma. Intensive scattering of CRs was substantiated by observing high degree of isotropy of the cosmic radiation.

Jokipii [Jokipii (1971)] noted that at first "diffusion of cosmic rays was an ad hoc hypothesis, based on intuition... The first observational evidence that diffusion was in fact an excellent approximation to CR motion was put forth by [Meyer *et al.* (1956)]. In this important paper, a study of the February 23, 1956, solar-flare event established that the observed time-intensity profile could be accounted for quantitatively by equation (4.19) with appropriate parameters."

Important features of CR transport were revealed from the solar modulation of the galactic CR intensity (see, e.g. [Potgieter (2013)]). This effect first reported by [Forbush (1954)] consists in significant global and temporal variations in galactic CR intensity antiphase with solar activity. The concept of a continuous solar wind [Parker (1958b)] clarified the picture. The interplanetary CR diffusion can be explained by scattering of charged particles by interplanetary magnetic field irregularities and the field is carried out with the solar wind [Jokipii (1971)]. Parker [Parker (1958a)] pointed at the radial convection of particles due to the irregular magnetic field frozen into the solar wind.

Kinetic equation of CR transport was proposed by [Krymskii (1964)] and [Parker (1965)]. In frames of consistent kinetic approach, it was derived by [Dolginov and Toptygin (1967)].

[Getmantsev (1962, 1963)] considered compound diffusion as a possible explanation for the isotropy of CRs in the galaxy. That diffusion is anisotropic, as a result of the presence of a nonfluctuating (average) part of the magnetic field. The presence of this part in the interplanetary field was discussed at first by [Axford (1965)] and [Parker (1965)]. Due to the divergence of the solar-wind velocity, [Krymskii (1964)], [Parker (1965)] and [Gleeson and Axford (1967)] pointed out an adiabatic change in CR energy.

[Jokipii (1966, 1967); Hasselmann and Wibberenz (1968)] related the CR diffusion tensor to the magnetic-field power spectrum. This important step indicated that the CR diffusion equation could be derived as a statistical limit of equations of motion [Jokipii (1971)].

Recently, important statistical features of SCR propagation were revealed [Perri and Zimbardo (2007, 2008, 2009b)]. Analyzing the events detected by Ulysses and Voyager-2, Perri and Zimbardo found parallel superdiffusive regime of propagation for electrons and protons. These data were interpreted in terms of anomalous diffusion propagators in [Perri and Zimbardo (2007, 2008, 2009b); Uchaikin *et al.* (2014); Litvinenko and Effenberger (2014)]. Here, we continue to study these processes using the methods of fractional kinetics.

[Perri and Zimbardo (2007, 2008, 2009b)] analyzed data sets of interplanetary shocks in the solar wind, the CIR-associated forward and reverse shocks detected by spacecrafts Ulysses and Voyager 2. Perri and Zimbardo proposed the procedure for computing the energetic particle profiles by means of a propagator formalism: namely, the propagator $G(x, t)$ determines the probability density function of the longitudinal displacement during time t. In [Perri and Zimbardo (2007)], authors substantiate that the source of energetic particles accelerated at interplanetary shock waves associated with corotating interaction regions can be described by a distribution function

$$f_{\text{sh}}(x, E, t) = f_0(E)\delta(x - V_{\text{sh}}t), \tag{4.20}$$

where $f_0(E)$ represents the distribution function of particles of energy E emitted at the shock front, V_{sh} is the shock wave speed. Substituting Gaussian propagator

$$G(x, t) = \frac{1}{\sqrt{4\pi Dt}} \exp\left(-\frac{x^2}{4Dt}\right) \tag{4.21}$$

and $f_{\text{sh}}(x, E, t)$ into the formula for the particle time profile

$$f(x, E, t) = \int_0^t dt' \int_{-\infty}^{\infty} dx' G(x - x', t - t') f_{\text{sh}}(x', E, t'), \tag{4.22}$$

they obtained the exponential decay function

$$f(0, E, t) = \frac{f_0(E)}{V_{\text{sh}}} \exp\left(-\frac{V_{\text{sh}}^2 |t|}{D}\right). \tag{4.23}$$

Schematic explaining the model is presented in Fig. 4.13. Here x is a distance upstream from the shock, D parallel diffusion coefficient. This result is consistent with the observed data for protons presented in [Perri and Zimbardo (2008)] (Fig. 4.14a). But the observed time profiles presented in [Perri and Zimbardo (2008)] (for electrons) and in [Perri and Zimbardo (2009b)] (for electrons and protons) cannot be approximated by this function (Fig. 4.14b).

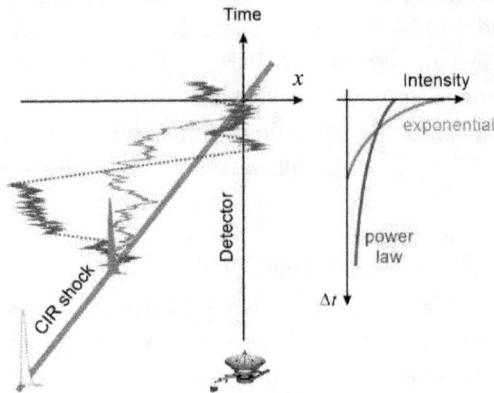

Fig. 4.13   Schematic explaining model (4.22).

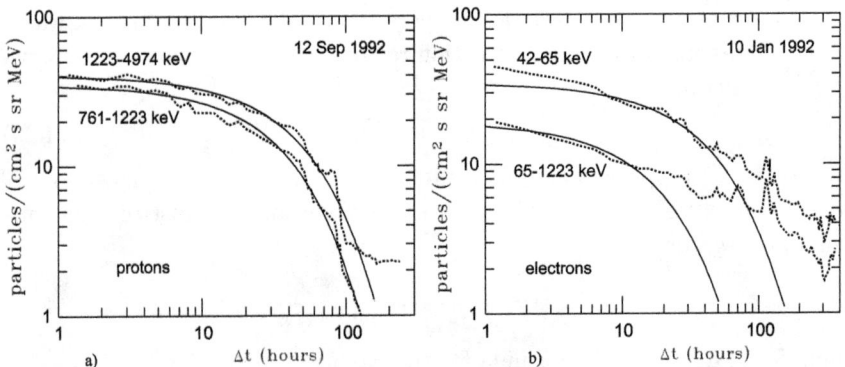

Fig. 4.14   Data on proton (a) and electron (b) fluxes digitized from Figs. 6 and 7 presented in [Perri and Zimbardo (2008)] (dotted lines). Energy ranges are indicated. The solid lines are exponential function plots corresponding to Gaussian propagators.

To describe superdiffusive regime, [Perri and Zimbardo (2007, 2008, 2009b)] take asymptotic power-law tails obtained by [Zumofen and Klafter (1993)] for Lévy walks:

$$G(x,t) = b\frac{t}{x^{\alpha+1}}, \tag{4.24}$$

where b is a constant and $1 < \alpha < 2$. Substituting propagator (4.24) and source function (4.20) into the formula for the particle time profile, they obtained a power law dependence

$$f(0, E, t) = \int G(0 - x', t - t')f_{\text{sh}}(x', E, t')dx'dt' \propto (\Delta t)^{1-\alpha}, \quad 1 < \alpha < 2,$$

that is consistent with tails observed in the data on the intensity of solar energetic particles (Fig. 4.15a).

The power-type decay of a propagator may serve as a sign of its belonging to the family of fractional differential equation solutions [Uchaikin (2013a)]. Taking into account that the fractional calculus has proved oneself to be an effective tool in galactic cosmic ray modeling [Uchaikin and Sibatov (2012a)]. We involve it in our subsequent consideration. As a result we find propagators of an asymmetric one-dimensional fractal walk with finite speed, which are consistent with (4.24). Using these propagators we calculate the particle time profiles which describe not only the tails of the observed kinetics, but the middle part as well, including the observed kink.

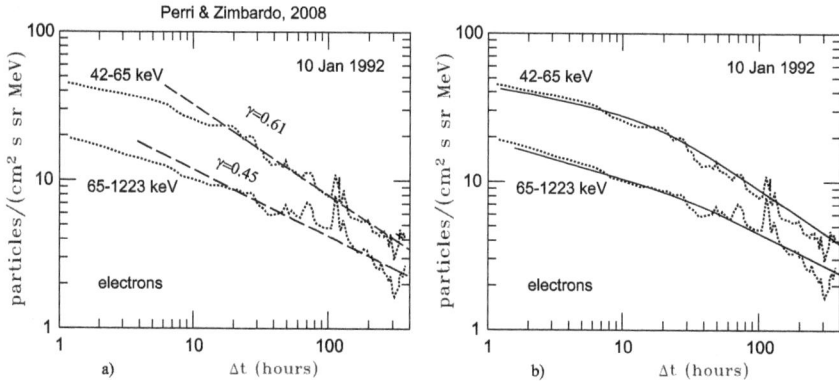

Fig. 4.15 Electron fluxes digitized from Figs. 6 of [Perri and Zimbardo (2008)] (dotted lines). Dashed lines correspond to asymptotic power law behavior obtained by [Perri and Zimbardo (2007)]. Solid lines in the right panel correspond to the numerical integration of Eq. (4.22), where propagators (7.57) derived in [Uchaikin *et al.* (2014)] are taken into account.

Interaction of electrons and protons with magnetic turbulence can lead to different diffusion regimes. Perri and Zimbardo [Perri and Zimbardo

(2009a)] write: "Due to the small Larmor radius, electrons can interact weakly with waves, causing a decrease in pitch angle diffusion and a faster parallel transport. It is interesting to notice that electron whistler inter-action can lead to a fast electron pitch angle diffusion, and that enhanced levels of whistler waves are frequently observed close to the shock ramp [Veltri and Zimbardo (1993)]. Yet, it appears that these whistlers are con-fined to a relatively narrow region around the shock, so that they are not influencing the electron propagation farther upstream. In addition, the su-perdiffusive transport of ions, here found, could be due to a lower level of magnetic turbulence upstream of the shock, owing to the higher heliocen-tric distance ($\sim$ 7 a.u.) ... Indeed, the turbulence level decreases with the distance from the Sun, as a consequence of solar wind expansion [Roberts *et al.* (1990); Zank *et al.* (1996)]. When the magnetic turbulence level goes down, the pitch angle scattering becomes weaker, so that the parallel velocity does not change sign very frequently; a persistent, long range cor-related velocity is one of the characteristic features of Lévy random walks and superdiffusion ...

These findings are likely to have a profound impact on the models of cosmic ray acceleration [Bell (1978)], as well as on the analysis of ener-getic particle propagation throughout the heliosphere [Shalchi *et al.* (2004); Verkhoglyadova and Le Roux (2005)]."

For $0 < \alpha < 1$ the pdf of particles performing Lévy walks along magnetic field line has the form (see section 4.5)

$$G(z,t) = (Kt)^{-1/\alpha} g\left((z - \beta vt)(Kt)^{-1/\alpha}; \alpha, \beta \right), \qquad (4.25)$$

where

$$K = \frac{cv}{m} \sin \frac{\pi(\alpha - 1)}{2}.$$

We use this propagator instead of (4.24) to calculate proton and electron fluxes by means of Eq. (4.22).

In Fig. 4.16, we compare solutions obtained with regular propagators and the observed time profiles of electron and proton fluxes. Data from [Perri and Zimbardo (2007, 2008, 2009b)] is quantized. Left panel represents the time relationship for electron fluxes. Points were restored by the authors of [Perri and Zimbardo (2007)] according to the information collected by the Ulysses spacecraft (January 22, 1993; 5 a.u.) for electrons with energies ranging from 42 up to 65 keV ($\times$) and from 65 up to 112 keV ($\circ$). Solid lines show power dependencies (11) with α=1.62 and α=1.47. Right panel represents the time relationship for proton fluxes. Points were restored by the authors of [Perri and Zimbardo (2007)] according to the information collected by the Ulysses spacecraft (January 22, 1993; 5 a.u.) for the ranges 761–1223 keV ($\triangle$) and 1233–4974 keV ($\diamond$), and according to the information collected by the Voyager 2 spacecraft (2006–2007; 8.7 a.u.) for the ranges 540–990 keV ($\circ$), 990–2140 keV ($\triangledown$), 2140–3500 keV ($\bullet$). Solid lines show the relationships (11) with $\alpha = 1.80$, 1.85, and 1.9.

We can see that the results from observations for electrons are in good agreement with superdiffusion solutions obtained using propagator (7.57) ($1 < \alpha < 2$; $\beta = 0$). The time profiles for protons obtained by Perri and Zimbardo based on data collected by the same spacecraft (Ulysses; 5 a.u.) are in this case described properly by damped exponential (4.23), meaning that protons mostly propagate in a diffusive manner in the above scales (5 a.u). However, the results obtained in [Perri and Zimbardo (2009b)] using data collected by the Voyager 2 spacecraft (2006–2007; 83.7 a.u) indicates that there were tails with powers of -0.8 to -0.9, meaning that the proton propagation was superdiffusive ($\alpha \approx 1.8$, 1.9). Figure 4.16 shows that solutions calculated with propagator (7.57) describe the experimental data better than the asymptotic relationships obtained by Perri and Zimbardo in [Perri and Zimbardo (2007, 2008, 2009b)].

The model that we developed on the basis of an equation with material fractional-order derivatives to describe the paths of cosmic ray charged particles with finite velocity of free motion [Uchaikin and Sibatov (2004)] is thus in good agreement with observed processes and can be used in the pre-asymptotic area to interpret the results from other experiments.

Authors [Zimbardo *et al.* (2006)] note: "Superdiffusive transport occurs, when particles during gyromotion are subjected to only weak variations of the magnetic field, so that pitch-angle diffusion is very slow (in other words, the magnetic moment is nearly conserved). Conversely, when the transverse variation of the magnetic field is stronger, and the pitch-angle diffusion is faster, leading to a Gaussian diffusion process. Clearly, pitch-angle diffusion can also occur because of magnetic field variations along $\mathbf{H}$, due to 0 the combined effect of parallel motion and gyromotion. On the other hand, an increase in the Larmor radius allows particles to be subjected to stronger variations along their orbits, so that pitch-angle diffusion is increased."

4.5 Tempered superdiffusion parallel to magnetic field

There are many papers [Giacalone and Jokipii (1999); Casse *et al.* (2001)] devoted to computation of the diffusion coefficient parallel and perpen-

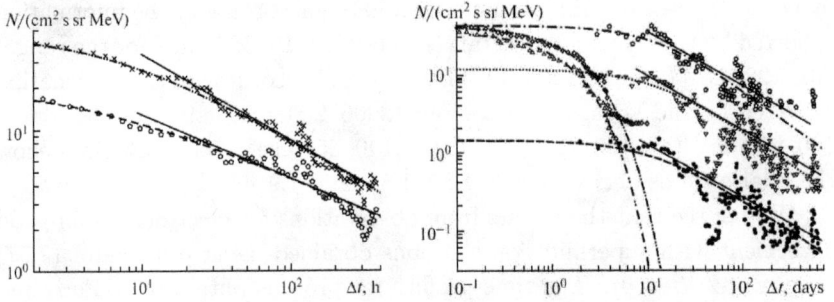

Fig. 4.16   Comparison of our solution (dashed line and dash-dotted line) and the results of the model in [Perri and Zimbardo (2007)] (solid line) and experimental data (points). Details in the text.

dicular to an average magnetic field $\mathbf{H}_0$. Different transport regimes are observed among which are normal diffusion, ballistic motion, super- and subdiffusion. It is ascertained that the transport of charged particles in turbulent magnetic field is defined by several parameters, such as turbulence level $\delta B/B_0$, turbulence anisotropy, and ratio of Larmor radius R_L to correlation length l_c.

[Matthaeus *et al.* (2003)] developed the theory of nonlinear guiding center which agrees with numerical calculations of perpendicular diffusion. This theory is based on the decorrelation hypothesis leading to diffusive motion of guiding centers along magnetic field lines, which undergo random walk in perpendicular direction due to developed turbulence. Perpendicular subdiffusion of particles was reported in [Chuvilgin and Ptuskin (1993); Qin *et al.* (2002b)]. For statistical description of such behavior, the compound diffusion model was developed in [Webb *et al.* (2006)]. The model assumes that guiding centers of CR particles move along magnetic field lines. This motion is not regular, direction of the particle velocity at constriction points, magnetic plug, and other irregularities, so the guiding center undergoes a one-dimensional random walk. In [Webb *et al.* (2006)] such motion is modeled by Brownian motion, the field line wandering is considered as normal diffusion. The model is characterized by two parameters, parallel and perpendicular diffusion coefficients. Perpendicular subdiffusion is interpreted in [Webb *et al.* (2006)] as a result of combined action of field lines wandering and diffusive back and forth motion along field lines. Similar conclusions about subdiffusion for slab turbulence were achieved by Shalchi and Kourakis [Shalchi and Kourakis (2007b); Shalchi (2010)] by means of analysis of fourth-order correlations. For the composite model of slab and 2D-turbulence with 80% of fluctuation energy in 2D spectrum, [Qin *et al.* (2002a)] obtained normal diffusion for perpendicular transport that is explained by strong exponential wandering of field lines for 2D turbulence. The influence of turbulence anisotropy on particle transport was studied in [Zimbardo *et al.* (2006); Pommois *et al.* (2007)]. For quasi-slab turbulence, parallel superdiffusion and perpendicular subdiffusion take place. For isotropic turbulence, there are parallel superdiffusion and perpendicular normal diffusion that confirms the results of [Qin *et al.* (2002a)]. Superdiffusive regime is also obtained in the composite turbulence model [Shalchi and Kourakis (2007b)]. [Pucci *et al.* (2016)] indicate very narrow spectral turbulence range used in [Zimbardo *et al.* (2006); Pommois *et al.* (2007)] and exclude this defect. In [Pucci *et al.* (2016)], turbulent magnetic field is simulated as a superposition of spatially localized fluctuations of magnetic field on different scales. The spectrum of generated turbulence is isotropic with adjustable spectral index.

[Pucci *et al.* (2016)] calculate time dependencies of diffusion coefficients $D_\perp(t)$ and $D_\parallel(t)$, pdf $\rho(\tau)$ of time between consecutive inversions of parallel velocity sign. The tempered power law form of $\rho(\tau)$, dependence of truncation parameter on the turbulence level, transition from linear dependence of $D_\parallel(t)$ to a constant value indicate the possible realization of the tempered Lévy statistics [Rosiński (2007); Baeumer and Meerschaert (2010)]. Below we associate the stochastic process of tempered Lévy walks to parallel transport of charged particles in turbulent magnetic field simulated in [Pucci *et al.* (2016)]. Probability density function of elementary displacements has the form

$$w(x) \propto |x|^{-1-\alpha} e^{-\gamma|x|}. \qquad (4.26)$$

[Pucci *et al.* (2016)] determine the diffusion coefficient using the second moment of particle displacement. The particular form of the equation from [Uchaikin (1998a)] for the second moment $m_2(t)$ of fractal walk with finite velocity of free motion has the form

$$m_2(t) = 2 \int_0^{vt} x P(x) dx + \int_0^{vt} m_2(t - x/v) p(x) dx. \qquad (4.27)$$

Here $p(x)$ is a pdf of free path length, $P(x) = \int_x^\infty p(x') dx'$ is a complementary distribution, v is a particle velocity. After Laplace transformation, one can obtain for $\tilde{m}_2(\lambda)$ [Uchaikin (1998a)]

$$\tilde{m}_2(\lambda) = \frac{2\tilde{w}(\lambda/v)}{\lambda[1 - \tilde{p}(\lambda/v)]},$$

where $\tilde{p}(\lambda)$, $\tilde{w}(\lambda)$ are Laplace images of free path length pdf and function $xP(x)$. Image $\tilde{w}(\lambda)$ is related to $\tilde{p}(\lambda)$ by expression

$$\tilde{w}(\lambda) = \mathcal{L}\{xP(x)\} = -\frac{d\tilde{P}(\lambda)}{d\lambda} = -\frac{d}{d\lambda}\frac{1-\tilde{p}(\lambda)}{\lambda} = \frac{1+\lambda\tilde{p}'(\lambda)-\tilde{p}(\lambda)}{\lambda^2}.$$

For $\tilde{m}_2(\lambda)$ one can write

$$\tilde{m}_2(\lambda) = \frac{2v^2[1+\lambda\tilde{p}'_\lambda(\lambda/v)-\tilde{p}(\lambda/v)]}{\lambda^3[1-\tilde{p}(\lambda/v)]} = \frac{2v^2}{\lambda^3} + \frac{2v^2\tilde{p}'_\lambda(\lambda/v)}{\lambda^2[1-\tilde{p}(\lambda/v)]}. \quad (4.28)$$

In case of distribution with power law tail $P(x) \propto x^{-\alpha}$:

$$\tilde{p}(\lambda) \sim 1 - c_1\lambda^\alpha, \quad \lambda \to 0, \quad 0 < \alpha < 1, \quad (4.29)$$

$$\tilde{p}(\lambda) \sim 1 - l\lambda + c_2\lambda^\alpha, \quad \lambda \to 0, \quad 1 < \alpha < 2. \quad (4.30)$$

For asymptotic ($\lambda \to 0$) transform of the second moment, we obtain

$$\tilde{m}_2(\lambda) \sim \frac{2v^2}{\lambda^3}(1-\alpha), \quad 0 < \alpha < 1,$$

$$\tilde{m}_2(\lambda) \sim \frac{2v^2}{\lambda^3}\frac{(\alpha-1)c_2(\lambda/v)^{\alpha-1}}{l-c_2(\lambda/v)^{\alpha-1}} \sim 2(\alpha-1)c_2l^{-1}v^{3-\alpha}\lambda^{\alpha-4}, \quad 1 < \alpha < 2.$$

Inverse Laplace transformation leads to

$$m_2(t) \sim v^2t^2(1-\alpha), \quad 0 < \alpha < 1,$$

$$m_2(t) \sim \frac{2(\alpha-1)}{\Gamma(4-\alpha)}c_2l^{-1}(vt)^{3-\alpha} \quad 1 < \alpha < 2.$$

For the case of tempered superdiffusion with exponentially truncated power law distribution of free path lengths,

$$\tilde{p}(\lambda) \sim 1 + c_\alpha\gamma^\alpha - c_\alpha(\lambda+\gamma)^\alpha, \quad 0 < \alpha < 1, \quad (4.31)$$

$$\tilde{p}(\lambda) \sim 1 - c_1\lambda + c_\alpha(\lambda+\gamma)^\alpha - c_\alpha\gamma^\alpha, \quad 1 < \alpha < 2. \quad (4.32)$$

Substituting (4.31) into (4.28) for case $0 < \alpha < 1$ leads to equation

$$\tilde{m}_2(\lambda) \sim \frac{2v^2}{\lambda^3} - \frac{2v\alpha c_\alpha(\gamma+\lambda/v)^{\alpha-1}}{\lambda^2[c_\alpha(\gamma+\lambda/v)^\alpha - c_\alpha\gamma^\alpha]} \sim \begin{cases} 2v^2(1-\alpha)\lambda^{-3}, & \gamma \ll \lambda/v; \\ v(1-\alpha)\gamma^{-1}\lambda^{-2}, & \gamma \gg \lambda/v. \end{cases}$$

After inverse Laplace transformation, we obtain

$$D_\parallel(t) = \frac{m_2(t)}{2t} \sim \begin{cases} v^2t(1-\alpha)/2, & vt \ll \gamma^{-1}; \\ v(1-\alpha)(2\gamma)^{-1}, & vt \gg \gamma^{-1}, \end{cases} \quad 0 < \alpha < 1. \quad (4.33)$$

For case $1 < \alpha < 2$ after substitution of (4.31) into (4.28), we obtain

$$\tilde{m}_2(\lambda) \sim \frac{2v^2}{\lambda^3} - \frac{2vc_1 - 2v\alpha c_\alpha(\gamma + \lambda/v)^{\alpha-1}}{\lambda^2[c_1\lambda/v + c_\alpha\gamma^\alpha - c_\alpha(\gamma + \lambda/v)^\alpha]}$$

$$\sim \begin{cases} \dfrac{2(\alpha-1)c_\alpha v^{3-\alpha}}{c_1\Gamma(4-\alpha)\lambda^{4-\alpha}}, & \gamma \ll \lambda/v; \\[3mm] \dfrac{v\alpha(\alpha-1)c_\alpha\gamma^{\alpha-2}}{\lambda^2(c_1 - \alpha c_\alpha\gamma^{\alpha-1})}, & \gamma \gg \lambda/v. \end{cases}$$

Inverse Laplace transformation with $l = c_1 - \alpha c_\alpha\gamma^{\alpha-1}$ leads to

$$D_\parallel(t) \sim \begin{cases} (\alpha-1)v(c_\alpha/c_1)(vt)^{2-\alpha}/\Gamma(4-\alpha), & vt \ll \gamma^{-1}; \\[2mm] \alpha(\alpha-1)v(c_\alpha/2l)\gamma^{\alpha-2}, & vt \gg \gamma^{-1}, \end{cases} \quad 1 < \alpha < 2. \tag{4.34}$$

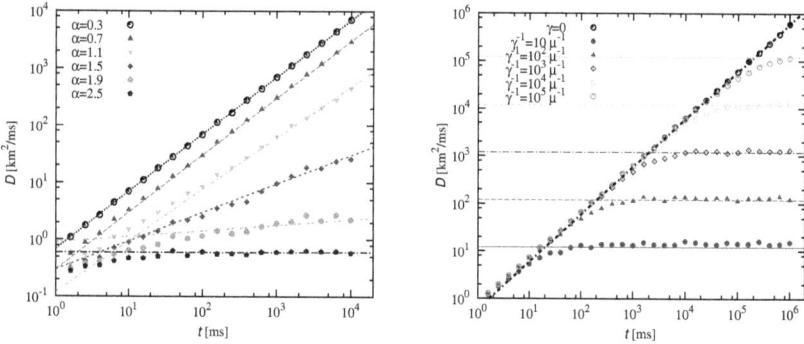

Fig. 4.17 Time dependence of the diffusion coefficient for Lévy walks with pure (left panel) and tempered (right panel) power law distributions of free path lengths. Curves correspond to (4.33).

Based on the results of simulation [Pucci *et al.* (2016)] for protons with energy 1 MeV, determining slope and plateau of $D_\parallel(t)$ we computed parameters of tempered superdiffusion using (4.33) (see Table 4.1) for several values of the turbulence level ($\delta b = \delta B/B_0$). Direct Monte Carlo simulation of the tempered Lévy walks with fitted parameters is in a good agreement with originally simulated dependence $D_\parallel(t)$ from [Pucci *et al.* (2016)] (see Fig. 4.18). Also, we found superdiffusion exponent and truncation parameter for different values of intermittency and spectral range width (Tables 4.2 and 4.3).

Table 4.1. Parameters of tempered superdiffusion dependent on turbulence level $\delta b = \delta B/B_0$ for 1 MeV proton, $p = 0.5$, $\xi = 1024$, $v = 1.41 \cdot 10^9$ cm/s.

δb	0.2	0.5	0.75	1.0
α	0.37	0.44	0.55	0.63
γ^{-1}, 10^{11} cm	48	7.1	3.8	2.8

Table 4.2. Parameters of tempered superdiffusion dependent on intermittency p for 1 MeV proton, $\delta b = 1.0$, $\xi = 16384$, $v = 1.41 \cdot 10^9$ cm/s.

p	0.5	0.7	0.8	0.9	0.95
α	0.49	0.53	0.58	0.60	0.63
γ^{-1}, 10^{11} cm	2.6	3.08	3.67	5.50	8.2

Table 4.3. Parameters of tempered superdiffusion dependent on spectral range width ξ for 1 MeV proton, $p = 0.5$, $\delta b = 1.0$, $v = 1.41 \cdot 10^9$ cm/s.

ξ	4	16	64	128	256	1024	16384
α	0.56	0.56	0.56	0.56	0.56	0.56	0.56
γ^{-1}, 10^{11} cm	145	22.5	4.5	3.0	2.4	2.4	2.4

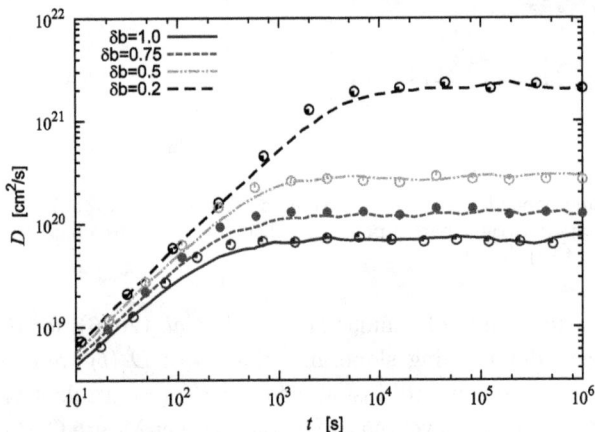

Fig. 4.18   Comparison of numerical results. Lines are time dependencies of parallel diffusion coefficient computed in numerical experiment [Pucci *et al.* (2016)], points represent result for tempered Lévy walk.

Let us list expressions for several asymptotic moments of tempered Lévy walks with $0 < \alpha < 1$, for the fourth moment,

$$m_4(t) \sim \begin{cases} (vt)^4(1-\alpha)(3-\alpha-\alpha^2)/3, \ vt \ll \gamma^{-1}; \\ 12v^2(1-\alpha)^2(2\gamma)^{-2}t^2, \quad vt \gg \gamma^{-1}, \end{cases}$$

the sixth one,

$$m_6(t) \sim \begin{cases} (vt)^6(1-\alpha)(15-8\alpha-8\alpha^2+2\alpha^3+2\alpha^4)/15, \ vt \ll \gamma^{-1}; \\ 120v^3(1-\alpha)^3(2\gamma)^{-3}t^3, \quad\quad\quad\quad\quad vt \gg \gamma^{-1}, \end{cases}$$

and the eighth one,

$$m_8(t) \sim \begin{cases} (vt)^8(1-\alpha)\frac{315-213\alpha-213\alpha^2+95\alpha^3+95\alpha^4-17\alpha^5-17\alpha^6}{315}, \ vt \ll \gamma^{-1}; \\ 1680v^4(1-\alpha)^4(2\gamma)^{-4}t^4, \quad\quad\quad\quad\quad\quad vt \gg \gamma^{-1}. \end{cases}$$

Thus, simulation results [Pucci *et al.* (2016)] of transport of protons with energy 1 MeV in turbulent magnetic field indicate tempered superdiffusive behavior with superdiffusion exponent $\alpha \in (0,1)$ and truncation parameter dependent on turbulence level, Larmor radius, intermittency and spectral width.

Chapter 5

From classic to fractional models of cosmic ray transport

This chapter gives a short review of popular models used for description of intragalactic transport of cosmic rays. Section 5.1 refers to the pioneer Fermi article, where the interstellar turbulent magnetic field was thought of as a set of randomly scattered clouds changing directions of motion of particles falling into them. There was a scheme like that which served as a basement for neutron transport theory, to creation of which Fermi also put his hand. Ginsburg and Syrovatsky reduced this model into diffusion process, similar to Brownian motion. An alternative class of models can be integrated by the term *collisionless* (Section 5.2). In Section 5.3, the influence of turbulent magnetic field on cosmic ray propagation is discussed. Section 5.4 describes the appearance of fractional derivatives in cosmic ray phenomenology. In Section 5.5, we review some results of the nonlocal (nonrelativistic) diffusion model based on bifractional diffusion equation. Percolation of cosmic rays is discussed in Section 5.6.

5.1   Ginzburg-Syrovatsky model

The charge component of cosmic rays (electrons, protons, nuclei) can be considered as a high-energy part of the cosmic plasma inheriting its main characteristic feature: turbulence. The turbulence is generated by chaotic evolution of interstellar magnetic field linked with a turbulent moving gas, dust and other interstellar medium components. They have a crucial effect upon the charge CR transport, but the inverse influence is a bit weaker. Thus, except of special regions (say, supernova explosions), one can formulate the CR propagation problem as a problem of propagation of charged particles through a random magnetic field with given stochastic properties.

The lack of detailed information on this medium stimulated development of very simple models of the process. The first model of this kind was offered in Fermi's paper on the nature of cosmic rays [Fermi (1949)]. Fermi proposed the hypothesis that "cosmic rays originate and are accelerated primarily in the interstellar space, although they are assumed to be prevented by magnetic fields from leaving the boundaries of the galaxy... Such fields have a remarkably great stability because of their large dimensions (of the order of magnitude of light years), and of the relatively high electrical conductivity of the interstellar space. Indeed, the conductivity is so high that one might describe the magnetic lines of force as attached to the matter and partaking in its streaming motions... The evidence indicates, however, that this matter is not uniformly spread, but that there are condensations where the density may be as much as ten or a hundred times as large and which extend to average dimensions of the order of 10 parsec ... Such relatively dense clouds occupy approximately 5 percent of the interstellar space" [Strömgren (1948)]. Fermi argued that the acceleration of a particle moving in the interstellar space was the result of its scattering in collisions with magnetized clouds.[1] In a short time after Fermi's publication, the models of random walk through an irregular magnetic field were discussed in [Cocconi (1951)] and [Terletskii and Logunov (1951)].

Five years after Fermi's article, Ginzburg wrote: "The motion of charged particles in the interstellar space resembles Brownian motion or motion of molecules in a gas. Indeed, due to the presence of the interstellar magnetic field, in the region where this field is quasihomogeneous, the trajectory of a particle winds around a magnetic field line and, upon averaging over the rotation period, is close to a straight line. However, on passing to a region with a different field direction, the trajectory changes and becomes a broken line as a whole. If the size of regions where the field direction noticeably changes is small compared to that of regions with a quasihomogeneous field, the particle motion can be treated as the motion of a molecule in a gas: the motion is free in the homogeneous field, and a change in the velocity direction at a boundary is similar to a collision with another molecule and can be usually assumed instantaneous. Hence, the size of the region with a quasihomogeneous field plays the role of the mean free path l. The mean free time is $\tau = l/v_0$, where v_0 is the translational velocity along the trajectory, which is by an order of magnitude equal to the usual velocity of the particle itself (and we therefore assume below that $\tau \sim l/v$, where v is the particle velocity). If magnetic fields do not change in time, this collision process leads only to the diffusion of particles and the 'mixing' of their velocities over directions but not to a change in the energy of the particles. It is known from the diffusion theory that the mean square distance

---

[1]Nowadays, the events of turbulent reconnection of magnetic lines claim the role of scatterers [Lazarian (2005)].

L propagated by a particle in a time t is
$$L = \sqrt{6Dt} \sim \sqrt{lvt},$$
where $D \sim lv/3$ is the diffusion coefficient. According to astronomical data, $l > 10^{19}$ cm in the interstellar medium, and for $v \sim c$ and $t \sim T \sim 10^{16}$ s (the proton lifetime), we obtain $L \sim 3 \cdot 10^{22}$ cm, which is of the order of the size of the Galaxy. Therefore, for $l < 10^{19}$ cm, protons, and all the more so nuclei, have no time to escape in great numbers from the Galaxy" ([Ginzburg (1953)] pp. 368-369).

Of course, it was clear from the very beginning that the intricate cosmic-ray transfer process cannot be fully described by the classical diffusion model. Ginzburg and Syrovatskii write in the first monograph on cosmic rays [Ginzburg and Syrovatskii (1964)]: "The high degree of isotropy of cosmic rays was one of the first indications that cosmic rays fall on Earth not directly from sources but after complicated motion and scattering in interstellar magnetic fields. This motion can be considered a 'diffusion' of cosmic rays in the interstellar space during which particles 'forget' about their initial direction of motion. However, the determination of the real nature of this diffusion is a quite challenging problem." Giving the model of the adiabatic motion of particles along field lines the due credit, the authors point out that the necessary condition for this motion (the radius of curvature being much smaller than the size of inhomogeneities of the magnetic field) is not satisfied everywhere, and its violation (in shock waves with a small front width or in regions with a zero magnetic field strength) eventually leads to the diffusion process. Sources are preferably concentrated in the Galactic disk and radial distribution of supernova remnants is usually taken as a source distribution. Fig. 5.1 depicts the most up-to-date information about distribution of visible matter in our Galaxy. It is reasonably to suppose that the distribution of supernovae is similar.

Despite such apparently intuitive and phenomenological rather than physical and mathematical, foundations, the diffusion direction in cosmic-ray physics exists and is still being developed. In this connection, we quote a remarkable note by Heisenberg [Heisenberg (1966)]: "A 'phenomenological' theory is understood as the formulation of regularities in the field of observed physical phenomena that does not attempt to reduce the described relations to the underlying general laws of nature through which they could be understood. Such phenomenological theories have always played a considerable role in the development of physics... . Of course, phenomenological theories are always developed where the observed phenomena cannot yet be reduced to the general laws of nature. The reason

Fig. 5.1 a) Artist's concept depicting the most up-to-date information about the shape of our galaxy (Milky Way). Credit: NASA/JPL-Caltech/R. Hurt (SSC/Caltech).

for this can be either an extreme complexity of these phenomena, which makes such a reduction impossible because of mathematical difficulties, or the lack of knowledge about these laws."

To account for a change in the energy spectrum of particles with the distance from their source due to ionization, synchrotron radiation, and additional acceleration by fluctuations of magnetic fields, shock waves, and supernova remnants, the energy-dependence was introduced the diffusion coefficient. In view of the observed decrease in the fraction of secondary nuclei with energy, the diffusion coefficient was approximated by a power-law function

$$D(E) = D_0 E^{\delta}, \qquad (5.1)$$

where E is the particle energy in GeV, with the exponent $\delta \simeq 0.3 - 0.7$, which was consistent, in particular, with the data on the cosmic-ray anisotropy [Berezinskii *et al.* (1984)].

Despite the anisotropic character of CR transport, it is surprising that isotropic diffusion equation describes propagation of galactic CRs so well. In many models of galactic CRs, isotropic diffusion (possibly with a galactic wind) is considered [Lerche and Schlickeiser (1982); Berezinskii *et al.* (1984); Webber *et al.* (1992); Bloemen *et al.* (1993)]. It is utilized in modern computational codes such as GALPROP and DRAGON. Justification of the isotropic model can involve fluctuations in the magnetic field related to instabilities, leading to the field line wandering out of the galactic plane and effective diffusion across the galactic disk [Jokipii and Parker (1969b); Jokipii (1973a)].

Blasi and Amato [Blasi and Amato (2012a,b)] considered diffusive propagation of CRs in the Galaxy accounting for spatiotemporal distribution of sources, diffusion in halo and spallation of nuclei. Using this model, they calculated energy spectrum, chemical composition and anisotropy of galactic CRs in the near-Earth space. Diffusion coefficient depends on energy according to power law $D(E) \propto E^\delta$. Large-scale distribution of supernova remnants is modeled in accordance with the observable distribution of pulsars accounting for spiral structure of the Galaxy. In the mentioned papers, authors used the equation of isotropic diffusion

$$\frac{\partial n_k(E, \mathbf{x}, t)}{\partial t} = \nabla[D_k(E)\nabla n_k(E, \mathbf{x}, t)] - \Gamma_k^{\mathrm{sp}}(E)n_k(E, \mathbf{x}, t)$$

$$+N_k(E)\delta(t - t_s)\delta^3(\mathbf{x} - \mathbf{x}_s), \tag{5.2}$$

where $n_k(E, \mathbf{x}, t)$ is a density of particles (nuclei) of k-th type with energy E in point $\mathbf{x}$ at time point t, $D_k(E)$ is a space-invariant diffusion coefficient, $\Gamma_k^{\mathrm{sp}}(E)$ is a spallation rate of nuclei of k-th type, $N_k(E)$ is an injection spectrum of the source exploded at time t_s. Values of k are chosen in the following way: $k = 1$ for hydrogen H, $k = 2$ for helium, $k = 3$ for CNO group, $k = 4$ for Mg-Al-Si group, $k = 5$ for Fe.

Diffusion coefficient, "determined from statistically reliable data (up to 100 GeV/nuclon) on secondary nuclei" [Ptuskin (2007)], depends on energy according to the power law $\propto R^\delta$, where $R = pc/Z$ is a magnetic rigidity, p momentum of a particle, Z nucleus charge, $\beta = v/c$, v particle velocity. Exponent $\delta = 0.6$ in pure diffusion model, and $\delta = 0.34$ accounting for diffusive reacceleration of particles by random magnetohydrodynamic waves in interstellar medium and Alfven velocity $V_a \approx 36$ km·s^{-1} [Ptuskin (2007)]. Popular dependencies for diffusion coefficients in different models for $R > R_0 = 3$ GV are listed in Table 5.1.

To calculate characteristics of CRs, one has to know the injection spectrum of sources and their isotopic composition. Boundary conditions depend on a model. One usually supposes $\psi = 0$ at halo boundaries, where particles exit into extragalactic space [Berezinskii *et al.* (1984)]. To define a galactic domain for CR propagation, the cylindrical model is often used [Berezinskii *et al.* (1984); Sveshnikova (2003); Blasi and Amato (2012a)]. The diffusion equation is solved under stationary condition, when one assumes $\partial \psi / \partial t = 0$. The spectrum in the source is modeled by a power law with an exponential cutoff

$$N_k(E) = \zeta E^{-\gamma} \exp\left[-\left(\frac{E}{E_{\max,k}}\right)\right],$$

where parameter $E_{\max,k}$ is determined by relation

$$E_{\max,k} = Z_k E_{\max,k}^H.$$

Here Z_k is a charge number, and $E_{\max,k}^H$ is chosen from fitting calculations to the observed energy spectrum of particles. The following charge numbers correspond to listed groups of nuclei

$$Z_1 = 1, \quad Z_2 = 2, \quad Z_3 = 7, \quad Z_4 = 13, \quad Z_5 = 26.$$

The solution of Eq. (5.2) is the Gaussian density,

$$G_k^{\text{free}}(\mathbf{x}, t; \mathbf{x}_s, t_s) = \frac{N_k(E)}{[4\pi D_k(E)\tau]^{3/2}} \exp[-\Gamma_k^{\text{sp}}(E)\tau] \exp\left[-\frac{(\mathbf{x} - \mathbf{x}_s)^2}{4D_k(E)\tau}\right].$$

Here, $\tau = t - t_s$.

The boundary conditions are usually formulated on the assumption that CRs leave the Galaxy coming up to the halo boundaries. In other words, at the boundary the particle flux directed into the Galaxy is zero. Neglecting radial sizes of the halo, with boundary conditions

$$G_k(x, y, z = \pm H, t; x_s, y_s, z_s, t_s) = 0,$$

one can write the solution using the method of images

$$G_k(\mathbf{x}, t; \mathbf{x}_s, t_s) = \frac{N_k(E)}{[4\pi D_k(E)\tau]^{3/2}} \exp[-\Gamma_k^{\text{sp}}(E)\tau] \exp\left[-\frac{(x - x_s)^2 + (y - y_s)^2}{4D_k(E)\tau}\right]$$

$$\times \sum_{n=-\infty}^{+\infty} (-1)^n \exp\left[-\frac{(z - z_n')^2}{4D_k(E)\tau}\right]; \quad z_n' = (-1)^n z_s + 2nH.$$

5.2 Collisionless Vlasov equation

Similarly to the spatial diffusion description, one can try to represent the process in the velocity or momentum subspace of the phase space. The direct analogy may slightly discourage us: the process of diffusion away from the initial point in the space looks as unlimited acceleration without visible causes... The reason is that in nonhomogeneity and anisotropy of this $\mathbf{v}-$ (or $\mathbf{p}-$) space with respect to the particle position. Indeed, the more velocity has the particle in some direction, the larger the resistance force seeking to return it to the origin of this frame. Before discussion of this process in details, we introduce the kinetic equation in the full phase space $(\mathbf{x}, \mathbf{p})$.

We restrict ourselves to consideration of such a region the size of which is essentially smaller than the mean free path of CR-particles with respect to Coulomb and nuclear collisions (ionization, spallation, irradiation), and only the interplanet or interstellar electromagnetic field effects on its trajectory with the Lorentz force $\mathbf{F}$. In this case, the motion of CR-particles can be described by using the collisionless Boltzmann equation (often called the *Vlasov equation*) for the joint spatial-momentum distribution density $f(\mathbf{x}, \mathbf{p}, t)$:

$$\frac{\partial f}{\partial t} + \mathbf{v} \cdot \frac{\partial f}{\partial \mathbf{x}} + \mathbf{F} \cdot \frac{\partial f}{\partial \mathbf{p}} = 0. \tag{5.3}$$

Neglecting a weak back contribution of CR-particles into the electromagnetic field, we will consider that $\mathbf{F}$ doesn't depend on f and Eq. (5.3) is a linear equation.

Due to the turbulent character of this force, it is usually treated as some of the regular (say, mean) component $\mathbf{F}_0$ and the turbulent (random) component $\mathbf{F}_1$:

$$\mathbf{F} = \mathbf{F}_0 + \mathbf{F}_1. \tag{5.4}$$

The second term generates irregular variations of $f(\mathbf{x}, \mathbf{p}, t)$ which in its turn becomes a *random field*. The modern CR-physics deals mainly with the pdf averaged over the statistical ensemble of the fluctuations,

$$\overline{f}(\mathbf{x}, \mathbf{p}, t) = \langle f(\mathbf{x}, \mathbf{p}, t) \rangle,$$

exactly this function should be found first of all.

However, inserting (5.4) into (5.3) and averaging the result over fluctuations, we don't get a closed equation for $\overline{f}$ because of the field-particle correlations term:

$$\frac{\partial \overline{f}}{\partial t} + \mathbf{v} \cdot \frac{\partial \overline{f}}{\partial \mathbf{x}} + \mathbf{F}_0 \cdot \frac{\partial \overline{f}}{\partial \mathbf{p}} = - \left\langle \mathbf{F}_1 \cdot \frac{\partial f}{\partial \mathbf{p}} \right\rangle. \tag{5.5}$$

This trouble was successfully overcame in [Dolginov and Toptygin (1966)], [Dorman and Kats (1977)]. In outline, it looks as follows.

Rewriting the initial equation in the form

$$\frac{\partial f}{\partial t} = \mathsf{L}_0 f + \mathsf{L}_1 f, \tag{5.6}$$

where

$$L_0 f = -\mathbf{v} \cdot \frac{\partial f}{\partial \mathbf{x}} - \mathbf{F}_0 \cdot \frac{\partial f}{\partial \mathbf{p}}$$

and

$$L_1 f = -\mathbf{F}_1 \cdot \frac{\partial f}{\partial \mathbf{p}}.$$

Differential equation (5.6) with initial condition $f_0 = f(\mathbf{x}, \mathbf{p}, 0)$ can be transformed into integral form

$$f(t) = f_0 + \int_0^t [L_0(\tau) + L_1(\tau)] f(\tau) d\tau.$$

Its formal solution is given by means of inverse operator $\Sigma(t)$,

$$f(t) = \Sigma(t) f_0,$$

which is found by successive iterations method. Omitting details (the reader can find them in the original article [Kichatinov and Matyukhin (1981)]), we represent the resulting equation for the averaged phase distribution in the tensor form:

$$\frac{\partial \overline{f}}{\partial t} + v_i \frac{\partial \overline{f}}{\partial x_i} + F_{0i} \frac{\partial \overline{f}}{\partial p_i} = \frac{\partial}{\partial p_i} \int_0^t K'_{ij}(\mathbf{x}, \mathbf{p}, t; \mathbf{x}(\tau, t), \mathbf{p}(\tau, t), \tau)$$

$$\times \left[\frac{\partial f}{\partial p_j} + \frac{\partial \Delta p_k(\tau, t)}{\partial p_j} \frac{\partial}{\partial p_k} + \frac{\partial \Delta x_k(\tau, t)}{\partial p_j} \frac{\partial}{\partial x_k} \right] \overline{f}(\tau) d\tau. \qquad (5.7)$$

In this equation, $(\mathbf{x}(\tau, t), \mathbf{x}(\tau, t))$ are the phase coordinates of the particle at time $\tau < t$ which is at point $(\mathbf{x}, \mathbf{p})$ at time t,

$$\Delta \mathbf{x}(\tau, t) = \mathbf{x}(\tau, t) - \mathbf{x}(t)$$

and

$$\Delta \mathbf{p}(\tau, t) = \mathbf{x}(\tau, t) - \mathbf{x}(t)$$

are corresponding increments, and

$$K'_{ij}(\mathbf{x}, \mathbf{p}, t; \mathbf{x}', \mathbf{p}', t') = \langle F_{1i}(\mathbf{x}, \mathbf{p}, t) F_{1j}(\mathbf{x}', \mathbf{p}', t') \rangle.$$

Solution of the equation essentially depends on the correlation length of magnetic field l_c. On assumptions that

(1) the change of particle momentum during the correlation time $\tau_c = l_c/v$ is relatively small,

(2) the characteristic scale of distribution change mach longer than l_c, and
(3) the observation time t is much longer than τ_c,

Eq. (5.7) reduces to the form commonly used for solving CR transport under the above condition:

$$\frac{\partial \overline{f}}{\partial t} + \mathbf{v} \cdot \frac{\partial \overline{f}}{\partial \mathbf{x}} + \mathbf{F}_0 \cdot \frac{\partial \overline{f}}{\partial \mathbf{p}} = \mathsf{K}\overline{f}, \qquad (5.8)$$

where

$$\mathsf{K}\overline{f}(\mathbf{x}, \mathbf{p}, t) = \frac{\partial}{\partial p_i} \left(K_{ij}(\mathbf{x}, \mathbf{p}) \frac{\partial \overline{f}}{\partial p_j} \right),$$

$$K_{ij} = \epsilon_{ikl}\epsilon_{jmn} v_k v_m (e/c)^2 \int_0^\infty B_{ln}(\mathbf{x}, \mathbf{v}\tau) d\tau \qquad (5.9)$$

and ϵ_{ikl} is the antisymmetrical Levy-Civita tensor. The integrand in Eq. (5.9)

$$B_{ln}((\mathbf{x}_1 + \mathbf{x}_2)/2, \mathbf{x}_1 - \mathbf{x}_2) = \langle H_{1l}(\mathbf{x}_1, t) H_{1m}(\mathbf{x}_2, t) \rangle$$

is the correlation tensor of magnetic field random component $\mathbf{H}_1$. The turbulence is assumed to be stationary, so B_{ln} doesn't depend on time. Its first argument $\mathbf{x}_1 + \mathbf{x}_2$ relates to slow spatial changes of the mean square magnetic field, whereas the second one links correlations dependent on mutual arrangement of two points.

Some authors have already used particular form of the general expression Eq. (5.6). Thus, authors of the article [Fedorov and Stehlik (2008)] have used Eq. (5.8) with

$$K'_{ij} = \int_0^\infty d\tau K_{ij}(\mathbf{x}, \mathbf{p}, t; \mathbf{x} - \mathbf{v}\tau, \mathbf{p}, t - \tau). \qquad (5.10)$$

This correlation function corresponds to the right-hand side of Eq. (5.8) in the form

$$\mathsf{K}\overline{f}(\mathbf{x}, \mathbf{p}, t) = \frac{\partial}{\partial p_i} \int_0^t d\tau K_{ij}(\mathbf{x}, \mathbf{p}, t; \mathbf{x} - \mathbf{v}\tau, \mathbf{p}, t - \tau) \frac{\partial \overline{f}(\mathbf{x} - \mathbf{v}\tau, \mathbf{p}, t - \tau)}{\partial p_j}.$$

$$(5.11)$$

The arguments in this kernel allow us to interpret it as probability of transition from $\mathbf{x} - \mathbf{v}\tau$ to $\mathbf{x}$ with a constant momentum. In other words, we encounter the walk model with trajectories consisting of straight line segments along which the particle moves with constant velocity $\mathbf{v}$. The connection points could be interpreted as collision points, but there is some difference: in the collision approximation the right-hand side of the kinetic equation looks as follows:

$$K\bar{f}(\mathbf{x}, \mathbf{p}, t) = \int d\mathbf{p}' w(\mathbf{p}' \to \mathbf{p})\bar{f}(\mathbf{x}, \mathbf{p}', t). \qquad (5.12)$$

Here $w(\mathbf{p}' \to \mathbf{p})d\mathbf{p}$ is the probability of collision with transition $\mathbf{p}' \to d\mathbf{p}$ per unit time. We assume that the integral kernel describes a short-time change of $\mathbf{p}$ on a pulse variation of field in a small-size region as a process of scattering envisaging averaging over the space-time variables and for this reason w doesn't depend on them. Below, we consider the link between the integral representation of the process and two differential forms arising from the Fokker-Planck equation.

5.3   Cosmic rays in turbulent magnetic field

Along with the isotropic model developed by Ginzburg and Syrovatsky [Ginzburg and Syrovatskii (1964)], the anisotropic diffusion model is widely used in local problems of galactic cosmic-ray transfer. This model was initially developed in theoretical studies of the motion of charged particles in quasi-homogeneous regions with a fluctuating magnetic field slightly different from a constant homogeneous field. The development of this model led to the separation of the diffusion of charged particles into the longitudinal and transverse components, each of which was described by a diffusion equation of the corresponding dimension with its own diffusion coefficient [Getmantsev (1963); Jokipii and Parker (1969b); Forman *et al.* (1974)]. The transverse diffusion was the first example of anomalous diffusion. The transverse diffusion anomaly was manifested not only in its slowness compared to the normal diffusion (which could be achieved by simply introducing a smaller diffusion coefficient) but also in a different expansion law for a diffusion packet and its different shape. Some authors believe that the local interpretation of such a composite model of anomalous diffusion (compound diffusion) can be extended to the entire galactic disc. For example, Hayakawa writes: "In this model, interstellar magnetic fields are assumed almost homogeneous along spiral sleeves. Particles are drifting along field lines and are reflected at mirror points... . Particles captured and kept on a field line continue to diffuse... in accordance with the chaotic motion of the field line... . Because the magnetic field is homogeneous only at the distance of a few kiloparsecs, we can assume that particles have escaped from the Galaxy if they have propagated a path longer than the field homogeneity length" [Hayakawa (1969)]. Nonhomogeneity of ISM appears also in existing of such structural forms as voids, clouds, magnetic traps, etc., where...

Dorman writes in [Dorman (1975)] about magnetic traps the following:

"Cosmic rays in the cosmos are confined, in fact, in magnetic traps of one scale or another, not propagating freely in space (except for cosmic-ray gamma quanta and neutrinos, for which traps are absent). Giant traps for cosmic rays exhibit a wide variety of properties, and the behavior of charged particles in them essentially depends on the particle energy. A trap in the vicinity of Earth formed by a magnetic field close to a dipole field is highly stable, and the lifetime of particles in it is long. At the same time, traps in the vicinity of chromospheric flares or in solar corpuscular fluxes of magnetized plasmas are much more transparent to particles and the particles escape from them similarly to how diffusion in irregular magnetic fields occurs. Traps of various types are also formed in the vicinity of usual stars, in particular, in the solar system and in supernova shells. On the other hand, the galaxy (the galactic disc together with the halo) also forms a certain type of trap a few thousand parsecs in size, which well confines moderate- and high-energy particles (with a lifetime of $\sim 10^7$ years) and is quite transparent to ultra-high-energy particles. It is quite possible that galactic clusters form even more gigantic traps for ultra-high-energy particles.

... it seems reasonable to treat any magnetic regions where the motion and lifetime of charged particles considerably differ from those in free space of the same volume as cosmic magnetic traps."

For such a large spread in the size of objects, it is impossible to imagine their uniform or even independent position in space. The distribution of visible matter in space (star clusters, galaxies, and galactic clusters) produces examples of hierarchic structures, which approximately preserve their inhomogeneity type upon varying scales in a broad range. The mathematical model of such inhomogeneities, which cannot be smoothed by scale transformations, uses *fractals* characterized by power-law correlations of spatial structures. At the end of the 20th century, a new avenue analyzing the structure of magnetic fields based on the fractal concept emerged in the astrophysics of the interstellar medium [Combes (2000)].

To elucidate the relation between the CTRW model and the real cosmic-ray transfer process in a galactic magnetic field, we consider a homogeneous (diffusion-wise) medium and divide it into cubic volumes (cells) (Fig. 5.2). We assign the coordinate $\mathbf{x}_i$ to each particle entering the ith cell centered at the point $\mathbf{x}_i$ at an instant t and leaving it at an instant $t+T$, neglecting the

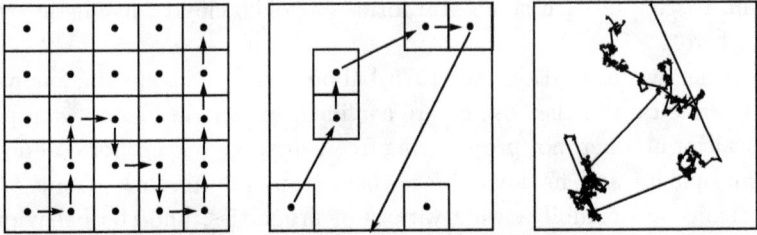

Fig. 5.2 Passing from (a) Brownian diffusion in a continuous homogeneous medium to (b) random walks in the medium alternating with voids and an example of the Lévy random walk trajectory obtained by the Monte Carlo method, with $\alpha = 1.67$ (c).

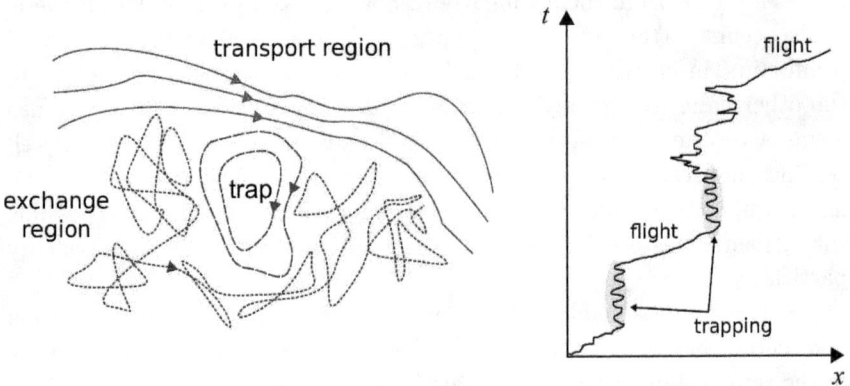

Fig. 5.3 Schematic of composite flow in plasma leading to CTRW according to [del Castillo-Negrete *et al.* (2004, 2005)].

motion of the particle inside the cell itself. After some (random) time T, the particle moves to one of the six neighboring cells, and the vector assigned to the particle moves from the center of the previous cell to the center of this new cell at the instant of the intersection of the face separating the cells. After a random time T', the particle moves to another neighboring cell, and so on. Thus, in a coarse-grain description, the particle coordinate moves jump-wise over three-dimensional lattice sites, staying in them for a random time. At a large scale, we then see a random walk, which is very close to Brownian motion (which it is, in fact).

However, a strong turbulent magnetic field does not necessarily exist in each cell. A considerable part of space between magnetic clouds is filled with weaker and quieter fields with magnetic field lines running smoothly over large distances. Charged particles in cosmic rays move along these lines and are sometimes captured in cloud traps, where they can be confined for a long time, forgetting their initial direction. To construct such a model, we remove part of the elements (cells), keeping others in their places. The passage from one cell to another is then no longer instantaneous, as was the case with crossing the face between neighboring cells, and the particle passes through 'almost empty' cells, propagating over random distances R. The time spent for these passages is proportional to the distance propagated in free space (the mean free path). The character of the entire process depends on the mutual arrangement of the remaining parts. If they are scattered homogeneously (in the statistical sense) and are independent of each other, as in the Bershadskii model [Bershadskii (1990)], we obtain a normal process with a larger diffusion coefficient. If they are arranged in a fractal manner, such that clusters and voids are observed at different scales (the method for constructing such distributions is shown in [Uchaikin *et al.* (1997)]), the particles can move over long distances in one run. In that case, the resulting process is determined by the order of the converging moments of a single run. We say that an anomalous diffusion process is a process of the *first kind* if the mean free path $\langle R \rangle = \infty$, and a process of the *second kind* if $\langle R \rangle < \infty$, and $\langle R^2 \rangle = \infty$. Below, we see a significant difference between these two types of anomalous diffusion.

We must note, however, that these considerations, despite their intentionally schematic and illustrative character, are in qualitative agreement with the picture of magnetic fields randomly spread in space, which was already proposed by Fermi. Quantitative agreement can probably be achieved by specifying (i) the appropriate distributions of clouds producing the corresponding distributions of mean free paths of particles between them and (ii) the laws of interaction of particles with individual clouds. In such a formulation, the problem can be considered in the framework of the standard multiple scattering theory, which, in particular, allows investigating the interaction of a particle with a cloud (the scattering cross section) and the transfer itself of particles in space (solving the kinetic equation) separately. In passing to the diffusion limit, the integro-differential kinetic equation is transformed into a differential diffusion equation, while the interaction cross section containing information on the mechanisms of this interaction (resonance interaction, scattering by Alfven and magnetosonic waves, etc.) is transformed into the diffusion coefficient, which then accumulates this information.

As regard the kinetics themselves, there are no grounds to assume that magnetic clouds are arranged homogeneously and move independently of each other. In that case, it would be reasonable to adopt an exponential distribution law for transits between scattering events. An example is given by an ideal gas, whose molecules do not interact with each other, which leads to the independence and exponential range distribution. But measurements of the electromagnetic radiation of the charged component of cosmic rays show that the interstellar region is not an ideal gas and is characterized by long-range power-law correlations, and this can be manifested in the higher probability of the long transits of particles propagating through voids. The mathematical model of such a process is already developed. This is Lévy motion: the random walk of a particle with an asymptotically power-law range distribution. An example of the trajectory of such a process is presented in Fig. 7c. We can see that the increase in the fraction of long free paths is accompanied by the increase in the fraction of short ones. This occurs due to the decrease in the probability of intermediate-length free paths. Now we can say that a trajectory consists of clusters of short free runs separated by long ones. The clusters of short free runs localized in space are capable of simulating the behavior of a particle in cells, which we mentioned above. At the same time, stable laws governing Lévy motion are directly and rigorously related to fractional derivatives [Uchaikin and Zolotarev (1999); Uchaikin (2003b)].

Following the evolution of the diffusion model as additional information is being gradually included, we infer that imitation possibilities of the model are already nearly exhausted. The reason for this is clear: the diffusion process is determined by the only parameter (except the space-time scales), the diffusion coefficient, and this single parameter (even if split into components, as in the case of compound diffusion) is insufficient. The natural way out, by replacing the diffusion coefficient with its random analog and subsequently averaging the equation and obtaining its averaged solution, was mathematically found only in the case of small fluctuations, which are of minor interest in our problems: in the turbulent interstellar space, a major role is played by large fluctuations alternating with different-scale voids and characterized by long-range power-law correlations. To describe the transfer of such a 'ragged' (fractal) medium, a special mathematical apparatus was required, and it was developed in the framework of fractional calculus.

The introduction of the method of fractional derivatives for solving a number of relevant problems [Lagutin *et al.* (2001b); Lagutin and Uchaikin (2003)] was stimulated by the use of the fractional differential technique in [Zolotarev *et al.* (1999); Uchaikin (1999)] and our experience, described in [Uchaikin and Zolotarev (1999)], of working with stable non-Gaussian distributions. The appearance of two new parameters, the spatial (α) and temporal (β) fractional-order derivatives, remarkably extended the family of solutions of the diffusion equation, formally preserving its form. The most important feature of new solutions is the power law of their asymptotic behavior, which is in excellent agreement with the known properties of the turbulent interstellar medium, the Fermi acceleration mechanism, and other processes affecting cosmic rays. At the same time, the fractional differential approach, unlike other nonlocal approaches, demonstrated a peculiar 'correspondence principle', incorporating normal diffusion as a particular case corresponding to $\alpha = 2$ and $\beta = 1$.

When the plasma moves with a high speed that the diffusion term in Eq. (3.3) can be neglected:

$$\frac{\partial \mathbf{H}}{\partial t} = \text{rot}\,[\mathbf{v}, \mathbf{H}]. \tag{5.13}$$

Let C be some closed loop moving together with the plasma, and S be a stretched surface. The magnetic flux through the surface changes and its increment during dt consists of two parts. The first of them, motion of the loop

$$d\Phi_1 = \int_S (\partial \mathbf{H}/\partial t)\,d\mathbf{S}dt$$

is caused by external sources.

The second part arises due to electric field induced at motion of the loop,

$$d\Phi_2 = -\int_S \text{rot}\,[\mathbf{v}, \mathbf{H}]\,d\mathbf{S}dt.$$

Combining both parts and taking Eq. (5.13) into account, we obtain

$$d\Phi = d\Phi_1 + d\Phi_2 = \int_S (\partial \mathbf{H}/\partial t - \text{rot}\,[\mathbf{v}, \mathbf{H}])\,d\mathbf{S}dt = 0.$$

Thus, the flux of magnetic field intensity through the material loop related to plasma is conserved. It follows from this, that the plasma within a magnetic flux tube always remains in that flux tube, as the plasma moves. Moreover, plasma elements that are linked by a magnetic field line will always remain on that magnetic field as the plasma and magnetic field line move. They say that the magnetic lines are "frozen into" the plasma (more exactly, the magnetic lines and the plasma are "frozen into each other").

Being stretched and bent by the turbulence, the fields resist deformation by means of the Lorentz force. This gives rise to the statistically steady state of fully developed MHD turbulence, in which, energy and momentum are transfered from larger scales to smaller scales and eventually dissipated. In principle, the picture is similar to the Kolmogorov model, but the system is more complicated, because it consists of two interacting subsystems: charged fluid and electromagnetic field.

Solving this problem has been started with the case of a plasma threaded by a straight uniform magnetic field of some external (i.e., large-scale) origin, whereas turbulent excitations are forced by weak wave-like disturbances running along the mean field. For this limit phenomenon known as *Alfvénic turbulence*, Kraichnan and (independently) Iroshnikov derived spectrum of isotropic turbulence in the form

$$F(k) \propto k^{-3/2}.$$

However, measurements of the solar wind performed in 70s-80s revealed that the turbulence was strongly anisotropic and that the spectral index was closer to $-5/3$ than to $-3/2$. A theory of anisotropic MHD turbulence proposed by [Goldreich and Sridhar (1995)] gives

$$F(k_\perp) \propto k_\perp^{-3/2}.$$

[Boldyrev (2005)] offered approximation

$$F(k_\perp) \propto k_\perp^{-(5+\alpha)/(3+\alpha)}.$$

Formally defining the corresponding longitudinal spectrum of fluctuations from the condition

$$F(k_\perp)dk_\perp = F(k_\parallel)dk_\parallel$$

with

$$k_\perp \propto k_\parallel^{(3+\alpha/2)},$$

he also found that

$$F(k_\parallel) \propto k_\parallel^{-2}.$$

Observe that this expression does not depend on α.

The MHD turbulence theory is a very complicated problem, solution of which is now in progress.

5.4 Appearance of fractional derivatives in CR phenomenology

Papers by Urch [Urch (1977a,b)] devoted to the study of the motion of charged particles in random magnetic fields can probably be considered forerunners of the appearance of fractional derivatives in the cosmic-ray phenomenology. Discussing the applicability of the Fokker-Planck equation under the assumption that the trajectories of particles propagating along the unperturbed trajectory over distances many times exceeding the correlation length are only slightly perturbed and a number of other conditions are satisfied (the gyroradius r'_g of particles due to the field perturbation is negligibly small compared to the correlation length L_c, of the field, the stochastic magnetic field consists of unpolarized Alfvén waves propagating along the main field $\mathbf{H}_0$, directed along the z axis, and the velocity v of particles greatly exceeds the Alfén wave velocity), Urch reaches the conclusion that for

$$v_z/v < \sqrt{L_c/r'_g}$$

the Fokker-Planck equation leading to Fick's law $J_x = -D\partial f/\partial x$ of normal diffusion is inapplicable. Urch performed calculations not related to the Fokker-Planck equation and found that the relation between the transverse component J_x of the particle current density and their concentration N in the given problem under consideration differs from the usual Fick law by the presence of the third derivative instead of the first one:

$$J_x = -D_\parallel D_L^2 \frac{\partial^3 f}{\partial x^3}.$$

In [Urch (1977a,b)], fractional derivatives were not mentioned, but Webb and coauthors [Webb *et al.* (2006)] later noticed that the Urch formula in conjunction with the continuity equation

$$\frac{\partial f}{\partial t} + \frac{\partial J_x}{\partial x} = 0$$

leads to an equation with the fourth derivative with respect to the coordinate,

$$\left[\frac{\partial}{\partial t} - D_\parallel D_L^2 \frac{\partial^4}{\partial x^4} \right] f(x,t) = 0.$$

Factoring the operator in the left-hand side of this equation,

$$\frac{\partial}{\partial t} - D_\parallel D_L^2 \frac{\partial^4}{\partial x^4} = \left(\sqrt{\frac{\partial}{\partial t}} + \sqrt{D_\parallel} D_L \frac{\partial^2}{\partial x^2} \right) \left(\sqrt{\frac{\partial}{\partial t}} - \sqrt{D_\parallel} D_L \frac{\partial^2}{\partial x^2} \right)$$

formally leads to a fractional differentiation operator (of the order $1/2$).

The next paper "in the vicinity of a fractional derivative" was [Webb et al. (2006)], where various compound-diffusion regimes were investigated in more detail. The description of one of these regimes resulted in an equation close to the one with a fractional derivative (see the details in [Chuvilgin and Ptuskin (1993)]).

Starting with the collisionless Vlasov equation for charged particles in an electromagnetic field, separating fluctuations from the means, and then using the quasilinear approximation for a weakly turbulent plasma with small-scale fluctuations [Jokipii (1973b); Schlickeiser (2002); Toptygin (1983); Tsytovich (1977)],

$$f_1 \ll \langle f \rangle,$$

Ptuskin and Chuvilgin [Ptuskin and Chuvilgin (1990); Chuvilgin and Ptuskin (1993)] considered the equation

$$\frac{\partial \langle f \rangle}{\partial t} + \mathbf{v} \nabla \langle f \rangle + \langle \mathbf{F} \rangle \frac{\partial \langle f \rangle}{\partial \mathbf{p}} = R,$$

whose right-hand side, being the usual collisional term describing scattering of particles by small-scale and small-amplitude fluctuations. The approximation of this term by the relaxation expression $-(f - \bar{f})/\tau$, where

$$\bar{f} = \bar{f}(\mathbf{x}, p, t) = \frac{1}{4\pi} \int_{4\pi} f(\mathbf{x}, p\mathbf{\Omega}, t) d\mathbf{\Omega},$$

leads to the equation

$$\frac{\partial \langle f \rangle}{\partial t} + \mathbf{v} \nabla \langle f \rangle + \frac{q}{c} [\mathbf{v}, \langle \mathbf{H} \rangle] \frac{\partial \langle f \rangle}{\partial \mathbf{p}} = -\frac{\langle f \rangle - \bar{f}}{\tau},$$

where τ is the characteristic time of scattering by a small-scale inhomogeneity, the angular brackets denote averaging over small-scale fluctuations, and the bar over f denotes time averaging.

Splitting the solution into parallel and perpendicular diffusion components and using the known expressions for corresponding coefficients

$$\kappa_\| = \frac{v^2 \tau}{3}, \quad \kappa_\perp = \frac{\kappa_\|}{1 + (\omega_H \tau)^2}, \quad \kappa_A = -\frac{\kappa_\| \omega_H \tau}{1 + (\omega_H \tau)^2}$$

(see [Toptygin (1983)]), the authors arrived at the following equation for the spatial distribution of the perpendicular displacements (Eq. (B11) in [Ptuskin and Chuvilgin (1990)])

$$\frac{\partial N_\perp(\mathbf{x}_\perp, t)}{\partial t} - D_\perp \Delta_\perp \left[N(\mathbf{x}, t) - \frac{1}{2} \int_1^\infty N(\mathbf{x}, t - yL^2/\kappa_\|) y^{-3/2} dy \right] = 0,$$

$$(5.14)$$

where L is the correlation length of field fluctuations.

Fractional derivatives are still absent here, but if we represent the expression in square brackets in the form

$$\left[N(\mathbf{x}, t) - \frac{1}{2} \int\limits_1^\infty N(\mathbf{x}, t - yL^2/\kappa_\|) y^{-3/2} dy \right]$$

$$= \frac{1}{2} \int\limits_1^\infty \left[N(\mathbf{x}, t) - N(\mathbf{x}, t - yL^2/\kappa_\|) \right] y^{-3/2} dy$$

$$= L \sqrt{\frac{\pi}{\kappa_\|}} \left\{ \frac{1}{\Gamma(-1/2)} \int\limits_{L^2/\kappa_\|}^\infty \left[N(\mathbf{x}, t - \tau) - N(\mathbf{x}, t) \right] \tau^{-3/2} d\tau \right\},$$

and, using the condition $t \gg t_d = L^2/\kappa_\|$, replace the lower integration limit by zero, the expression in braces becomes a fractional semi-derivative (in the Marchaud form, but identically equal to a fractional derivative in the Riemann-Liouville form we used; see [Uchaikin (2013a)]). This was done in [Ramaty and Lingenfelter (1971)], where the transverse diffusion equation was represented in the fractional differential form

$$\frac{\partial N_\perp}{\partial t} = D_\perp \frac{\partial^{1/2}}{\partial t^{1/2}} \Delta_\perp N_\perp(\mathbf{x}_\perp, t) + N_\perp(\mathbf{x}_\perp, 0)\delta(t),$$

which, after applying the operator $(\partial/\partial t)^{-1/2}$, yields the equation

$$\frac{\partial^{1/2} N_\perp}{\partial t^{1/2}} = D_\perp \Delta_\perp N_\perp(\mathbf{x}_\perp, t) + N_\perp(\mathbf{x}_\perp, 0)\delta_{1/2}(t). \tag{5.15}$$

[Ptuskin and Chuvilgin (1990)] point out an important feature of their equation for the distribution of transverse displacements, preserved in its fractional differential version: "The presence of integrals in the equation means that particles have a good 'memory' during compound diffusion, the values of the function N at the instant t being dependent on the values of N at preceding instants. Unlike the usual diffusion, compound diffusion is not a Markov process."

We note, however, that although these equations were mathematically obtained for the asymptotic behavior of the distribution (formally, for $t \to \infty$), their applicability is also restricted at long times, because, as mentioned above, a number of physical assumptions no longer correspond to the process. [Ptuskin and Chuvilgin (1990)] begin the derivation of the corresponding equations with the comment: "We assume that a random magnetic field is small compared to the mean homogeneous magnetic field, i.e., $A \ll 1$. This means that the field lines weakly deviate from the direction of the mean field $\mathbf{H}_0 = \text{const}$. The compound diffusion proceeds in the plane perpendicular to $\mathbf{H}_0$, and the length parameter along a field line S can be approximately replaced by displacement along the direction of $\mathbf{H}_0$." It is clear that the increase in the effect of strong large-scale inhomogeneities during the development of diffusion makes this process isotropic, and the correctness of the assumptions used in its description and therefore of the equations derived is rapidly lost. For this reason, long-term correlations represented by the function $y^{-3/2}$ in the integral term in Eq. (5.14) are then shortened, the integral no longer balances the preceding term, its role decreases with time, and the transverse diffusion equation in the large-time limit takes the normal form[2]:

$$\frac{\partial N(\mathbf{x}_\perp, t)}{\partial t} - D_\perp \Delta_\perp N(\mathbf{x}_\perp, t) = 0.$$

This confirms the conclusion in [Zybin and Istomin (1985)] that the motion of charged particles in a random magnetic field eventually becomes normal diffusion (this conclusion was made assuming that field fluctuations are isotropic and have a small scale, the kinetic operator is Hermitian, and the first approximation of the perturbation theory with respect to the ratio of the particle free path to the field correlation length is valid).

However, it is reasonable to assume that calculations of the transverse diffusion performed in the perturbation theory approximation should not be extended to longer times, because this method exhausts its possibilities as the cumulative effect of the perturbations builds up. For example, the calculations in [Zybin and Istomin (1985)] showed that the direct solution of the compound diffusion problem, which can be obtained in a simple model, also gives the normal diffusion in transverse directions in the long-time asymptotics. [Zybin and Istomin (1985)] also showed that this conclusion can be obtained in the perturbation theory by changing the sequence of averaging procedures. This remarkable fact emphasizes that conclusions based on the perturbation theory should be derived carefully.

Another important aspect of the propagation of galactic cosmic rays is the convection transfer mechanism caused by large-scale motions of a medium as a whole with a convection velocity $\mathbf{u}(\mathbf{x}, t)$. "The large-scale motions of a medium can be random, and then on the average at scales greatly exceeding the main turbulence scale L, the motion of particles in

---

[2]If the first moment (mean time) of the correlation function is finite [Uchaikin (1999)].

some cases (for $D \ll uL/3$) is reduced to diffusion with the effective turbulent diffusion coefficient of the order of $uL/3$" [Berezinskii *et al.* (1984)]. The most important difference between turbulent diffusion and molecular diffusion is the nonlocal (specifically, spatially nonlocal) character of the former: the presence of vortex formations at different scales gives rise to long-range spatial correlations of the velocity field.

In the conclusion in [Zybin and Istomin (1985)], the authors consider the simple problem of Brownian motion of a particle in the field of random velocities and again demonstrate the diffusive character of the asymptotic regime. This is quite consistent with the results obtained long ago in [Roberts (1960)] devoted to the same problem, where it was also shown that this process at shorter times is described by an integro-differential equation like the one derived in [Ptuskin and Chuvilgin (1990)]. That we are dealing here with different time regions is confirmed, in particular, in the second part of [Webb *et al.* (2009)], where the Parker equation [Parker (1965)] describing the motion of cosmic rays in plasmas is analyzed taking convection, diffusion, and transfer into account. [Webb *et al.* (2009)] also showed that the transverse displacement equation, having a fractional differential form at small times, asymptotically becomes the normal diffusion equation at large times. In other words, the solution of the fractional differential subdiffusion equation describes the pre-asymptotic behavior of the process, which is often called the intermediate asymptotic regime [Barenblatt and Zel'dovich (1971)].

The equation involved an integral of the solution over the time interval preceding the observation instant, which can be interpreted as a peculiar effect of 'magnetic traps', characteristic of plasma dynamics [Balescu (2000)]. At the same time, estimates in [Berezinskii *et al.* (1984)] based on a comparison of the anisotropy $\delta \sim 10^{-3}$ and the mean free path $L \lesssim 10^{21}$ cm in the disc with the size of the disc itself confirmed the old assumption in [Chuvilgin and Ptuskin (1993)] that "cosmic rays cannot freely propagate along the disc but should also efficiently shift in the transverse direction. Such a motion can be caused by the mixing and entanglement of lines of force themselves, which carry the frozen relativistic cosmic-ray gas away to the disc boundaries. Thus, something similar to diffusion should obviously take place."

The first use of bifractional approach to description of the anomalous CR diffusion in isotropic approximation is encountered in [Ragot and Kirk (1997)]. Authors of this article point out that, although galactic magnetic fields are located predominantly in the galactic plane (see [Zweibel and

Heiles (1997)]) and charged particles with small Larmor radii (compared to mean free paths) move along these lines, only slightly diffusing in transverse directions, the observed angular distribution of particles has a surprisingly high isotropy degree. The authors explain this by large-scale fluctuations of the interstellar fields and by exponential divergence of magnetic field lines. An important mechanism of active mixing of cosmic rays is also their acceleration at the leading edges of shock waves from supernovae.cosmic tsunamis that shake the particles off their field lines and bring them to a chaotic regime [Dendy *et al.* (1995); Duffy *et al.* (1995)]. Under the action of these and other factors, particles that initially moved along 'their' field lines lose the connection with them after some correlation time and enter the isotropic diffusion regime. The turbulent character of the interstellar medium, which is manifested in the alternation of weakly irregular magnetic fields with randomly dispersed islands (clouds, regions) of strong fluctuations, affects only the distributions of free paths and residence times of particles in different regions. [Ragot and Kirk (1997)] generalized the standard diffusion model to the CTRW model with power-law distributions of free paths and lifetimes in traps corresponding and wrote the mean square of the resulting distribution at the observation instant t in the form

$$\int |\mathbf{x}|^2 G(\mathbf{x}, t) d\mathbf{x} = \frac{1}{2\pi i} \frac{1}{2\pi} \int_{\Gamma} d\lambda \int d\mathbf{k} \int d\mathbf{x} |\mathbf{x}|^2 e^{\lambda t - i\mathbf{k}\mathbf{x}} \widetilde{G}(\mathbf{k}, \lambda), \qquad (5.16)$$

where

$$\widetilde{G}(\mathbf{k}, \lambda) = \frac{B\lambda^{\beta-1}}{B\lambda^{\beta} + C|\mathbf{k}|^{\alpha}} \qquad (5.17)$$

in our notation. The last expression is consistent with expression (1.21) representing fractional differential equation (1.18) in Fourier-Laplace variables. The authors justify the choice of representations (1.22) and (1.23) by the existence of 'stability islets' (traps) in the plasma capturing particles for a long time, thereby slowing down the diffusion process [White *et al.* (1993)], and by the property of field lines to perform rapid (compared to diffusion) and long-range flights, preserving their direction [Jokipii and Parker (1969a); Kadomtsev and Pogutse (1978)]. The competition between these two processes is reflected in the distribution obtained.

Two remarks concerning [Ragot and Kirk (1997)] are in order[3]. First, expression (5.16) makes no sense for $\alpha \neq 2$, because the integral diverges and the packet width should be described differently. However, the authors

---

[3]That paper was written fairly long ago, and it would not be necessary to point out some of its inconsistencies if they were not repeated in other later papers.

consider only the case $\alpha = 2$, and no objections can be raised here. The second remark concerns the range of the parameter β (denoted by α in their paper). The authors solved the problem for synchrotron radiation of electrons with $\beta = 0.5$, 1, 1.5, and pointed out that in contrast to [Ptuskin and Chuvilgin (1990); Chuvilgin and Ptuskin (1993); Balescu (2000)], they extended the range of β to the region $\beta > 1$, and this extension can be regarded as the passage from subdiffusion to superdiffusion.

At first glance, such an interpretation is quite reasonable: $\beta < 1$ corresponds to subdiffusion and $\beta = 1$ to normal diffusion, and therefore $\beta > 1$ should correspond to super-diffusion, and the authors had no doubts about it. However, a reason to doubt existed: it suffices to inspect waiting time pdf in [Ragot and Kirk (1997)], which has the form

$$q(t) \sim \frac{1}{\tau} \frac{\beta}{\Gamma(1-\beta)} \left(\frac{t}{\tau}\right)^{-\beta-1} \quad , \quad t \to \infty. \tag{5.18}$$

Obviously, a function with such an asymptotic behavior can never be a probability density for $\beta > 1$, simply because it is negative ($\Gamma(1 - \beta) < 0$). However, the issue is not that $q(t)$ has a power-law asymptotic behavior, because nothing prevents us from taking it, say, in the form

$$q(t) \sim \frac{\beta'}{\tau} \left(\frac{t}{\tau}\right)^{-\beta'-1} \quad , \quad t \to \infty, \tag{5.19}$$

where β' is any positive number, but that expression (5.18) is obtained from expansion (1.22), which is valid only for $\beta' = \beta \le 1$. If $1 < \beta' \le 2$, then the expansion

$$\widetilde{q}(\lambda) \sim 1 - \langle T \rangle \lambda + B' \lambda^{\beta'}, \quad \lambda \to \infty$$

holds instead of (1.22). For small λ corresponding to long times, the term with $\lambda^{\beta'}$ is asymptotically small compared to the linear term and can be omitted. Hence, for $\beta' > 1$ and $\alpha = 2$ we obtain (and the authors of [Ragot and Kirk (1997)] should have obtained) normal diffusion, rather than superdiffusion.[4] The latter would only follow for $\alpha < 2\beta$. However, the question can arise: what about the case $\alpha > 1$? Should a term proportional to a gradient appear in this case, in addition to the Laplacian? Indeed, this term should appear, but $\langle \mathbf{x} \rangle = 0$ for isotropic diffusion, and this term safely

---

[4]The type of anomalous diffusion is determined by the exponent $\gamma = \beta/\alpha$ in the expansion law of the diffusion packet $\Delta \propto t^\gamma$: subdiffusion corresponds to $\gamma < 1/2$ and superdiffusion, to $\gamma > 1/2$. It is better to call the process with $\gamma = 1/2$, $\alpha \neq 2$ a *quasi-normal diffusion*, because the shape of the diffusion packet in this case differs from the normal shape.

disappears. The duration of the interval T cannot be negative; therefore, assuming $\langle T \rangle = 0$, we automatically assume $T = 0$, i.e., we eliminate the traps. As the particle velocity is infinite, the particle immediately escapes to infinity in the absence of traps, and we do not observe the time sweep of the process.

If for some reason we wish to see a fractional time derivative of an order $\beta \in (1,2)$ in the equation, then, as follows from the previous expansion, this derivative should be introduced together with the first derivative:

$$\langle T \rangle \frac{\partial G(x,t)}{\partial t} + B \frac{\partial^\beta G(x,t)}{\partial t^\beta} + C\Delta_x^{\alpha/2} G(x,t) = \langle T \rangle \delta(t)\delta(x) + B\delta_\beta(t)\delta(x),$$

$$1 < \beta < 2.$$

Now everything is in place. Only we should bear in mind that the long-term asymptotic form of the solution already determines an equation with a lower-order time derivative, i.e., the asymptotic solution is still characterized by $\beta = 1$.

5.5   Nonlocal diffusion model: physical basement

The knee problem

In our first paper written in the framework of the fractional differential model [Lagutin *et al.* (2000)], we discussed the nature of the 'knee' in the energy spectrum of primary cosmic radiation at $E \sim 3 \cdot 10^{15}$ eV (Fig. 5.4). Since the discovery of the break in energy spectrum of Galactic CRs by [Kulikov and Khristiansen (1959)], origin of this 'knee' is an outstanding problem in astroparticle physics [Hörandel (2004)].

The most important data on energy spectrum, the mass composition and the arrival direction distribution of high-energy cosmic rays are obtained from analysis of Extensive Air Showers (Fig. 5.5) [Kampert and Watson (2012)]. CR impacting molecules and atoms in the Earth's atmosphere produce EAS of subatomic particles. Surface detector arrays and optical telescopes are used to detect this cascade.

During sufficiently long time, it has been suspected that the steepening of the cosmic-ray spectrum around 3 PeV has something to do with the upper end of galactic sources of cosmic rays [Peters (1961)]. An experimental confirmation of this assumption could be found in a change of CR composition. Acceleration and propagation both depend on magnetic fields and hence on magnetic rigidity or total momentum divided by charge of

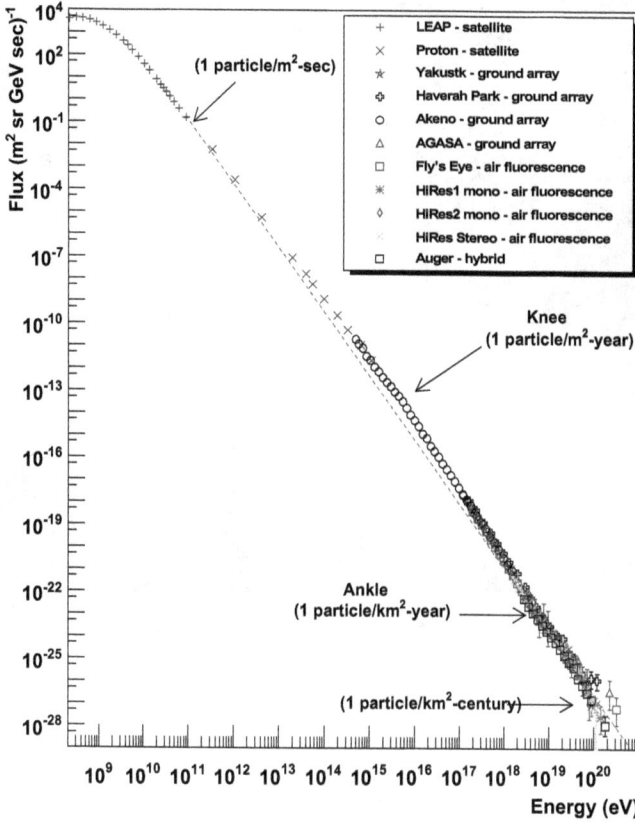

Fig. 5.4 Energy spectrum of cosmic rays. The data can be described by a power-law with spectral breaks at 4 PeV, referred to as the knee, a second knee at 400 PeV and the ankle at 1 EeV. Prepared from the spectrum presented by W.F. Hanlon (University of Utah) at www.physics.utah.edu.

the particle being accelerated. This brought to mind on the two-component hypothesis: for example, a major component due to diffusive shock acceleration of the ISM and another smaller component due to a recent local SN nearby [Erlykin and Wolfendale (1997)]. But realization of this scheme needed a 'fine-tuning' in order to avoid 'kinks' or 'bumps' in the junction region of the spectrum.

Among other possible mechanisms discussed at that time (to 2001, when our first work with fractional transport model was published), we should

Fig. 5.5 Artwork of cosmic rays impacting molecules and atoms in the Earth's atmosphere producing extensive air showers of subatomic particles. Credit: Mark Garlick / Science Photo Library.

mention assumptions on the maximum energy attained during the acceleration at supernova remnant shock waves [Stanev *et al.* (1993); Berezhko and Ksenofontov (1999); Erlykin and Wolfendale (2001)], leakage of particles from the Galaxy during the propagation process [Swordy (1995); Ptuskin *et al.* (1993)], interactions of CRs with the background light [Karakula and Tkaczyk (1993); Candia *et al.* (2002)] or neutrinos [Wigmans (2003)] before arriving at the Earth, or exotic interaction of CRs in the atmosphere where undetectable particles are produced and missing the detection [Nikolsky and Romachin (2000); Kazanas and Nicolaidis (2003)].

Listing these (and other) works in review [Hörandel (2006)], the author included our first work [Lagutin *et al.* (2001b)] using fractional order diffusion equation into the last group. However, the main idea of this article was not at all associated with the process of leakage of particles from the drawer. It originated from the phenomenological account of acceleration (by saving the energy dependence of the diffusion coefficient) combined with asymptotically power-law distribution of free paths due to the fractal structure of the interstellar medium. The knee phenomenon appears here as a result of non-Gaussian Lévy statistics, in turn, associated with fractional order derivatives. We explain this idea below.

Derivation of the equation

The fractional differential character of the diffusion equation was substantiated in the following way.

Based on the standard diffusion equation

$$\frac{\partial N}{\partial t} = D\Delta N(\mathbf{x}, t) + S(\mathbf{x}, t),$$

this knee is related to the decrease in the confinement efficiency of high-energy particles in the Galaxy, which in turn requires a long lifetime ($10^7 - 10^8$ years) of the proton-nuclear component in the system and the presence of remote sources [Berezinskii *et al.* (1984)]. But replacing the diffusion equation with the superdiffusion equation involving the fractional Laplacian,

$$\frac{\partial N}{\partial t} = -D_\alpha(-\Delta)^{\alpha/2}N(\mathbf{x}, t) + S(\mathbf{x}, t) \qquad (5.20)$$

showed the steepening of the spectrum even without a special assumption about the leakage of particles, if the sources of particles are explosions of supernovae nearest to the Solar System. In this case, the knee appears due to the fractional Laplacian, caused by a power-law distribution of free paths, which can in turn be interpreted as a result of averaging the exponential distribution of free paths over multiscale (fractal) fluctuations of the interstellar magnetic field (see, e.g., [Kulakov and Rumyantsev (1994); Elmegreen (1998)]).

Recall that Eq. (5.20) was derived as a long-time asymptotic equation for CTRJ-process [Uchaikin and Zolotarev (1999); Zolotarev *et al.* (1999)][5]. In 1999, V. Uchaikin met A. Lagutin and introduced him to these equations and their solutions. A. Lagutin drew attention to the fact that the power asymptotics of this 'anomalous' propagator can, at least qualitatively, reproduce the knee in the energy spectrum if to use the power type representation for the growth of the diffusion coefficient with energy taken at that time. In the first joint work [Lagutin *et al.* (2000)], V. Uchaikin gave

---

[5]One-dimensional version of this equation was derived and solved two years earlier in [Saichev and Zaslavsky (1997)].

a less formal derivation of this equation performed in natural (space-time) variables. It was obtained from integro-differential equation

$$\frac{\partial f_\epsilon}{\partial t} + \sigma f_\epsilon(\mathbf{x}, t) = \sigma \int f_\epsilon(\mathbf{x}', t) W\left(\frac{\mathbf{x} - \mathbf{x}'}{\epsilon}\right) \frac{d\mathbf{x}'}{\epsilon^3} + S_\epsilon(\mathbf{x}, t),$$

where ϵ is an auxiliary parameter. This equation can also be written in the form

$$\frac{\partial f_\epsilon}{\partial t} = \sigma \int [f_\epsilon(\mathbf{x}', t) - f_\epsilon(\mathbf{x}, t)] W\left(\frac{\mathbf{x} - \mathbf{x}'}{\epsilon}\right) \frac{d\mathbf{x}'}{\epsilon^3} + S_\epsilon(\mathbf{x}, t).$$

The equation in the first form describes the process of independent instantaneous jumps separated by random time intervals and distributed in accordance with the exponential law with the mean $1/\sigma$. The kernel W of the integral operator plays the role of the distribution density of the displacement vector in such a jump, and should therefore be integrable. For the equation represented in the second form, requirements imposed on the kernel W are relaxed, because the factor given by the difference of the solution values at nearby points serves as a regulator, and the integral can converge even if the kernel diverges. Assuming that the asymptotic form of the kernel for large ranges ('Lévy flights') is described by a power-law function, as is typical for fractal structures,

$$W(\mathbf{x}) \sim A r^{-3-\alpha}, \quad r \to \infty,$$

and introducing the notation $A' = A\sigma$, $t = \epsilon^\alpha s$, $N(\mathbf{x}, t) = \lim_{\epsilon \to 0} f_\epsilon(\mathbf{x}, t\epsilon^{-\alpha})$, $S(\mathbf{x}, t) = \lim_{\epsilon \to 0} \epsilon^\alpha S_\epsilon(\mathbf{x}, t)$, we obtain the equation for the three-dimensional isotropic Lévy motion:

$$\frac{\partial N(\mathbf{x}, t)}{\partial t} = A' \int \frac{N(\mathbf{x}', t) - N(\mathbf{x}, t)}{|\mathbf{x}' - \mathbf{x}|^{3+\alpha}} d\mathbf{x}' + S(\mathbf{x}, t), \quad 0 < \alpha < 1.$$

The range of α indicated here is determined by the convergence condition for the integral at $\mathbf{x}' \sim \mathbf{x}$. We assume that $N(\mathbf{x}, t)$ is a differentiable function of the coordinates, and therefore, as $|\mathbf{x} - \mathbf{x}'| \to 0$, we have

$$|N(\mathbf{x}', t) - N(\mathbf{x}, t)| \propto |\mathbf{x}' - \mathbf{x}|$$

and the integral converges for $\alpha < 1$. This integral operator can be continued to the domain of larger α by several methods [Uchaikin (2013a)]. The regularization by calculating the finite (Hadamard's) part of the integral brings it to the form

$$J = \int \frac{[N(\mathbf{x}', t) - N(\mathbf{x}, t)]_2}{|\mathbf{x}' - \mathbf{x}|^{3+\alpha}} d\mathbf{x}', \quad 1 < \alpha < 2,$$

where $[...]_2$ denotes a second-order difference.

In all these cases, the equation for the Fourier transform with respect to spatial variables,

$$\frac{\partial \widetilde{N}(\mathbf{k}, t)}{\partial t} = -A_\alpha |\mathbf{k}|^\alpha \widetilde{N}(\mathbf{k}, t) + \widetilde{S}(\mathbf{k}, t)$$

contains a term proportional to the fractional power α of the wave vector $|\mathbf{k}|$. Because $-|\mathbf{k}|^2$ is the Laplace operator transform, this term can be represented as

$$|\mathbf{k}|^\alpha \widetilde{N}(\mathbf{k}, t) \equiv (|\mathbf{k}|^2)^{\alpha/2} \widetilde{N}(\mathbf{k}, t) \Leftrightarrow (-\Delta)^{\alpha/2} N(\mathbf{x}, t),$$

and the equation is then written in the form (5.20).

From the physical standpoint, the fractional Laplacian can be regarded as the result of some averaging of the diffusion operator with a random diffusion coefficient $\tilde{D}$:

$$\langle \nabla(\tilde{D} \nabla N(\mathbf{x}, t)) \rangle \mapsto -K_\alpha (-\Delta)^{\alpha/2} \langle N(\mathbf{x}, t) \rangle.$$

It seems that the general derivation of this relation from the fractal structure of a medium where diffusion occurs does not exist, but there is a good example of a particular process of the propagation of excitations in plasma by resonance radiation. When averaging the transfer equation with an exponential range distribution (asymptotically equivalent to the standard diffusion equation) over the Lorentzian frequency distribution, the integral transfer operator is indeed transformed into the fractional Laplacian [Uchaikin (2013a)], and the equation exactly coincides with the one written above.

Physical justification for the fractional approach

The fractional-differential diffusion equation mainly differs from the standard diffusion equation in the range distribution: a power-like distribution in the first case and an exponential distribution in the second case. To elucidate the problem of cosmic-ray propagation in random magnetic fields, we refer again to the book [Ginzburg and Syrovatskii (1964)]: "Assume that the motion occurs only along tubes of force, but these tubes themselves are randomly entangled, for example, consist of rectilinear segments with the mean length l, any angle between the directions of neighboring segments being equally probable. Then the diffusion approximation can be fully applied to analyze the problem of averaging the spatial distribution of particles (cosmic rays) over large enough regions." What do we know about the lengths of these segments? Of course, it is natural to assume that they are random. As a continuous random quantity, this length is characterized by a distribution density $p(\xi)$, $\xi > 0$. The mention of the mean length of a rectilinear segment can be taken as an implicit assumption of the existence (i.e., convergence) of the improper integral

$$\int_0^\infty p(\xi)\xi d\xi = \langle R \rangle = l.$$

But this is not sufficient to provide the diffusive random walk of particles. The second moment of this distribution must also exist,

$$\int_0^\infty p(\xi)\xi^2 d\xi = \langle R^2 \rangle.$$

The standard diffusion coefficient is expressed in terms of this moment. These questions do not occur in the standard kinetics because the initial range distribution is assumed exponential, and all the moments of an exponential distribution are finite. We can say, of course, that this distribution is 'derived', i.e., obtained as a solution of the first-order differential equation $dp/d\xi = -\sigma p(\xi)$, but a close inspection of this derivation shows that it is based on the assumption that the random numbers of atoms on the segment $[0, \xi]$ and the adjacent element $d\xi$ are mutually independent. In classical kinetics, a particle moves in an ideal gas of noninteracting and unrelated atoms. In the case under study, the role of such atoms, in 'collisions' with which the cosmic-ray particles change the direction of their motion, is played by the ends (or, it is better to say, breaking points) of the rectilinear parts of magnetic field lines. Can we agree without any doubt that two such points lying at the ends of the same segment are mutually independent? Apparently not. But rejecting this hypothesis of the independence, we thereby cast doubt on the validity of the exponential distribution. This is not yet problematic because many other distributions exist with two finite first moments; however, the beginning has already been made. Any such distribution in the long-time limit brings us again to classical diffusion, but now the question already arises: What is there behind the convergence of the second moment? What do we sacrifice when we accept the assumption about the convergence of the second moment? And it turns out that we sacrifice the entire class of stable laws with a power-law asymptotic behavior, in problems with plasmas, turbulence, and random kinetics, which we are dealing with in studying cosmic rays. The difference between transfer processes with exponential and power-law range distributions is the same as that between molecular and turbulent diffusion, and cosmic-ray transfer in the Galaxy is the turbulent diffusion. The principal feature of the latter process is its nonlocality.

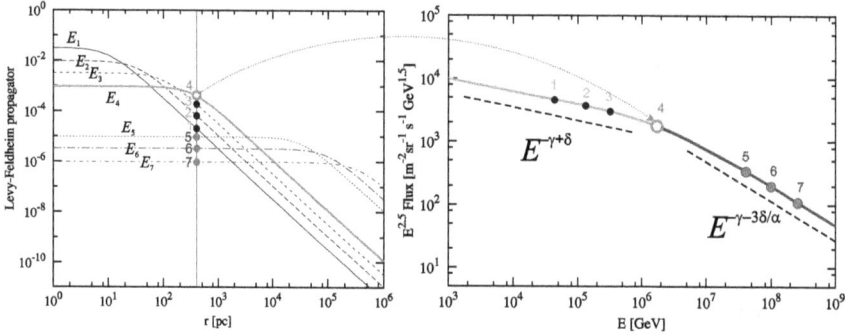

Fig. 5.6   The 'knee' in frames of the LU-model.

Of all the measurements performed in cosmic ray physics, only data on the energy spectra cover more than ten orders of magnitude, whereas the range of variations of other parameters is considerably smaller. This suggests that the effect of replacing the normal spatial distribution by distributions with power-law asymptotic forms is manifested, first of all, in energy spectra. However, it is obvious that a simple introduction of the energy spectrum $S(E)$ of a source into the transfer

$$\frac{\partial N}{\partial t} = -D_\alpha(-\Delta)^{\alpha/2} N(\mathbf{x}, t, E) + \delta(\mathbf{x})\delta(t)S(E), \qquad (5.21)$$

is not sufficient for estimating this effect: the source spectrum translates into the observed spectrum

$$N(\mathbf{x}, t, E) = [D_\alpha t]^{-3/\alpha} \Psi_3^{(\alpha)}([D_\alpha t]^{-1/\alpha} r)S(E),$$

without changes, because particles with different energies move in space with the same "diffusion coefficient". The situation changes if we introduce the energy dependence of this coefficient: in this case, different energies correspond to different values of the dimensional variable $\xi = [D_\alpha(E)t]^{-1/\alpha} r$. The choice of a power-law dependence means that the parameter $\xi = [D_{0\alpha} t E^\delta]^{-1/\alpha} r \equiv \xi_1 E^{-\delta/\alpha}$ of a high-energy particle lies near the top of the stable density (the left asymptotic region, for which we conventionally assume that $\xi < \xi_0$), whereas low-energy particles correspond to large values of ξ lying at the periphery of the spatial distribution $\Psi_3^{(\alpha)}(\xi)$ (the right asymptotic region, $\xi > \xi_0$). In the first case,

$$N(\mathbf{x}, t, E) \sim S_0 r^{-3} E^{-p} \left[\xi^3 \Psi_3^{(\alpha)}(0) \right] \propto E^{-p-3\delta/\alpha}, \quad E \to \infty,$$

for any admissible $\alpha \in (0, 2]$. In the second case $(E \to 0)$, a power-law asymptotic behavior is observed only for anomalous diffusion $(\alpha < 2)$:

$$N(\mathbf{x}, t, E) \propto \begin{cases} E^{-(p+3\delta/2)} \exp(-\xi_1^2/4E^\delta), & \alpha = 2; \\ E^{-(p-\delta)}, & \alpha < 2. \end{cases}$$

This was the reason in [Lagutin *et al.* (2000, 2001b)] to pass from the usual diffusion equation to the equation with fractional Laplacian (5.20), which allows relating this Laplacian to the fractal properties of the medium due to which the free paths of particles acquire the power-law form. The difference between the exponents of power-law asymptotic forms for low and high energies manifested for $\alpha < 2$ was interpreted as an indication of a 'knee' (see Fig. 5.6).

Choice of numerical parameters

Using experimental data on the position of the intermediate knee region $(E_<, E_>)$ and the exponent of the spectrum for $E < E_<$ and $E > E_>$, the main parameters $(D_{0\alpha}, \delta)$ of the model and the exponent p of particle generation in a source as a function of the exponent α were found. In [Lagutin *et al.* (2000)], calculations were originally performed for a source located at a distance r from the observation point and acting with a constant intensity for a time interval τ_S preceding the observation. In this case,

$$N(\mathbf{x}, t, E) = S_0 E^{-p} [D_{0\alpha} E^\delta]^{-3/\alpha} \int_{\max\{0, t-\tau_S\}}^{t} \Psi_3^{(\alpha)} \left(r[D_{0\alpha} E^\delta \tau]^{-1/\alpha} \right) \tau^{-3/\alpha} d\tau,$$

and the exponent of the observed spectrum changes from $p - \delta$ for $E \leq E_<$ to $p + \delta$ for $E \geq E_>$. The results of calculations showed that the best agreement with experimental data on the spectra of protons and nuclei and the total spectrum of all particles was achieved for $\alpha \approx 5/3$, $E_{\text{rad}} = (E_> + E_<)/2 = 3 \cdot 10^4 A$ GeV, per nucleon, $\delta = 0.25$, $p = 2.9$, $r \sim 200$ pc, $\tau_0 \sim 10^5$ years. The exponent of the spectrum observed in the kink region then changes from 2.65 to 3.15 (Fig. 5.4), which does not contradict the hypothesis that the sources of cosmic rays could be the explosions of the nearest supernovae during the last 100,000 years.

Bifractional equation

In [Lagutin and Uchaikin (2001, 2003)], we passed to the equation of a more general type, containing, along with the fractional Laplacian, the fractional time derivative to take the influence of magnetic traps into account (which, without a doubt, exist in the galactic medium):

$$\frac{\partial N}{\partial t} = -D_\alpha(E)\frac{\partial^{1-\beta}}{\partial t^{1-\beta}}(-\Delta)^{\alpha/2}N(\mathbf{x}, t, E) + S(\mathbf{x}, t, E). \qquad (5.22)$$

As the family of these equations includes Eq. (5.21) as a particular case ($\beta = 1$), we sacrificed nothing, but simply extended the class of solutions: instead of the one-parameter family of solutions expressed in terms of stable densities $\Psi_3^{(\alpha)}$, we obtained the two-parameter family of solutions determined by the fractionally stable densities $\Psi_3^{(\alpha,\beta)}$ [Uchaikin (2003a)]. For a point-like instantaneous source, the solution has the form

$$N(\mathbf{x}, t, E) = S_0 E^{-p}[D_{0\alpha}E^\delta t^\beta]^{-3/\alpha}\Psi_3^{(\alpha,\beta)}\left(r[D_{0\alpha}E^\delta t^\beta]^{-1/\alpha}\right),$$

whence it follows that

$$N(\mathbf{x}, t, E) \approx \begin{cases} S_0 D_{0\alpha}t^\beta r^{-3-\alpha}E^{-p+\delta}, & E < E_<; \\ S_0[D_{0\alpha}t^\beta]^{-1}r^{-3+\alpha}E^{-p-\delta}, & E > E_>. \end{cases}$$

Thus, as the energy increases, the spectrum steepness increases, which is manifested in the increase in the absolute value of the exponent of the spectrum by 2δ after passing through the interval $(E_<, E_>)$. In the case of a source acting with a constant intensity for a finite time, the spectrum steepness also increases, but the exponent changes by $(1 + 1/\beta)\delta$. We note that in both cases, the exponents characterizing the exponential behavior of the spectrum outside the knee region are independent of α, but a noticeable effect is observed in the intermediate region: the smaller α is, the broader the transition region, and the transition occurs more smoothly.

The choice of numerical values of these parameters was discussed in [Lagutin *et al.* (2001d)]: "To estimate the parameter β, we used results from paper [Cadavid *et al.* (1999)], where the anomalous diffusion of solar magnetic elements was studied. The authors showed that the distribution of the lifetime in a trap in asymptotics takes the form of the Lévy distribution with the spectral exponent $\beta \approx 0.8$. Assuming that the capture mechanism is characterized by a certain self-similarity, we can expect the same value of β at all scales under study. For this reason, we used the value $\beta = 0.8$ in our calculations. Assuming then that $\eta_{E\ll E_0} \sim 2.63$ and $\eta_{E\gg E_0} \sim 3.24$, we finally obtain $p \approx 2.9$ and $\delta \sim 0.27$.

To determine the next important parameter, the anomalous diffusion coefficient $D_{0\alpha}$, we used experimental data on the anisotropy of $10^3 - 10^4$ GeV particle fluxes within the framework of the scheme proposed by Osborne and coauthors in 1976 [Osborne *et al.* (1976)] and Dorman and coauthors in 1985 [Dorman *et al.* (1985)]. In particular, we found that $D_{0\alpha} \approx (1-4) \cdot 10^{-3}$ pc$^{1.7}$year$^{-0.8}$ $\alpha = 1.7$ and $\beta = 0.8$ for the three nearest sources.

In the model considered here, only one parameter α $(1 < \alpha < 2)$, related to the fractal structure of the medium, was found by fitting. Test calculations of cosmic-ray spectra showed that the best fit of experimental data was achieved for $\alpha \approx 1,7$".

In conclusion, the authors of [Lagutin *et al.* (2001d)] note that for the parameters presented above, the results of calculations are in agreement not only with the experimental increase in the spectrum steepness but also with the mass content in the energy region $10^2 - 10^5$ GeV per nucleon if the source content is p=72%, He=18%, CNO=5%, Ne-Si=3%, and Fe=2%.

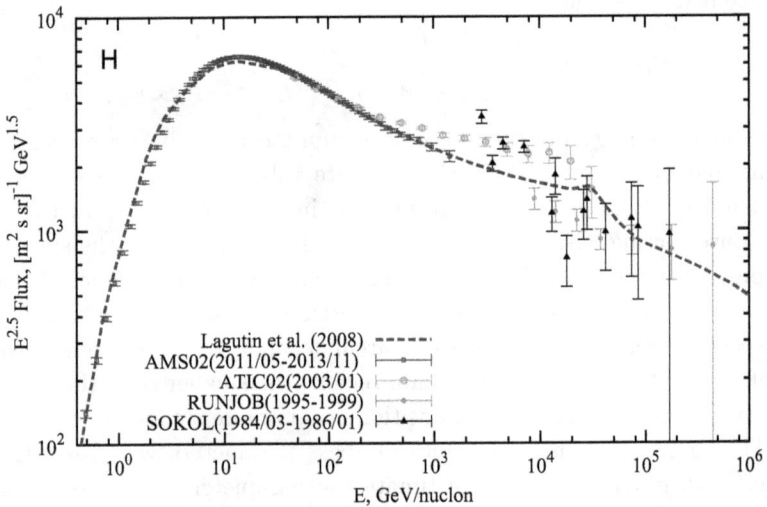

Fig. 5.7 Comparison of proton spectrum calculated in frames of the bifractional diffusion model [Lagutin *et al.* (2008)] with experimental data.

For $\alpha = 1$, the stable density can be written in a simple analytic form (the three-dimensional Cauchy distribution), which was used in [Lagutin *et al.* (2001d)] in solving Eq. (5.20) with the source

$$S(\mathbf{x}, t, E) = S_0 E^{-p} \delta(\mathbf{x}) 1_+(t).$$

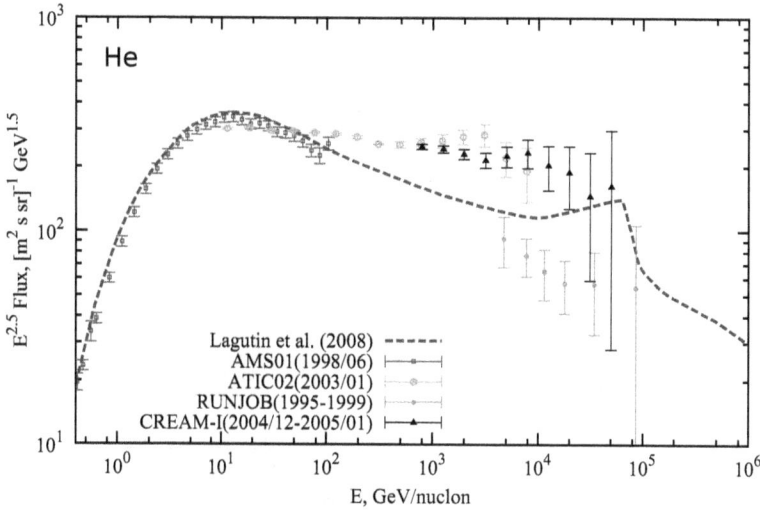

Fig. 5.8 Comparison of He spectrum calculated in frames of the bifractional diffusion model [Lagutin *et al.* (2008)] with experimental data.

As a result, the simple expression

$$N(\mathbf{x}, t, E) = \frac{S_0 E^{-p-\delta}}{2\pi D_{0,1} r^2 t} \frac{(E/E_{rad})^{2\delta}}{(E/E_{rad})^{2\delta} + 1} 1_+(t), \quad \alpha = 1, \quad \beta = 1,$$

was obtained with $E_{rad} = [r/(D_{0,1}t)]^{1/\delta}$, which allows passing from one set of asymptotic expressions to another. In particular, we can estimate the energy gap $\Delta E = E_> - E_<$, separating these two asymptotic forms. For a 20% agreement margin (the difference between the exact spectrum and its power-law asymptotic form), we obtain the energy gap somewhat smaller than two and a half orders of magnitude, which does not contradict the experimental data in general. Figures 5.7, 5.8 and 5.9 demonstrate comparison of energy spectra calculated in frames of the bifractional model [Lagutin *et al.* (2008)] with experimental data.

The stationary solution

In [Lagutin *et al.* (2001a)], a problem with a constant (in time) point-like source was considered. The solution of the stationary equation

$$D_\alpha(E)(-\Delta)^{\alpha/2} N(\mathbf{x}, E) = S_0 E^{-p} \delta(\mathbf{x}),$$

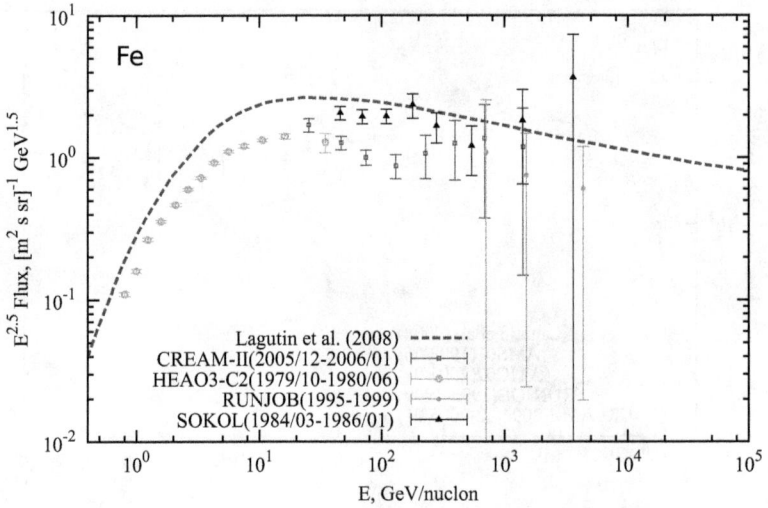

Fig. 5.9 Comparison of Fe spectrum calculated in frames of the bifractional diffusion model [Lagutin *et al.* (2008)] with experimental data.

following from (5.20) and found by using the known Mellin transformation of the three-dimensional stable density

$$\int_0^\infty \Psi_3^{(\alpha)}(r) r^{s-1} dr = \frac{2^s \Gamma(s/2) \Gamma((3-s)/\alpha)}{\alpha (4\pi)^{3/2} \Gamma((3-s)/2)}$$

under the same assumption about the energy dependence of the diffusion coefficient, has the form

$$N(\mathbf{x}, E) = S_0 E^{-p} [D_\alpha(E)]^{-3/\alpha} \int_0^\infty \Psi_3^{(\alpha)} \left(r [D_\alpha(E)t]^{-1/\alpha} \right) dt$$

$$= \frac{2^{-\alpha} S_0}{\pi^{3/2} D_{0\alpha} r^{3-\alpha}} \frac{\Gamma((3-\alpha/2)/2)}{\Gamma(\alpha/2)} E^{-p-\delta}.$$

We emphasize that this expression is an exact solution of the stationary equation, although this diffusion equation itself is approximate, representing the asymptotic regime of the process at large distances, and therefore the use of its solution at small distances is risky. In conclusion, the authors of [Lagutin *et al.* (2001a)] made an attempt to obtain the two-parameter (with $\beta \neq 1$) stationary solution as the limit (for $t \to \infty$) of the distribution of particles from a source switched on at the instant $t = 0$:

$$N_{\mathrm{st}}(\mathbf{x}, E) = \lim_{t \to \infty} N(\mathbf{x}, t, E)$$

$$= \frac{S_0 E^{-p}}{[D_\alpha(E)]^{3/2}} \int\limits_0^\infty \Psi_3^{(\alpha,\beta)} \left(r[D_\alpha(E)\tau^\beta]^{-1/\alpha} \right) \tau^{-3\beta/\alpha} d\tau.$$

However, this integral diverges, and they were forced to cut off the time integral at a large upper limit (10^{10} years), which would make sense if the limit existed. The divergence is explained by the fact that a fraction of the particles confined in traps escape from them after some time and continuously make up the total flux; however, due to the infinite mean lifetime in a trap, the equilibrium between particles captured in traps and those leaving them is not established. Mathematically, the matter is that the Riemann-Liouville fractional derivative used in this model vanishes when applied to a constant only for an integer order β, whereas a *fractional-order derivative of a constant is not zero* [Uchaikin (2013a)]:

$$\frac{\partial^\nu C}{\partial t^\nu} = \frac{Ct^{-\nu}}{\Gamma(1-\nu)}.$$

Therefore, for a stationary (time-independent) distribution of particles in a medium to exist, i.e., for the condition $\partial N_{st}/\partial t = 0$ to be satisfied, as follows from Eq. (5.22), the source density must satisfy the equation

$$S(\mathbf{x}, t, E) = D_\alpha(E)(-\Delta)^{\alpha/2} \frac{\partial^{1-\beta}}{\partial t^{1-\beta}} N_{st}(\mathbf{x}, E)$$

$$= \left[D_\alpha(E)(-\Delta)^{\alpha/2} N_{st}(\mathbf{x}, E) \right] \frac{t^{\beta-1}}{\Gamma(\beta)}.$$

Such a behavior of sources can be explained by the fact that they themselves are traps of the same type as others, emitting particles at a rate decreasing in accordance with a power law, which seems more natural than a source perpetually emitting particles at a constant rate.

Other results and other parameters

Using the results in [Lagutin *et al.* (2001d)], researchers in Lagutin's group calculated the energy spectra and mass composition by separating the particle flux into contributions from particles in a direct (unscattered) flux from near ($r < 1$ kpc) and distant ($r > 1$ kpc) sources, represented by three terms in the expression

$$J_i = \frac{v_i}{4\pi} \left[C_{0i} E^{-p+\delta/\beta} + \sum_{j:\ r_j<1} N_i(r_j, t_j, E) + C_{1i} E^{-p-\delta/\beta} \right],$$

$$i = \mathrm{p}, \mathrm{He}, \mathrm{CNO}, \mathrm{Ne} - \mathrm{Si}, \mathrm{Fe}.$$

The constants C_{0i}, and C_{1i} were determined from the same experimental data with which the results of calculations were compared. By introducing the correction for the solar modulation of galactic rays, the authors of [Lagutin *et al.* (2001a)] satisfactorily reproduced the energy spectra and the mass composition of components with different energies.

The same model was used in [Lagutin *et al.* (2001c)] for calculating the electron and positron spectra. The sources of high-energy ($E \geq 100$ GeV) electrons and positrons observed in the Solar System were shown to be comparatively young local sources (the distance is no more than 200 pc and the age is 10^5 years).

The fraction of positrons obtained from these calculations was in agreement with experimental data and the exponent $p_e = 2.95$ of the spectrum of the source for electrons and positrons proved to be close to the exponent $p_\mathrm{p} = 2.9$ found previously for protons. The authors believe that this suggests that the acceleration mechanism of these particles is the same. The energy losses of relativistic electrons were taken in the form

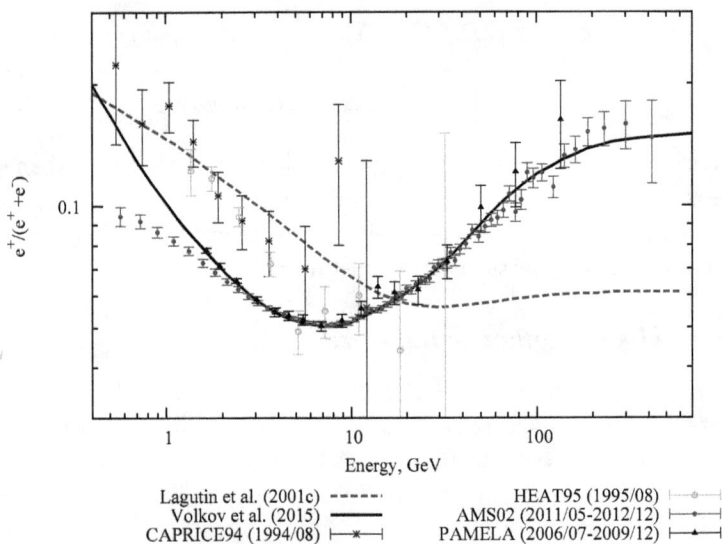

Fig. 5.10   Fraction of positrons in the electron-positron component of primary cosmic rays. Symbols are different experimental results, the dashed curve is calculated in the LU model ($\alpha = 1.7$, $\beta = 0.8$) [Lagutin *et al.* (2001c)] and solid line represents the result of the bifractional differential model [Volkov *et al.* (2015)].

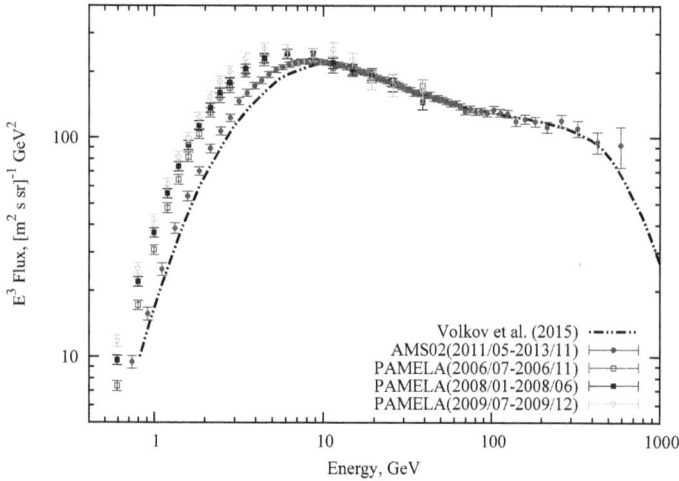

Fig. 5.11 Comparison of observed (by PAMELA and AMS-02) spectrum of electrons with the result of the fractional differential model [Volkov *et al.* (2015)].

$$-\frac{dE}{dt} = b(E) = b_0 + b_1 E + b_2 E^2,$$

(corresponding to a homogeneous medium, however), whereas the inhomogeneities of the magnetic field were assumed fractal. Such an approach assumes the absence of a correlation between the magnetic field and matter in the interstellar region, which contradicts the conventional concept of magnetic field lines "frozen" into plasma. In the case of complete correlation, the term with energy losses should enter the equation as a part of the material derivative operator raised to a fractional power.

The calculations by Lagutin *et al.* with the parameters $\alpha = 1.7$, $\beta = 0.8$ (which we refer to as the Lagutin-Uchaikin (LU) model for convenience) were continued in [Lagutin and Tyumentsev (2004)], and it seems that no serious disagreements with experiments Caprice94, HEAT95 and some others were observed (Fig. 5.10, dashed line). However, Fig. 5.10 shows that this theoretical result is not in good agreement with new more accurate measurements by PAMELA and AMS-02. In [Volkov *et al.* (2015)], authors presented new calculations of the energy spectra of CR electrons, positrons and positron fraction within the same fractional approach with $\alpha = 1.4$, $\beta = 1$, $p = 2.85$, $\delta = 0.27$ (Fig. 5.10, solid line). See also Figs. 5.11 and 5.12.

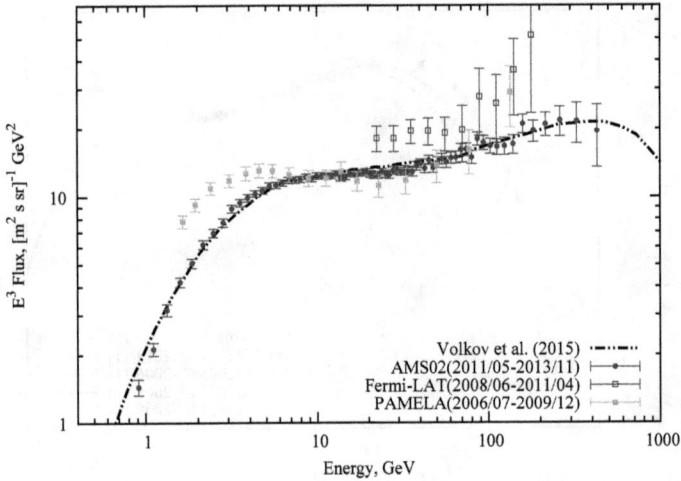

Fig. 5.12 Comparison of observed (by PAMELA and AMS-02) spectrum of positrons with the result of the fractional differential model [Volkov *et al.* (2015)].

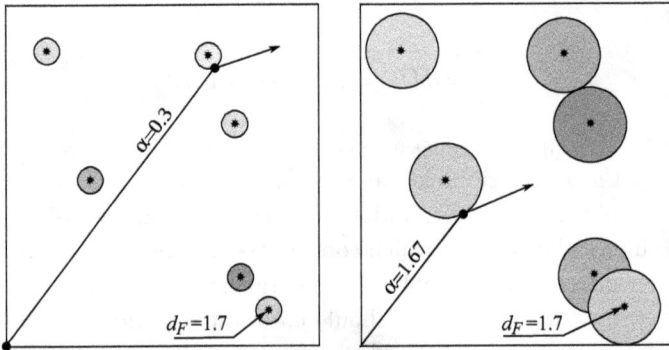

Fig. 5.13 Difference in the range distribution for the same distribution of centers and different radii of spheres simulating magnetic clouds: (a) LT model, (b) LU model.

However, the authors of [Erlykin *et al.* (2003)] performed calculations with $\alpha = 0.5, 1.0$ and 1.5 and concluded that the value $\alpha = 1.0$ provides the best fit of the experimental data and at the same time is consistent with the Kraichnan spectrum $F(k) \propto k^{-\xi}$ known from turbulence theory (the parameter α is related to the index ξ as $\alpha = (3 - \xi)/2$; we have $\xi = 3/2$ and $\alpha = 3/4$ for the magnetic energy). In [Erlykin and Wolfendale (2013)], the value $\alpha = 1$ is retained only for the periphery of the Galaxy, while $\alpha = 0.8$ is used for the inner part. Such an approach seems dubious, because it uses

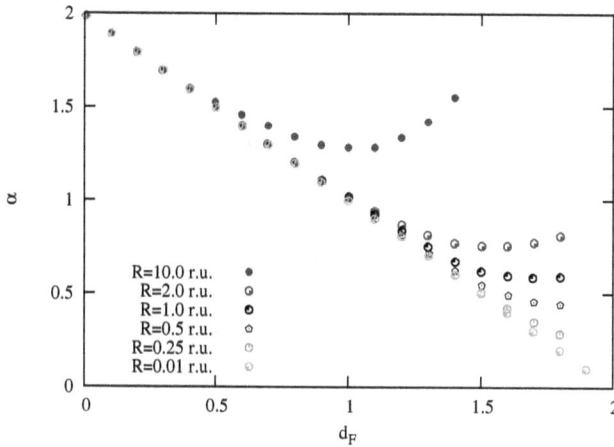

Fig. 5.14 Exponent α of the range distribution (digitized from [Lagutin *et al.* (2004)]) as a function of the fractal dimension of the distribution of the centers of spheres and their radii $R = 0.01$, 0.25, 0.5, 1.0, 2.0, and 10.0 rel. units.

propagators obtained for a homogeneous infinite medium with a unified exponent α.

Moving further in this direction, Lagutin and Tyumentsev [Lagutin and Tyumentsev (2004)] took a new value of the key parameter $\alpha = 0.3$ (we call this variant the LT version of the fractional differential model of galactic cosmic ray transfer) and performed extensive calculations with this parameter in 2004-2010 (which are available in proceedings of cosmic ray conferences). The motivation for choosing this value of α was the finding in [Uchaikin (2004b)] that the fractional exponent α in the range distribution does not coincide with the fractal dimension d_F of the medium. In addition, it was desirable to match mean free paths with the known parameters of the real interstellar medium. The first calculations of this type were performed in our paper [Uchaikin and Korobko (1998)]. Lagutin *et al.* simulated the free paths of particles in a medium with randomly distributed spherical targets and wrote in [Lagutin *et al.* (2004)] that because "for media with a fractal dimension $1 < d_F < 2$,

$$\alpha \approx 2 - d_F, \tag{5.23}$$

we find $\alpha = 0.3$ for the galactic medium with $d_F = 1.7$ [Duffy *et al.* (1995)]. We set the exponent β equal to 0.8, as in [Lagutin and Uchaikin (2003)]. Another important parameter of the model, the anomalous diffusion coefficient $D_{0\alpha}$, can be estimated by comparing the position of the knee in the observed cosmic-ray spectrum with the position of the "breaking point" in fractionally stable distributions $\Psi_3^{(\alpha,\beta)}(\xi)$. Because the breaking point of $\Psi_3^{(0,3;0,8)}(\xi)$ is observed at $\xi \approx 2.3$, we obtain

$$r(D_{0\alpha}E^\delta t^\beta)^{-1/\alpha} \approx 2.3.$$

Assuming that near sources are also involved in the formation of the knee in the energy spectrum, we find for $r \approx 10^2$ pc and $t \approx 10^5$ years that

$$D_{0\alpha} \approx (3-5) \cdot 10^{-6} \text{ pc}^{0.3}/\text{y}^{0.8}.$$

For such $D_{0\alpha}$ and the parameters δ, α, and β chosen above, a unique relation exists between r and t for the sources providing the knee in the spectrum at $E_{\text{fr}} \approx 3 \cdot 10^6$ GeV."

This conclusion is incorrect, because expression (5.23) is valid only for extremely small elements forming a fractal. The size of magnetic inhomogeneities of the galactic medium does not belong to this type: they occupy 5% of the volume (according to old estimates made in [Strömgren (1948)], which, however, were confirmed by subsequent astronomical measurements: for example, the volume fraction of the interstellar space not filled with hydrogen was estimated as 95% in [Lagutin and Uchaikin (2001)]), which becomes more than 30% on passing to the linear scale. That formula (5.23) is invalid for such scales can be clearly seen, in particular, from [Elmegreen (1998)], where, based on the experimental studies of the cloudy structure of interstellar hydrogen, the value $d_F = 2.3$ was obtained, which is obviously incompatible with expression (5.23). We also recall that the fractal dimension alone does not characterize a fractal-like structure completely: the size of its elements and their concentration in space are determined not by the exponent d_F in the fractal formula $V_F(R) = CR^{d_F}$ but by the coefficient $C = V_F(1)$ determining the volume fraction of a unit-radius ball filled with a fractal. An increase in the size of magnetic clouds, preserving the same fractal dimension d_F, obviously reduces the range distribution, thereby increasing the exponent α. This effect is also clearly seen in Fig. 5.14 taken from [Lagutin et al. (2004)].

Samples of random walk trajectory for three propagation models under discussion are presented in Fig. 5.15. Parameters are indicated in the caption. The ordinary diffusion model and LNU-model manifest trajectories looking quite acceptable in this scale while LT [Lagutin and Tyumentsev (2004)] trajectories are absolutely unexpected: we see a small cluster of points where the particle wastes its life time and then instantly crosses the Galaxy and disappears. Such long rectilinear instantaneous travels contradict to physical principles and our knowledge about ISM property.

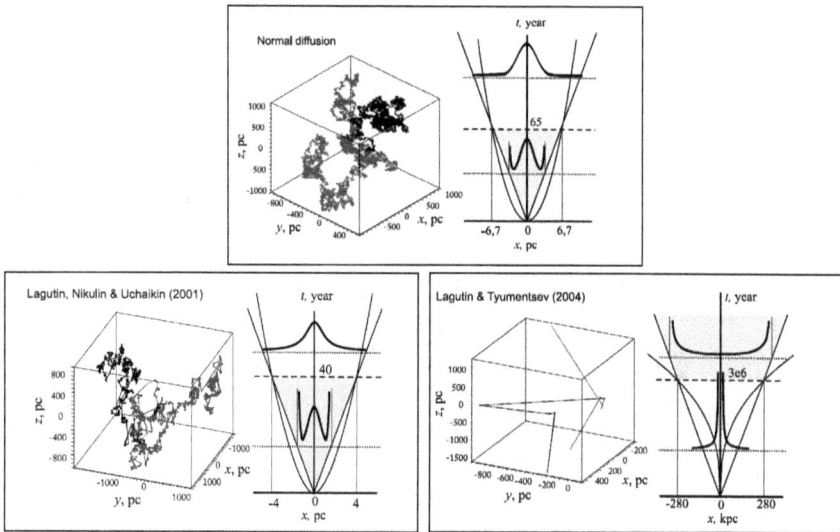

Fig. 5.15   Examples of trajectories in three models: normal diffusion, LNU-model ($\alpha =$ 1.7, $\beta = 0.8$, $D_0 = 2.4 \cdot 10^3$ pc$^{1.7}$/year$^{0.8}$, $D = D0R^{0.27}$, $E = 10^6$ GeV), and LT-model ($\alpha = 0.3$, $\beta = 0.8$, $D_0 = 4 \cdot 10^6$ pc$^{0.3}$/year$^{0.8}$, $D = D_0R^{0.27}$, $E = 10^6$ GeV). The left panels present schematically the diffusion packet spreading laws, shadowed regions correspond to regions with determining influence of finiteness of velocity. equation.

The original LU model has been used by other authors. The authors of [Ketabi and Fatemi (2009)] used the stationary solution for $\alpha = 1.8$ in Monte Carlo simulations of cosmic-ray diffusion from a supernova (in the Erlykin-Wolfendale model) and obtained good agreement with the observed characteristics. Based on these results, they concluded that the source of cosmic rays is a supernova. The authors of [Kermani and Fatemi (2011)] in fact repeated our first calculations and concluded again that the value $\alpha = 1.65$ provides the best fit of the energy spectrum and radial gradient in the vicinity of the Solar System, and the admissible range of α values is $1.6 - 1.9$, whereas the value $\alpha = 2$, corresponding to classical diffusion, is unacceptable. [Doostmohammadi and Fatemi (2012)] is entirely devoted, in fact, to estimating the exponent α. The authors studied the transfer of cosmic rays with the energy in the range from 10^{12} to 10^{19} eV from a supernova with the energy fraction converted to cosmic rays from 0.01 to 0.1 and the supernova age from 10^4 to 10^7 years for different $\alpha \in [0.5; 2]$ and concluded that the propagation of cosmic rays in the Galaxy is governed by anomalous diffusion of the second kind (with $\alpha = 1.7$) and is not described by the normal diffusion model ($\alpha = 2$). Therefore, this conclusion also rejects the LT hypothesis about anomalous diffusion with $\alpha = 0.3$ (diffusion of the first kind).

After critical paper [Uchaikin (2010b)], the authors of the LT version decided to return to the original model. In 2011, they published calculations with $\alpha = 1.1$ [Raikin and Lagutin (2011)] and presented the results of Monte Carlo simulations with $\alpha = 1.7$ at the European Cosmic Ray Symposium in 2012 [Lagutin and Tyumentsev (2013)].

Table 5.1. Parameters of the fractional differential model.

α	β	p	δ	D_0 pc$^\alpha$/year$^\beta$	References (year)
1.7	0.8	2.9	0.27	$1 - 4 \cdot 10^{-3}$	[Lagutin *et al.* (2001b)]
0.3	0.8	2.86	0.27	$3 - 5 \cdot 10^{-6}$	[Lagutin and Tyumentsev (2004)]
0.7	1	2.6	0.27	$2 \cdot 10^{-5}$	[Lagutin *et al.* (2008)]
1.1	0.8	2.85	0.27	$1 \cdot 10^{-4}$	[Raikin and Lagutin (2011)]
1.7	0.8	2.9	0.27	$2, 4 \cdot 10^{-3}$	[Lagutin and Tyumentsev (2013)]

The coefficient D_0 suffers the greatest changes under these variations of α (Table 5.1). However, the final results (the spectrum and composition) were practically unchanged (Table 5.1). Obviously, a change in parameter α was compensated by the corresponding change in the diffusion coefficient. However, as a change, the dimension of the diffusion coefficient also changes, and we cannot quantitatively estimate these changes or estimate the difference between velocity and acceleration values. It is possible that the coefficients C_i and C_{0i} play an important role here, and the results can be changed by varying these coefficients. At the same time, these manipulations drastically change the space-time shapes of particle trajectories, which was neglected by the authors of the LT version. If the dynamics of the process described by a fractional differential equation are not consistent with the reality, simulations become a multiparametric approximation of the known experimental results.

We also note that the parameter $\gamma = \beta/\alpha$ characterizing the expansion law of a diffusion packet in the LU model turns out to be suspiciously close to the classical value $1/2$. Hence, if we initially attempt to construct a model in which the expansion of a diffusion packet is consistent with the standard theory, but the shape is self-similar and has a power-law asymptotic behavior required for the description of the observed knee, then, for $\beta = 0.8$, we arrive precisely at the LU model ($\beta/\alpha \approx 0.5$) rather than at the LT version ($\beta/\alpha \approx 2.66$). Obviously, the diffusion packet dynamics in the LT version in the large-time asymptotic regime contradicts the physical reality, because the packet expansion velocity increases infinitely, despite the natural restriction imposed on the velocities of particles in the packet.

Boundary effects

Apart from particles associated with solar flares, the cosmic radiation comes from outside the solar system. The high energy part of it includes all stable charged particles and nuclei with lifetimes of order 10^6 years or longer. This indicates an important role that the boundary effects can play due to the presence of the galactic disk and halo (see [Schlickeiser (2009); Artmann *et al.* (2011); Litvinenko *et al.* (2015)]). CRs are produced in sources distributed in thin galactic disk with radius $R_d \approx 20$ kpc. The diffusion domain is the cylinder modeling the magnetized Galactic halo of height $H \approx 4$ kpc. Its boundaries are assumed to be absorbed: CRs reaching these boundaries leave the Galaxy and never return. For the diffusion and telegraph equations, the boundary conditions are well established [Masoliver *et al.* (1992, 1993)]. However, the need to take into account the turbulent nature of ISM makes the problem more complicated.

The ordinary Laplacian is a local operator, having the same form independently of boundaries, but the fractional-order Laplacian is a nonlocal operator depending on the medium beyond boundaries. For this reason, the definition based on the Fourier transform cannot be applied to the fractional Laplacian acting in a bounded medium. The statement of such a problem should be accompanied with a specification of the desired function values throughout an outer region. So, we have to return to the integral representation of the operator. The random flight interpretation can help in specifying the conditions but some subtle points such as distinction between first-passage and first-arrival times or between free and reflecting boundary conditions appear. [Zoia *et al.* (2007)] have investigated the matrix representation of the one-dimensional fractional Laplacian and solved numerically in connection to the first-passage problem (the Lévy-flights under absorbing boundary conditions) and to the long-ranged interfaces with no constraints at the ends (the free boundary conditions).

[Krepysheva *et al.* (2006)] analyzed the symmetric Lévy flights restricted to a semi-infinite domain by a reflective barrier. They show that the introduction of the boundary condition induces a modification in the kernel of the nonlocal operator. The operators $-(-\Delta)^{\alpha/2}$ and $-(-\Delta)^{\alpha/2}_{\text{refl}}$ differ in the kernels, but the difference becomes small when $x + \xi$ is large. Nevertheless, omitting the term $(x + \xi)^{1-\alpha}$ we would get a decreasing integral with respect to x, whereas the total amount of the diffusing matter should be preserved.

[Rafeiro and Samko (2005)] have introduced a version of the fractional Laplacian for a bounded domain as a generalization of the Marchaud formula for one-dimensional fractional derivatives on interval (a, b), $-\infty < a < b \leq \infty$, to the multidimensional case of functions defined on domain $G \subset \mathbb{R}^d$. In other words, they derived the Riesz fractional derivative of the zero continuation of $f(\mathbf{x})$ from G to the whole space $\mathbb{R}^d$.

[Guan and Ma (2006)], investigating the reflected symmetric α-stable processes, introduced *regional fractional Laplacian*. For more detail, the reader can refer to [Hu and Kallianpur (2000); Bogdan and Byczkowski (2000); Song and Vondracek (2003); Denisov *et al.* (2008); Jeng *et al.* (2010)]. Better understanding of the Laplacian in a bounded domain can be achieved on the base of the nonlocal operator theory [Vázquez (2004); Gunzburger and Lehoucq (2010); Du *et al.* (2012)].

5.6 Percolation of cosmic rays

In inhomogeneous media with irregular, porous, large-grain, and winding structures, the slowed down diffusion (subdiffusion) of a liquid can be observed (percolation). There are several models of such a process leading to fractional differential equations [Uchaikin (2013b)].

One of them involves a periodic lattice each of whose sites can be either occupied or free with a certain probability. A set of neighboring free points forms a cluster. Lines (paths) connecting these points can be either conducting (with their ends coming out to the cluster surface) or blind. We take one conducting line and stretch it along a straight line, accurately directing other lines coming out of its sites perpendicular to it. Those lines can, in turn, branch or, on the contrary, break; without going into further details, we assume for simplicity that they are unbranching infinite lines. In this way, we obtain a 'comb' (Fig. 5.16).

We now trace a liquid particle (a point mass) entering the main line. For this, we specify the law of motion of the particle along the line; let it be the ordinary diffusion. Having arrived at the first site, the particle passes to a side branch. According to the diffusion laws, even in the case of an infinite (more exactly, a semi-infinite) branch, the particle will necessarily (i.e., with probability 1) return, but the distribution of the return time has a power tail $\mathsf{P}(T > t) \propto (t/\tau_0)^{-\beta}$ with the exponent $\beta = 1/2$ and the

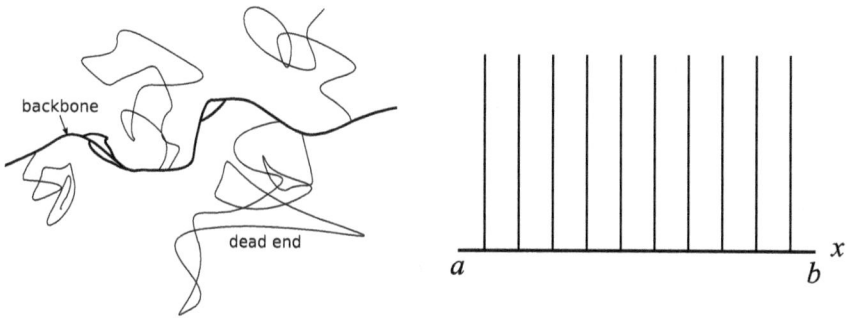

Fig. 5.16 Schematic of a percolation cluster and its comb model.

characteristic time τ_0. Following only the coordinate x, we can say that the particle stopped at this point for some time (captured in a trap), then continued to diffuse along the x axis with the coefficient D_x, then was again captured in the same or neighboring trap and remained in it for a different time, etc. In the limit of small distances between traps, such that

$$\tau_1 D_x / \tau_0^\beta \to D > 0,$$

where τ_1 1 is the mean diffusion time between hitting the traps, this process is described by the integral equation

$$f(x, t) = \delta(x) + \frac{1}{\Gamma(\beta)} \int_0^t (t - \tau)^{\beta - 1} D \Delta f(x, \tau) d\tau, \quad \beta = 1/2,$$

whose kernel reflects the delay in the diffusion of the coordinate x caused by the residence of the particle outside this axis.

We recall Cauchy's integral formula representing the n-fold integral in terms of the ordinary integral,

$$I^n f(t) = \frac{1}{(n - 1)!} \int_0^t (t - \tau)^{n-1} f(\tau) d\tau, \quad n = 1, 2, 3, \ldots$$

and allowing an analytic continuation to fractional (and even complex) values of the exponent n:

$$I^\beta f(t) = \frac{1}{\Gamma(\beta)} \int_0^t (t - \tau)^{\beta - 1} f(\tau) d\tau, \quad \beta > 0.$$

Introducing the Riemann-Liouville fractional derivatives

$$\frac{\partial^\beta}{\partial t^\beta} f(t) = \frac{\partial}{\partial t} l^{1-\beta}(t) = \frac{1}{\Gamma(1-\beta)} \frac{\partial}{\partial t} \int\limits_0^t \frac{f(\tau)d\tau}{(t-\tau)^\beta}, \quad 0 < \beta < 1,$$

and applying them to the integral equation written above, we obtain the fractional differential subdiffusion equation

$$\frac{\partial^\beta f(x,t)}{\partial t^\beta} = D\Delta f(x,t) + \delta(x)\delta_\beta(t),$$

where $\delta_\beta(t) = t^{-\beta}/\Gamma(1-\beta)$.

We make some remarks. First, the 'fractional delta function' is so called because it is a continuation of the known definition of the delta function as the derivative (in the generalized sense) of the Heaviside function

$$1_+(t) = \begin{cases} 0, \, t < 0; \\ 1, \, t > 0, \end{cases}$$

namely,

$$\delta_\beta(t) = \frac{d^\beta 1_+(t)}{dt^\beta}.$$

Second, a solution of the subdiffusion equation has the meaning of a probability density, and $\delta_\beta(t)$ ensures that the integral of the solution is constant in x (normalization). Finally, the fractional character of the derivative is here a consequence of the infinite length of side branches. If their length is limited, we again have the first-order time derivative instead of a fractional derivative in the x-diffusion equation, i.e., the usual diffusion equation (albeit with the diminished diffusion coefficient).

The comb model was introduced in [White and Barma (1984); Weiss and Havlin (1986)]. The fractional diffusion equation to describe transport on the continuous comb has been obtained by [Arkhincheev and Baskin (1991)]. The fractional diffusion on the continuous combs was studied in [Lubashevskii and Zemlyanov (1998); Baskin and Iomin (2004); Iomin (2011); Sibatov and Morozova (2015); Sandev *et al.* (2016a,b)]. The advection-diffusion in comb can be described in the framework of the CTRW concept [Metzler and Klafter (2000)]. Diffusion in a teeth could be considered as a trapping event characterizing by a random waiting time, propagation occurs due to motion along the backbone [Sandev *et al.* (2016a)]. Generalizations of the comb model consider trap-limited diffusion [Sandev *et al.* (2016a)], multiple trapping [Sibatov and Morozova (2015)] or fractional Brownian motion [Zahran *et al.* (2003)] on combs, fractal comb-like geometry [Iomin (2011); Sandev *et al.* (2016b)], multidimensional generalization [Arkhincheev (2010)] and others.

The problem of the percolation of a liquid through a porous medium (which might look like a very particular problem) attracted the attention of researchers in different fields (including cosmic-ray physics) because the percolation process turned out to be critical. A signature of this important property is the existence of percolation threshold, the minimum density of free sites of the lattice above which the liquid percolates over the entire infinite lattice and below which the liquid occupies only a finite region of the medium. For densities close to the threshold density, percolation occurs over a fractal set, and this process is governed exclusively by the laws of criticality, irrespective of the macroscopic properties of the medium [Bakunin (2008); Isichenko (1992)]. A deep connection between the percolation model and cosmic electrodynamics (the multiscale interaction of fields and currents in the distant Earth's magnetotail, self-organization processes in magnetized plasmas, the evolution of large-scale magnetic fields in the solar photosphere and interstellar space, and the construction of a self-consistent model of the turbulent current sheet) is demonstrated in the remarkable review by [Zelenyi and Milovanov (2004)].

Chapter 6

Acceleration of cosmic rays

One of the fundamental processes in space physics is acceleration of charged particles. Energetic electrons and ions are accelerated in supernova explosions, in ISM, on the Sun, on the termination shock of the Solar system, in interplanetary space, in the Earth's magnetosphere. This chapter considers modification of the models of acceleration process in case of anomalous diffusion of cosmic rays. In section 6.1, a short overview of acceleration mechanisms is given. Section 6.2 describes two cases of momentum diffusion operators in the kinetic equation necessary to consider an acceleration process. Superdiffusive and subdiffusive shock acceleration, or modifications of first order Fermi acceleration, are considered in sections 6.3 and 6.4, respectively. The model of anomalous diffusive reacceleration or generalized second order Fermi acceleration is described in section 6.5. In last section of the chapter, stochastic acceleration in Earth's magnetotail is considered.

6.1 Acceleration mechanisms

An important feature of a collisionless cosmic plasma is inherent processes leading to the generation and acceleration of fast charged particles with energies much higher than the thermal energy [Berezhko and Krymskii (1988)]. The first model of CR acceleration was proposed by [Fermi (1949)]. Two classes of Fermi acceleration are distinguished. Both work in a collisionless environment, otherwise collisions will lead to losses neutralizing the process of acceleration. The Fermi acceleration of the first order occurs in shock waves. As a rule, before and behind the front of the shock wave, there are magnetic inhomogeneities, which are capable of reflecting charged particles. Multiple reflection between two inhomogeneities will lead to a significant acceleration of the particles. The diffusion-like transport of charged particles trapped near magnetized shock front allows them to repeatedly cross the shock front [Krymskii (1977)]. The mechanism proposed by Fermi gives that each pair of successive intersections increases the energy of the particle in proportion to the energy already achieved. After multiple intersections, CR particle achieves a high energy allowing it to leave the trapping region. The resultant energy spectrum of the particles has a power form with an exponent of 2. The acceleration magnitude is proportional to the first order of the shock wave velocity. Only particles whose energy is greater than the energy of plasma will be accelerated, and low-energy particles cannot cross the shock front, they are magnetized. There is no acceleration from zero energy, the original particles must be injected into the region of the shock front. The probability of leaving the shock increases with growing energy. So, the number of particles decreases as the energy increases approximately in a power-law manner. Acceleration is proved to be very effective, and the spectrum of accelerated particles is quite hard: $\propto E^{-2}$ up to the maximum achievable energy $E_{\max}$ of accelerated particles. It is necessary to take into account the inverse effect of CRs on the medium, which leads to a modification of the shock wave and the appearance of a smooth long segment, the so-called prefront, in addition to the usual thermal front. Such a modification, in turn, affects the spectrum of CRs. Thus, it is necessary to account for the inverse influence of CRs on the medium and to use a self-consistent solution.

The second-order Fermi acceleration describes the acceleration of CRs in collisions with moving magnetic clouds. A particle acquires energy when it meet a cloud moving opposite, or loses energy when it collides with a cloud running away. The probability of collision with the incoming mirror is higher than with the running mirror, so in general the particles are accelerated. The increment of energy is proportional to the square of the average velocity of the magnetic clouds (from here, the name *second order*). An alternative name for the second-order Fermi acceleration is stochastic acceleration or *reacceleration*. Two kinds of Fermi acceleration are depicted schematically in Fig. 6.1.

Consider the second order Fermi acceleration in more detail. Let E_i be an initial energy of a CR "scattering" elastically on a magnetic cloud moving with velocity $V \ll c$. The variables, energy and angle (E_i, θ_i) characterize a particle before entering the cloud, and variables (E_f, θ_f) correspond to a CR exiting the cloud. These variables are shown in Fig. 6.1 (right panel). Using Lorentz transformation between the laboratory and cloud system, one could connect energies of a CR entering the cloud (see, e.g. [Kachelriess (2008)])

$$E_i' = \gamma E_i (1 - \beta \cos \theta_1), \quad \beta = V/c, \quad \gamma = 1/\sqrt{1 - \beta^2},$$

First order Fermi acceleration   Second order Fermi acceleration

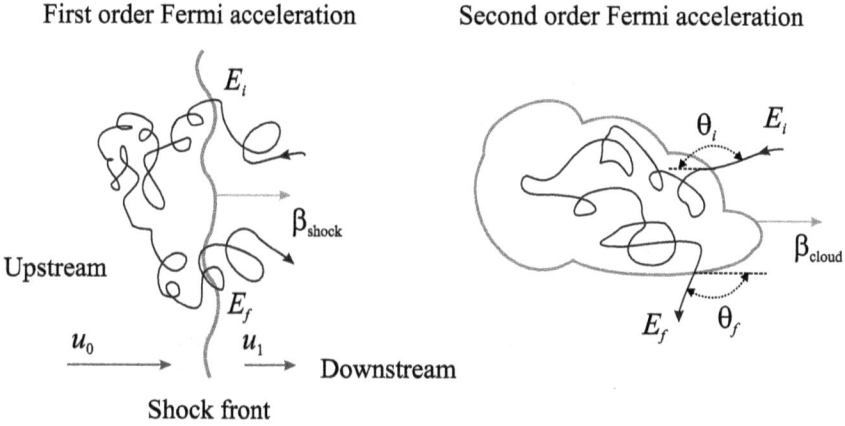

Fig. 6.1   Fermi acceleration mechanisms.

and energies after exiting the cloud

$$E_f = \gamma E'_f (1 + \beta \cos \theta'_2).$$

Scattering on magnetic inhomogeneities is collisionless, and the cloud is quite massive. So, energy is conserved in the cloud system, $E'_f = E'_i$. As a result one could obtain energy gain after single "scattering",

$$\xi = \frac{E_f - E_i}{E_i} = \frac{1 - \beta \cos \theta_i + \beta \cos \theta'_f - \beta^2 \cos \theta_i \cos \theta'_f}{1 - \beta^2} - 1.$$

To obtain the mean gain, we have to average $\cos \theta_i$ and $\cos \theta_f$. Because of multiple scattering of CR on magnetic irregularities in the cloud, the exit direction is assumed to be isotropically distributed, and we have $\langle \cos \theta'_f \rangle = 0$. The "scattering" rate is proportional to the relative velocity $v - V \cos \theta_i$. If we consider ultrarelativistic CRs, $v \approx c$, the collision rate is

$$\frac{dn}{d\Omega_i} \propto (1 - \beta \cos \theta_i).$$

Now, one could obtain the average value of $\langle \cos \theta_i \rangle$,

$$\langle \cos \theta_i \rangle = \frac{\int \cos \theta_i \frac{dn}{d\Omega_1} d\Omega_i}{\int \frac{dn}{d\Omega_1} d\Omega_i} = -\frac{\beta}{3}.$$

Substituting results for $\langle \cos \theta_i \rangle$ and $\langle \cos \theta_f \rangle$ into formula for energy gain, we have

$$\langle \xi \rangle = \frac{1 + \beta^2/3}{1 - \beta^2} - 1 \approx \frac{4}{3} \beta^2.$$

This acceleration is not very effective, because the energy gain per single "scattering" is only of second order of the small parameter $\beta = V/c$.

Approximately, first and second orders Fermi acceleration could be considered as some multiplicative process with increment $\Delta E/E = \xi$ [Kachelriess (2008)]. After n cycles of acceleration, the energy is

$$E_n = E_0(1 + \xi)^n.$$

So, the number of cycles needed to reach energy E_n is expressed as

$$n = \ln(E_n/E_0)/\ln(1 + \xi).$$

If escape probability p_{esc} per acceleration cycle is constant, then the probability to stay in the acceleration region after n acceleration events is $P(k > n) = (1 - p_{esc})^n$. So, the probability for the particle to have energy higher than E is

$$P(\varepsilon > E) = \sum_{m=n}^{\infty} (1 - p_{esc})^m = \frac{(1 - p_{esc})^n}{p_{esc}} \propto \frac{1}{p_{esc}} \left(\frac{E}{E_0} \right)^{\gamma},$$

where

$$\gamma = \ln \left(\frac{1}{1 - p_{esc}} \right) / \ln(1 + \xi) \approx p_{esc}/\xi, \quad \xi \ll 1, \quad p_{esc} \ll 1.$$

So, both first and second orders Fermi acceleration lead to a power-law energy spectrum of accelerated particles. The power law behavior is valid up to some maximal energy limit existing due to the finite life-time and finiteness of number of cycles, due to the energy dependent escape probability p_{esc} increasing with ε, and due to energy losses that balance the energy gain at some level.

There exist another acceleration mechanisms of charged particles in astrophysical plasma. In a crossed electric and magnetic field, charged particles move in the direction normal to the magnetic and electric field vectors. If the magnetic field is inhomogeneous and there is a gradient of the magnetic field in the direction of motion, the particle energy increases (so-called, $\mathbf{E} \times \mathbf{B}$ *drift acceleration*).

The increment of particle energy in the *betatron mechanism* is related to conservation of the first adiabatic invariant $\mu = E/B$ at increasing magnetic field strength. The increment of the magnetic flux on the area ringed by the trajectory of the particle leads to the generation of an electric field directed along the trajectory of the particle. Betatron acceleration works in all cases when a particle is transferred to a region with a stronger magnetic field.

If the particle moves along a curved line of force or is transferred to another force line with a change in the direction of the magnetic field vector, it will be accelerated by the centrifugal force if the direction of the force coincides with the direction of the particle transport. This is so-called *centrifugal acceleration.*

Inductive acceleration mechanisms include a variety of mechanisms of acceleration by an induction electric field. As a rule, electric field produced due to rapid changes in the magnetic field accelerates the charged particles due to transfer to an area with a higher **H**. At a certain configuration of the magnetic field, when the field lines with opposite direction are close to each other, the reconnection can occur (see Fig. 6.2). As a result, induction electric field appears, which accelerates the particles. The reconnection is the main mechanism in explaining the acceleration of solar CRs, and it is involved in the theory of magnetospheric substorms.

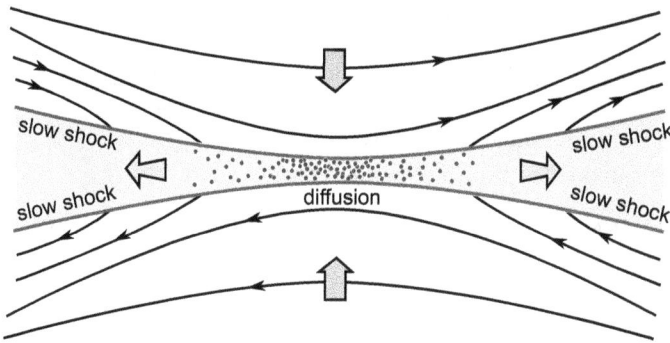

Fig. 6.2   To the inductive acceleration mechanism.

6.2   Two kinds of momentum diffusion operator

If we consider the passage of a CR-particle through a homogeneous on average turbulent medium and do not interested in its coordinates, $\overline{f}$ does not depend on $\mathbf{x}$, $\overline{f}(\mathbf{x}, \mathbf{p}, t) = f(\mathbf{p}, t)$, and Eq. (5.8) becomes

$$\frac{\partial f}{\partial t} + \mathbf{F}_0 \cdot \frac{\partial f}{\partial \mathbf{p}} = \mathsf{K} f(\mathbf{p}, t). \tag{6.1}$$

As in the spatial diffusion, reducing the integro-differential kinetic equation to the momentum diffusion equation is performed through the expansion of the integrand in the collisional term. There exist a few versions of such expansion. One of them is based on assumption that the *absolute value of the momentum change* $|\Delta\mathbf{p}| = |\mathbf{p} - \mathbf{p}'|$ is small, as for instance in the elastic collision of a heavy incident particle with a light atom (we denote this case A). Another version (case B) assumes smallness of *change of absolute value of momentum* $\Delta p = |\mathbf{p}| - |\mathbf{p}'|$ whereas the momentum direction is allowed to undergo an arbitrary change, as when a light particle collides with a heavy atom. Assumption on isotropic scattering leads to the following fractional generalizations of the FP-equation (case A):

$$\frac{\partial f(\mathbf{p}, t)}{\partial t} = \Delta_{\mathbf{p}}(K(\mathbf{p})f(\mathbf{p}, t)), \tag{6.2}$$

where

$$K(\mathbf{p}) = (1/2) \int (\Delta \mathbf{p})^2 w(\mathbf{p} \to \mathbf{p} + \Delta \mathbf{p}) d\Delta \mathbf{p}$$

is the diffusion coefficient in the momentum space. In the CR physics, the energy analogue of Eq. (6.2) is well-known:

$$\frac{\partial N(E, t)}{\partial t} = \frac{\partial [a_1(E)N(E, t)]}{\partial E} + \frac{\partial^2 [a_2(E)N(E, t)]}{\partial E^2} \tag{6.3}$$

(see Eq. (14.2) in [Ginzburg and Syrovatskii (1964)]).

At the same time, Eq. (6.2) with diffusion term

$$\Delta_{\mathbf{p}}(K(\mathbf{p})f(\mathbf{p}, t)) = (\Delta_{\mathbf{p}}K(\mathbf{p}))f(\mathbf{p}, t) + 2(\nabla_{\mathbf{p}}K(\mathbf{p}))\nabla_{\mathbf{p}}f(\mathbf{p}, t) + K(\mathbf{p})\Delta_{\mathbf{p}}f(\mathbf{p}, t)$$

essentially differs from another equation of the diffusion type (case B, see Eq. (9.57) in [Dorman (1975)])

$$\frac{\partial f(\mathbf{p}, t)}{\partial t} = \nabla_{\mathbf{p}}(K(\mathbf{p})\nabla_{\mathbf{p}}f(\mathbf{p}, t)). \tag{6.4}$$

The latter is an exact copy of spatial diffusion equation with a variable diffusion coefficient (with $\mathbf{p}$ instead of $\mathbf{x}$).

Underline once more: the reason for the difference lies in different assumption using in their derivation: Eq. (6.2) was derived on assumption that the collision instantaneous throws over the representative point $\mathbf{p}'$ of the particle to another $\mathbf{p}$, perhaps being far from it, but the lengths $|\mathbf{p}'|$ and $|\mathbf{p}|$ of these vectors are close to each other, whereas deriving Eq. (6.4) was performed under condition that $\mathbf{p}(t)$ is continuous and even differentiable, so $\mathbf{p}'$ and $\mathbf{p}$ are close to each other.

Equations (6.2)–(6.4), added by source-, Coulomb collision- and nuclear collision terms, lie in the base of the standard set of mathematical tools for description of CR stochastic acceleration in interstellar and even interplanetary turbulent electromagnetic fields.

The most popular in CR-physics equation originated from the Fokker-Planck model is the *Parker equation*. During the propagation through the heliosphere, CR particles are modulated by the solar wind and heliospheric magnetic field (HMF). Modulation of the CR is a result of action of four primary processes: convection by the solar wind, diffusion on irregularities of HMF, particles drifts in the non-uniform magnetic field and adiabatic cooling. Using instead of energy or momentum variable the *rigidity* R, Parker derived the kinetic equation for the omnidirectional distribution function $f(\mathbf{x}, R, t)$ in the form [Parker (1965)]

$$\frac{\partial f}{\partial t} = \frac{\partial}{\partial x_i}\left(K_{ij}^S \frac{\partial f}{\partial x_j}\right) - (u_i + U_i)\frac{\partial f}{\partial x_i} + \frac{R}{3}\left(\frac{\partial U_i}{\partial x_i}\right)\frac{\partial f}{\partial R}, \quad (6.5)$$

where $\mathbf{u}$ is the guiding-center drift velocity, $\mathbf{U}$ the solar wind velocity and K_{ij}^S denotes the symmetric part of the CR diffusion tensor. For deeper understanding of the Parker equation, we can recommend to read Moraal's article [Moraal (2013)]. He modified the equation by passing from R to p in the last term:

$$\frac{\partial f}{\partial t} = \frac{\partial}{\partial x_i}\left(K_{ij}^S \frac{\partial f}{\partial x_j}\right) - (u_i + U_i)\frac{\partial f}{\partial x_i} + \frac{1}{3p^2}\frac{\partial}{\partial p}\left(p^3 u_i \frac{\partial f}{\partial x_i}\right)\frac{\partial(p^3 f)}{\partial p}. \quad (6.6)$$

B. A. Tverskoy [Tverskoy (1968)] has used the velocity-diffusion model for description of interaction of charged particles with turbulent pulsations in short-wave region of spectrum, accelerating the particles. He showed that such acceleration is a universal property of plasma turbulence, belonging to fundamental features of the processes under consideration:

a) the resonant nature of particle scattering on waves;
b) the continuity of the turbulent energy flow in the space of wave numbers;
c) increasing $K(v)$ with v as $v^{\nu-1}$, where ν is the index of the turbulent pulsations spectrum.

On assumption of the hydromagnetic character of the interplanetary turbulence, and that its basic scale is much larger than Larmor radius of the particles under acceleration, the author has obtained for the isotropic velocity distribution

$$\frac{\partial f(\mathbf{v}, t)}{\partial t} = \frac{K_1}{v^2}\frac{\partial}{\partial v}\left(v^{\nu+1}\frac{\partial f(\mathbf{v}, t)}{\partial v}\right). \quad (6.7)$$

Its solution under initial condition $f(\mathbf{v}, 0) \propto \delta(\mathbf{v})$ is of the form

$$f(\mathbf{v}, t) = \text{const} \cdot t^{-3/(3-\nu)} \exp\left\{-\frac{v^{3-\nu}}{(3-\nu)^2 K_1 t}\right\}. \quad (6.8)$$

Observe that under $\nu = 1$ the solution reproduces the Maxwellian spectrum, and under $\nu = 2$ the proton spectrum from solar flashes e^{-v/v_0}. The values

$\nu > 2$ are not of interest because in this case the acceleration is determined by the Fermi mechanism, not by cyclotron resonance.

[Fedorov and Stehlik (2008)] found the solution to the equation

$$\frac{\partial f}{\partial t} - \frac{1}{p^2}\left(p^2 K_1 p^\gamma \frac{\partial f}{\partial p}\right) = \frac{1}{p^2}\delta(p - p_0)\delta(t) \qquad (6.9)$$

in the form

$$N(p,t) = \frac{\exp[-\frac{1+\gamma}{2}\ln(pp_0)]}{(2-\gamma)(mc)^3 t}\exp\left(-\frac{p^{2-\gamma} + p_0^{2-\gamma}}{(2-\gamma)^2 t}\right) I_\nu\left(\frac{2\exp[\tilde{\gamma}\ln(pp_0)]}{(2-\gamma)^2 t}\right).$$

Here p is expressed in mc, t in $(mc)^2/D_1$, and $\tilde{\gamma} = (2-\gamma)/2$.

6.3   Superdiffusive shock acceleration

According to the modern point of view, the CR spectrum in the range $1 - 10^6$ GeV is produced by shock acceleration in the Galaxy. Its power-law shape $N(E) \sim E^{2.7}$ is believed to be produced by strong shocks, probably of young supernova remnants (SNR) [Blandford and Ostriker (1980)]. The shock wave (a surface of discontinuity, which moves inside the medium, with pressure, density, temperature and velocity experiencing a jump) is generated due to motion of the shell released after explosion in the surrounding ISM.

The kinetic equation proposed by [Parker (1965); Dolginov and Toptygin (1966)] and written in previous section is widely used for studying energetic particle transport and diffusive shock acceleration [Bell (1978); Lee and Fisk (1982); Drury (1983); Jones and Ellison (1991); Berezhko and Ksenofontov (1999)]. [Lagage and Cesarsky (1983)] used it to study the acceleration of CRs at supernova remnant shocks, [Zank et al. (2000)] described acceleration and transport of solar energetic particles at coronal mass ejection driven shocks, [Zank et al. (2015)] applied it to study acceleration and transport of CRs at the solar wind termination shock.

Consider the infinite planar shock in the frame of reference at rest with the shock, the particle source is situated at the shock, $Q = \Phi_0 \delta(x)$. The kinetic equation in this case could be reduced to one-dimensional form (see e.g. [Moraal (2013)]),

$$\frac{\partial f}{\partial t} + u\frac{\partial f}{\partial x} = \frac{\partial}{\partial x}\left[D_{xx}\frac{\partial f}{\partial x}\right] + \frac{p}{3}\frac{\partial u}{\partial x}\frac{\partial f}{\partial p} + \frac{1}{p^2}\frac{\partial}{\partial p}\left(p^2 D_{pp}\frac{\partial f}{\partial p}\right) + Q. \quad (6.10)$$

The constant flow speed is u_1 upstream of the shock, $x < 0$, and u_2 downstream of the shock, $x > 0$, with both u_1, $u_2 > 0$ (e.g., [Jones and Ellison

(1991)]). Then the divergence of the fluid velocity, $\partial u/\partial x$, is different from zero only at the shock, in agreement with the fact that shock acceleration is due to shock compression [Lee and Fisk (1982)].

$$u(x) = \begin{cases} u_1, & x < 0; \\ u_2, & x > 0. \end{cases}$$

For shock acceleration, first order Fermi acceleration is thought to be much faster than second order Fermi acceleration, and the term $p^{-2}(\partial/\partial p)(p^2 D_{pp}\partial f/\partial p)$ can be neglected (e.g., [Drury (1983)]).

Consider the stationary case, that means that the shock time evolution is slow compared to both the CR acceleration time and the energy loss processes. For simplicity, a spatially independent diffusion coefficient D both upstream and downstream of the shock is often considered (e.g., [Lee and Fisk (1982)]). Solving the steady-state 1d equation ($\partial f/\partial t = 0$) with a source at the shock and neglecting by the momentum diffusion term containing D_{pp},

$$D\frac{\partial^2 f}{\partial x^2} - u\frac{\partial f}{\partial x} + \frac{1}{3}\frac{du}{dx}p\frac{\partial f}{\partial p} + \delta(p - p_0)\delta(x) = 0,$$

one could obtain the steady-state density upstream and downstream of the shock [Berezhko and Krymskii (1988)],

$$f(x,p) = \frac{3}{u_1 - u_2}\left(\frac{p}{p_0}\right)^{-\gamma_1} \begin{cases} \exp(u_0 x/D), & x < 0, \\ 1, & x > 0, \end{cases} \quad p > p_0.$$

One can see from this solution that the distribution is a power function of momentum, the exponent of which depends only on degree of compression of the shock wave. This latter value for strong shock waves lies within narrow limits, which makes the power-law spectrum of particles very stable to changing conditions.

[Zimbardo *et al.* (2017)] used fractional advection-diffusion equation to obtain steady-state distributions of particles upstream and downstream of the shock in the case of superdiffusion,

$$\frac{\partial f}{\partial t} + u\frac{\partial f}{\partial x} = -k_\alpha\left(-\frac{\partial^2}{\partial x^2}\right)^{\alpha/2} f + \Phi_0\delta(x). \tag{6.11}$$

They write "When describing the transport process by a differential equation in phase space, the neglect of the derivative with respect to pitch angle requires that the particle distribution function $f(x,p,t)$ be nearly isotropic (e.g., [Arthur and Le Roux (2013)]). This requires that enough scattering be present, and we point out that the Lévy walk model allows for a non-negligible probability of very long free paths, but also a very high probability of very short free paths [Perri and Zimbardo (2015)], which means frequent pitch angle scattering." So, they consider the case of isotropic superdiffusion and used the one-dimensional equation (6.11). This is not quite correct, because the equation with fractional Laplacian for isotropic three-dimensional Lévy walk or Lévy flights could not be factorized to a set of 1d equations with 1d fractional Laplacians as in the case of Brownian motion (see section 1.3). Rather, this equation is more suitable for parallel shock acceleration (6.3). Moreover, superdiffusive motion of charged particles is predicted for parallel motion in turbulent magnetic field [Casse *et al.* (2001); Pucci *et al.* (2016)].

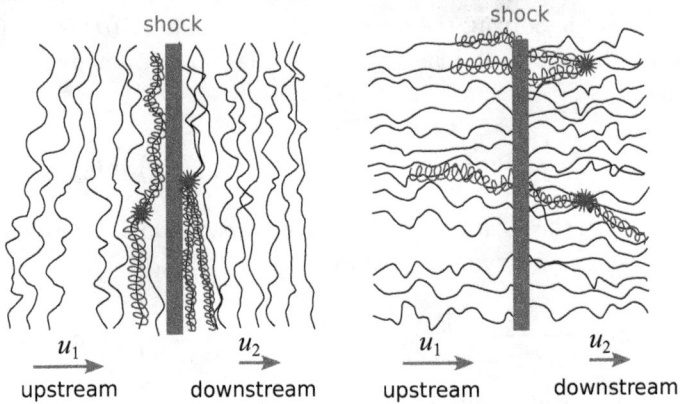

Fig. 6.3   Perpendicular and parallel diffusive shock acceleration.

Considering steady-state solutions for the superdiffusive case, they arrive at the equations

$$u_1 f - k_{\alpha,1} \, {}^{\alpha-1}_x \mathsf{D}_0 \, f = 0$$

for $x < 0$ (upstream) and

$$u_2 f - k_{\alpha,2} \, {}^{\alpha-1}_0 \mathsf{D}_x \, f = \Phi_0$$

for $x \geq 0$ (downstream).

To study the fractional diffusion-advection equation in the presence of a shock, [Zimbardo *et al.* (2017)] proposed to use the left and right Caputo fractional derivatives in space. It is well known that analytical solutions of these equations are fractional generalizations of the exponential function expressed through the Mittag-Leffler special functions (see Fig. 6.4). When considering the case of a planar shock with the shock as the source of energetic particles, the Mittag-Leffler functions correspond to a stretched exponential close to the shock, and to a power law profile far upstream of the shock. The latter result is in agreement with those obtained from a probabilistic approach to superdiffusion by Perri and Zimbardo (2007, 2008), and from a time asymptotic solution obtained by Litvinenko and Effenberger (2014).

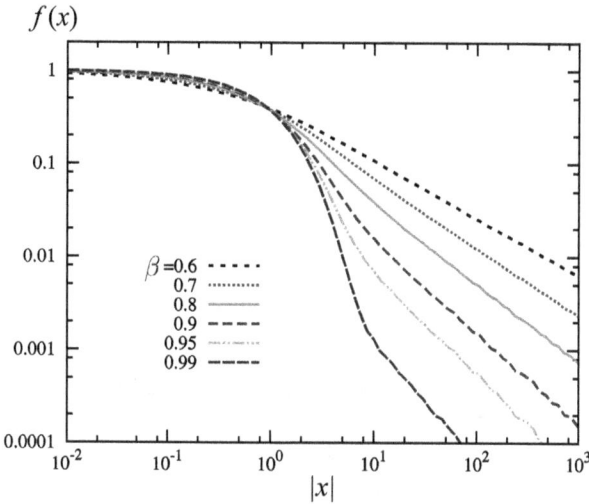

Fig. 6.4 Plot of fractional exponential function $f(x) = E_\beta(-|x|^\beta)$ in log-log scale for different β values.

6.4 Subdiffusive shock acceleration

[Lagutin (2010)] have considered the subdiffusive case of particle motion during the acceleration on shock wave front. The propagation of particles in the regions upstream and downstream of the shock is described by an equation with a fractional time derivative generalizing the kinetic transport equation for the case of turbulent motion of non-Markovian type. [Lagutin (2010)] starts with the subdiffusion equation

$$\frac{\partial n}{\partial t} = {}_0D_t^{1-\beta}\left\{\kappa\frac{\partial^2 n}{\partial x^2} - \frac{\partial}{\partial x}[u(x)n]\right\}$$

and associates the corresponding kinetic equation

$$\frac{\partial f}{\partial t} = {}_0D_t^{1-\beta}\left\{\kappa\frac{\partial^2 f}{\partial x^2} - u\frac{\partial f}{\partial x} + \frac{1}{3}\frac{du}{dx}p\frac{\partial f}{\partial p}\right\} + S(x,p,t).$$

He rewrites the equation in a more convenient form

$${}_0D_t^{\beta}f = \kappa\frac{\partial^2 f}{\partial x^2} - u\frac{\partial f}{\partial x} + \frac{1}{3}\frac{du}{dx}p\frac{\partial f}{\partial p} + {}_0I_t^{1-\beta}S(x,p,t).$$

For delta-shaped shock front, we have

$${}_0D_t^{\beta}f = \kappa\frac{\partial^2 f}{\partial x^2} - u\frac{\partial f}{\partial x} + \frac{1}{3}(v_2 - v_1)\delta(x)p\frac{\partial f}{\partial p} + S_0(p)\delta(x)\frac{t^{1-\beta}}{\Gamma(2-\beta)}.$$

Using Laplace transformation and solving the boundary problem, [Lagutin (2010)] obtained distribution for large times in the form

$$\tilde{f}_0(p,\lambda) = \frac{\lambda^{\beta-1}}{\lambda}\exp\left(-\tau^{\beta}\lambda^{\beta}\right)A\left(\frac{p}{p_i}\right)^{-1-3u_1/u_2}\begin{cases}\exp\left(\frac{u_1 x}{2\kappa_1}\right), & x < 0;\\[2mm]\exp\left(-\frac{u_2 x}{2\kappa_2}\right), & x > 0,\end{cases}$$

where

$$\tau = \left[\frac{3}{u_1 - u_2}\int_{p_i}^{p}\frac{dp'}{p'}\left(\frac{\kappa_1}{u_1} + \frac{\kappa_2}{u_2}\right)\right]^{1/\beta}$$

is considered as a characteristic time of acceleration from initial momentum value p_i to p.

So, [Lagutin (2010)] established that at stationary injection of particles into the acceleration process near the shock front, the non-Markovian nature of particle propagation affects only the rate of acceleration and does not modify the spectrum.

6.5 Anomalous second order Fermi acceleration

Using simple mathematical tools, Fermi displayed that interaction of a charged particle with chaotically moving magnetic clouds in interstellar space brings to its acceleration. Neglecting statistical fluctuations, one can obtain that the energy increment dw is proportional to energy itself w and consequently the function $w(t)$ is exponential, namely

$$w(t) = mc^2 e^{b^2 t/\tau},$$

where m is the particle mass, c speed of light, $b = u/c$, u is the characteristic speed of a magnetic cloud, and τ is the mean time between consecutive collisions of the particle with magnetic clouds. The variable t means here the time of acceleration. Obviously, it is a random quantity. Denoting its random value by T and assuming following Fermi that its distribution is exponential,

$$P(T > t) = e^{-t/\theta},$$

we arrive at his well-known formula:

$$P(w(T) > E) = P(mc^2 e^{b^2 T/\tau} > E) = P\left(T > \frac{\tau}{b^2} \ln\left(\frac{E}{mc^2}\right)\right)$$

$$= \exp\left(-\frac{\tau}{b^2\theta} \ln\left(\frac{E}{mc^2}\right)\right) = \left(\frac{E}{mc^2}\right)^{-\gamma}, \quad \gamma = \frac{\tau}{b^2\theta}.$$

Obviously, $P(w(T) > E) = 1$ for $E < mc^2$.

This elegant result did not become a solution of the problem mainly because of one reason: the life-times θ for different cosmic particles (protons, alpha-particles, other nuclei) are very different, meanwhile the observed exponents γ are close to each other.

After a CR particle has left the SNR site and is propagating in the Galaxy, it may encounter another SNR, and be accelerated repeatedly, and this process may continue up to escape the particle from the galaxy. As it is accepted to think that the probability to encounter a SNR is proportional to its volume. It turns out to be significant only for old SNR, hence with weak shocks. Indeed, observations of the ratio of primary to secondary cosmic rays suggest such a secondary acceleration process, but at the same time constrain its amount [Eichler (1980); Cowsik (1986)]. [Wandel et al. (1987)] have performed calculations which confirmed the observed constraints on the secondary acceleration. As a result, the authors proposed that the main cause of this constraint is the escape of the cosmic rays from the galaxy. Another possible explanation of this effect is based on taking into account the nonhomogeneity of the acceleration event distribution in time.

The calculations in [Wandel et al. (1987)] were based on the integro-differential equation

$$\frac{\partial f(\mathbf{p}, t)}{\partial t} = \mu A f(\mathbf{p}, t) + f_0(\mathbf{p})\delta(t), \tag{6.12}$$

where $f(\mathbf{p}, t)$ is the pdf of position of a cosmic ray particle in the momentum space at time t and

$$A f(\mathbf{p}, t) = \int w(\mathbf{p} \leftarrow \mathbf{p}') f(\mathbf{p}', t) d\mathbf{p}' - f(\mathbf{p}, t) \tag{6.13}$$

is the acceleration integral with transitions density $w(\mathbf{p} \leftarrow \mathbf{p}')$ (we omitted in Eq. (6.12) terms describing energy losses as less significant in comparison with the accelerator).

The physical meaning of this model is quite visible: successive inter-actions (collisions) of a charged particle with strong shock waves in the supernova remnants mentioned by Berezhko and Krymskii in their review [Berezhko and Krymskii (1988)] are considered as instantaneous jumps of the point represented the particle, from one point of the momentum space to another. The momenta $\Delta \mathbf{p}_i$ acquired by the particle in these collisions are random and the point

$$\mathbf{p} = \mathbf{p}_0 + \Delta \mathbf{p}_1 + \Delta \mathbf{p}_2 + \Delta \mathbf{p}_3 + \dots,$$

representing the particle in the momentum space moves away from the injection point (momentum) $\mathbf{p}_0$.

The increment of the momentum in the multiplicative model is propor-tional (in the statistical case) to the absolute value of the momentum p' of the particle coming into interaction,

$$\Delta \mathbf{p} = p' \mathbf{q}, \tag{6.14}$$

$$\int\limits_{|\Delta \mathbf{p}| > p} w(\Delta \mathbf{p}; \mathbf{p}') d\Delta \mathbf{p} \propto (p/p')^{-\gamma}, \; p \to \infty. \tag{6.15}$$

Under the assumption that the distribution of the proportionality vector $\mathbf{q}$ is independent of $\mathbf{p}'$ and isotropic,

$$W(\mathbf{q}; \mathbf{p}') d\mathbf{q} = (1/2) V(q) dq d\xi, \; \xi = \cos(\mathbf{q}, \mathbf{p}),$$

kinetic equation (6.12) can be written in the form

$$\frac{\partial}{\partial t} f(p, t) = \mu \left\{ \int\limits_{-1}^{1} \frac{d\xi}{2} \int\limits_{0}^{\infty} \frac{f\left(p / \sqrt{1 + 2\xi q + q^2}, t \right)}{\left(\sqrt{1 + 2\xi q + q^2} \right)^3} V(q) dq - f(p, t) \right\}$$

$$+ f_0(p) \delta(t), \tag{6.16}$$

representing a new model of a distributed reacceleration, more pre-cisely, a new modification of the model proposed in [Blandford and Os-triker (1980)], [Wandel *et al.* (1987)]. This model called *multiplicative walk* [Uchaikin (2010a)] in order to distinguish it from the *additive walk* model. As in the work, cited above, the energy of the accelerated parti-cle is expressed in terms of the product of independent random variables, rather than their sum. In order to make this model closer to real processes of reacceleration, e.g., in the case of the intersection of shock fronts in the remnants of supernova, we assume that

$$V(q) = \gamma q^{-\gamma - 1}, \quad \gamma > 1.$$

However, observation data indicate a slower dependence $\langle p(t) \rangle$ than obtained by solution of the above equations.

Besides a lot of other questions concerning the creating and/or amplifying of magnetic fields (MF) by turbulence (see in [Moffatt (1978); Krause and Rädler (2016)]), another question is, how does MHD turbulence influence the transport of heat and cosmic rays [Dennis and Chandran (2005); Cho and Lazarian (2006)]. MHD turbulence is an important agent for particle acceleration as was pointed first by Fermi [Fermi (1949)].

The second-order Fermi mechanism (stochastic acceleration of particles by scattering with randomly moving magnetized clouds) has application in a wide range of astrophysical objects including the solar wind (SW) and solar flares [Miller *et al.* (1990); Chandran (2003)], cluster of galaxies [Dogiel *et al.* (2007)], the Galactic center [Liu *et al.* (2004)], etc. Usually, this mechanism was applied in cases, where protons are accelerated from a thermal distribution.

Another relating stochastic acceleration mechanism is connected with the created magnetic field in the helical MHD turbulence [Fedorov *et al.* (1992); Hnatich *et al.* (2001); Shakhov and Stehlik (2008)]. In fact, due to an anisotropic helical turbulence the large scale electric field is generated (α-effect) [Krause and Rädler (2016)]. The electric field is directed along regular magnetic field (MF) and it can efficiently accelerate charged particles in the turbulent plasma. The equation describing such acceleration mechanism (α-acceleration) was firstly derived in [Kichatinov (1983)]. The stochastic particle acceleration by anisotropic helical turbulence was investigated on the frame of kinetic equation for charged energetic particle distribution function in the paper [Fedorov *et al.* (1992)].

The further transformation of the system is associated with the specification of the time interval distributions $q(t)$. This distribution is usually (one can always say) taken in the exponential form $q(t) = \mu e^{-\mu t}$ and therefore the master equation begins with the first-order time derivative $\partial f(\mathbf{p}, t)/\partial t = \ldots$. This means that the process is assumed to be Markovian. However, the real distribution of the time intervals between collisions is unknown. For example, one can assume that it is of power law, $Q(t) \propto t^{-\alpha}$, rather than exponential. This is in agreement with the self-similar pattern of turbulent motions and with its power-type laws. The hypothesis of the fractal character of the interstellar medium [Kulakov and Rumyantsev (1994)] also provides power-law distributions. This concerns the behavior of magnetic field lines in interstellar space. They are usually represented as relatively smooth lines, which somewhere rest at the magnetic traps, intersect each other, sharply change their directions, and performing a "diffusion dance" in time. The leading centers of particles moving along spiral trajectories move along these lines. If these smooth sections become invisible and chaotic patterns of the structure become prevalent with the expansion of the field of view and the corresponding decrease in the scale, then this is an asymptotically homogeneous medium where the mean free paths can be simulated by a usual exponential function. If the expanding field of view includes increasing straight segments replacing the segments becoming small due to a decrease in the scale so that the structure remains (qualitatively) unchanged, this is a fractal structure. Under these conditions, the exponential distribution of free paths characteristic of a strongly mixing medium cannot be expected, but cannot be completely rejected as well. The best compromise would be a family of distributions including both exponential and power-law distributions. Fortunately, such a family exists: it is a set of the functions

$$Q_\alpha(t) = \mathrm{Prob}(T > t) = E_\alpha(-\mu t^\alpha), \quad \alpha \in (0,1],$$

where $E_\alpha(z) = \sum\limits_{n=0}^{\infty} z^n/\Gamma(\alpha n + 1)$ are the Mittag-Leffler functions. If $\alpha = 1$, $Q_\alpha(t)$ is the exponential function. When $\alpha < 1$, $Q_\alpha(t)$ is a fractional exponential function having a power-law asymptotic behavior $t^{-\alpha}$, $t \to \infty$. The corresponding density $q_\alpha(t)$ satisfies a fractional differential equation; as a result, Eq. (6.14) takes the fractional time-differential form:

$$_0\mathsf{D}_t^\alpha f(\mathbf{p}, t) = \mu A f(\mathbf{p}, t) + f_0(\mathbf{p})\delta_\alpha(t). \tag{6.17}$$

The time series of accelerating collisions forms a *fractional Poisson process of order* α [Uchaikin et al. (2008)], with $\alpha \to 1$ becoming an ordinary Poisson process which underlies classical kinetic equation. Investigations performed in [Uchaikin et al. (2008)] indicate a qualitatively new property of this process: the average number of collisions increases more slowly ($\propto t^\alpha$) than in the usual case ($\propto t$) and relative fluctuations of the number of accelerations in the limit $t \to \infty$ do not disappear, but tend to a limiting distribution depending on α.

Solutions of Eq. (6.17), obtained by Monte Carlo simulation, are presented in Fig. 6.5. All of them are characterized by power law tails. For smaller α, acceleration is slower.

Let us consider the equation for the spectral function in two extreme cases. In the first case, $\gamma > 2$, the second moment of the momentum increment proportional to E^2 exists, and this case corresponds to the classical diffusion with variable coefficients:

$$
{}_0\mathsf{D}_t^\alpha n(E,t) = \frac{\partial[a_1 E n(E,t)]}{\partial E} + \frac{\partial^2[a_2 E^2 n(E,t)]}{\partial E^2} + n_0(E)\delta_\alpha(t).
$$

In the second case, we suppose $\gamma << 2$, so that only the term q^2 may be retained in the radicand in Eq. (6.17):

$$
{}_0\mathsf{D}_t^\alpha n(E,t) = \mu \left\{ \int\limits_1^\infty \gamma q^{-\gamma-1} n(E/q,t)dq/q - n(E,t) \right\}
$$

$$
+ n_0(E)\delta_\alpha(t). \tag{6.18}
$$

We do not claim that this approximation is very good, but namely this acceleration operator was used in [Wandel *et al.* (1987)] for some calculations. Solving Eq. (6.18) with the method of the Mellin Laplace transforms, the following expression is obtained for the case of the mono-energetic source $(n(E) = \delta(E - E_0))$:

$$
N_\alpha(E;\tau) = \frac{\mu \tau^\alpha \gamma}{(1 + \mu \tau^\alpha)^2} \left(\frac{E}{E_0} \right)^{-1-\gamma/(1+\mu\tau^\alpha)} \frac{1}{E_0}. \tag{6.19}
$$

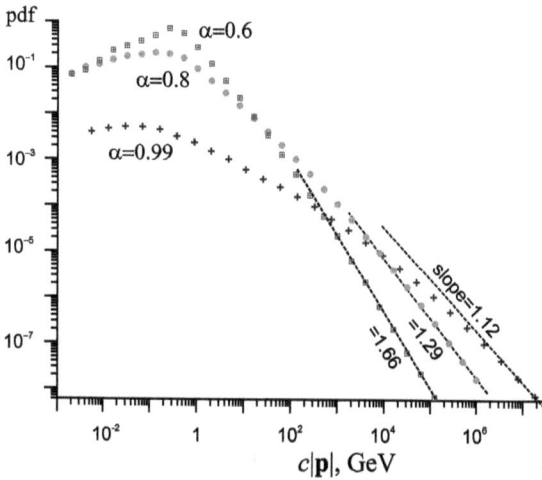

Fig. 6.5 Probability density function of the momentum magnitude for three values of α. The points correspond to the Monte Carlo simulation results, the lines are power law asymptotics with exponents indicated in figure.

Although Eq. (6.19) was derived under the assumptions that strongly simplify the real situation and has a qualitative sense, it compactly presents the effect of *all three sources of fluctuation acceleration*: fluctuations of the age of the particle (parameter τ), fluctuations of the number of the acceleration events (α μ), and fluctuations of the energy acquired in a single event (γ). Representing the scaling parameter μ in the form $\mu = \tau_A^\alpha$, where τ_A is the characteristic time interval between the events of the acceleration of particles in the remnants of various supernovae (recall that τ is the mean lifetime with respect to nuclear collisions), the absolute value of the exponent of the integral spectrum can be written in a clearer form $\gamma' = 1 + \gamma/[1 + (\tau/\tau_A)^\alpha]$. At $\alpha = 1$, and $\mu\tau \gg 1$, we arrive at the Fermi formula (2) with $a = \mu/\gamma$.

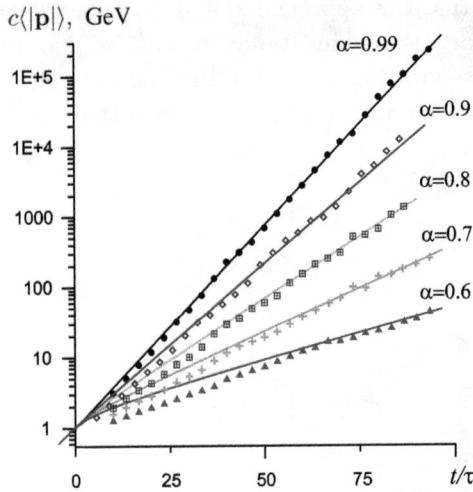

Fig. 6.6 Dependence of the mean momentum on time for different values of α. Points correspond to the Monte Carlo simulation results, lines are functions (6.20). Here $\gamma = 3$, $\mu K = 2/15$ for all curves.

Using Eq. (6.17) one can find the average momentum variation with time. Multiplying Eq. (6.17) by $|\mathbf{p}|$ and integrating over the momentum space, we arrive at

$$_0D_t^\alpha \langle|\mathbf{p}(t)|\rangle = \mu K(\gamma)\langle|\mathbf{p}(t)|\rangle + \langle|\mathbf{p}(0)|\rangle\delta_\alpha(t).$$

For the initial condition $\langle|\mathbf{p}(0)|\rangle = 1$, the solution of this equation is expressed through the one-parameter Mittag-Leffler function [Uchaikin (2013a)]:

$$\langle|\mathbf{p}(t)|\rangle = E_\alpha(\mu K t^\alpha). \tag{6.20}$$

The results of Monte Carlo simulation for different values of α are presented in Fig. 6.6. Analytical solutions (6.20) are in good agreement with numerical results (there are no adjustable parameters). In all simulations presented here, $\gamma = 3$, $\mu K = 2/15$.

Note that the operator describing acceleration at shock waves of supernovae remnants does not reduce to the fractional differential form, but holds in the integral form. Equation (6.17) should be supplemented by terms presenting energy losses, exist of particles from the Galaxy, and nuclear interactions (fragmentation).

6.6 Stochastic acceleration in Earth's magnetotail

Considering stochastic acceleration processes in the Earth magnetotail [Milovanov and Zelenyi (2002)] (see Fig. 4.11), emphasized that this consideration was applied to the test particles and does not include the inverse effect of the hot plasma on the magnetic field turbulence. They interpreted this phenomenon as the self-interaction of the turbulence in the nonlinear saturation regime. The self-interaction appears in the generation of magnetic fluctuations by the particles accelerated in the stochastic inductive electric field. This limits the particle energy gain from the inductive fields and may have an impact on the energy distribution in the turbulent system. In case of isotropic turbulence, inclusion of self-interactions leads to the equation

$$\frac{1}{K}\frac{\partial^\gamma \psi}{\partial t^\gamma} = \frac{1}{v^2}\frac{\partial}{\partial v}\left[v^\gamma\frac{\partial \psi}{\partial v}\right] - Rv^2\left[\int\limits_{\sim 0}^{v} v'^3\psi(v',t)dv'\right]^2$$

with $R = 64\pi^3 e^2 Q/K$, where constant Q is the characteristic interaction amplitude.

Thanks to the presence of the last term, a *nonequilibrium stationary solution* $\psi_0(v)$ now exists and can be found as an asymptotic solution of the equation at $t \to \infty$. Taking into account that

$$\frac{1}{K}\frac{\partial^\gamma \psi_0(v)}{\partial t^\gamma} = \frac{1}{K}\frac{\psi_0(v)t^{-\gamma}}{\Gamma(1-\gamma)},$$

one obtains the nonlinear equation

$$\frac{1}{v^2}\frac{\partial}{\partial v}\left[v^\gamma\frac{\partial \psi_0(v)}{\partial v}\right] = Rv^2\left[\int\limits_{\sim 0}^{v} v'^3\psi_0(v')dv'\right]^2.$$

Searching its solution in the inverse power form,

$$\psi_0(v) \propto v^{-\beta}, \quad v \to \infty,$$

the authors come to the following link between slope β and parameter γ:

$$\beta = 14 - \gamma.$$

The kinetic energy spectrum in nonrelativistic case, considered in the cited article, bocomes

$$\psi_0(E) \propto E^{-\beta/2} = E^{7-\gamma/2}.$$

The quantity $\beta/2$ is related to parameter κ used in works [Hasegawa *et al.* (1985); Christon *et al.* (1991)] via relation

$$\beta = \kappa + 1.$$

In view of $\gamma \in (0,1)$ (but not in (0,2), as assumed in [Milovanov and Zelenyi (2002)]), we observe a weaker influence of the time operator fractionality on the κ,

$$5.5 < \kappa < 6,$$

instead of $5 < \kappa < 6$ obtained in the cited work.

Nevertheless, it is still in agreement with the magnetotail particle population survey by [Christon *et al.* (1991)] who concluded that for both ions and electrons a most probable value of κ is between 5 and 6.

Chapter 7

Nonlocal relativistic diffusion model

In this chapter, we describe recently developed fractional model of cosmic ray propagation. It is called the nonlocal relativistic diffusion (NoRD) model. The model accounts for the turbulent character of the interstellar medium and the relativistic principle of the speed limitation. Motivation and assumptions are presented in section 7.1. In section 7.2 a special attention is paid to the energy spectrum and the knee problem. In sections 7.3 and 7.4, in frames of the NoRD model, we consider some specific problems such as CR anisotropy inversion and back-scattering from a fractal half-space. The obtained numerical calculation results do speak well for the NoRD-model as compared with the traditional one based on integer-order operators.

7.1 Relativistic speed-limit requirement and CR transport

Motivation

Because of the lack of detailed information on the structure of the interstellar medium, a consistent analysis of the cosmic rays propagation in the Galaxy was undertaken on the basis of normal (parabolic) diffusion equations [Ginzburg (1953)]. An important reason to use this approach is high level of isotropy of the observed flux and chemical composition, indicating that path traversed by the particles many times higher than estimated distances to the CR sources. Their path should present tortuous crooked lines. The best-known natural type of such a movement is the Brownian motion, which determines the type of diffusion differential equation modeling the propagation process.

Today it is obvious that this choice is not quite adequate to the process under consideration. There are at least three reasons for this conclusion. The model of Brownian motion of a molecule assumes that positions of other molecules with which it collides are random and independent (in the probabilistic sense) of each other, as in a hypothetical ideal gas. In reality it is not so: the irregularities of the magnetic field leading to a sharp change in direction of particle motion are linked by field lines, generating correlation effects and requiring introduction of nonlocal operators for their description. The second reason is the turbulence of the interstellar medium, its self-similar heterogeneity, power laws of which lead to integro-differential operators of fractional order. The third reason is the special (different from the hydrodynamic turbulence) intermittency of interstellar medium: clots entangling trajectories of particles are interchanged by large-scale voids, where high-energy particles move along segments close to a straight line. In these situations, relativistic limitation on particle velocity must be taken into account. In this chapter, we discuss the propagators constructed on these principles and the results of calculations that demonstrate the importance of factors enumerated above.

The transport of CRs is an extremely complex process dependent on changeable in space and time interstellar medium characteristics which are known to us only in outline. This is why we are forced to use more or less simplified models for description of the process. The standard diffusion model is a very crude approximation which gives only qualitative estimates of only some aspects of the process. Suffice to say that the diffusion model of CR propagation in space violates the relativistic principle of velocity limitation. In spite of the conventional diffusivity representation $D \propto vl$, the diffusion equation relates to the limit case with the infinite velocity v and zero free path l. For this reason, the diffusion model is incapable of describing the CR transport near boundary separating regions with different properties. Trajectories of this process called the Brownian motion are nowhere differentiable which makes them infinitely far from their physical prototypes (Fig. 7.1a). In particular, the length of any section of such trajectory between two arbitrary points is infinitely long.

Because of their self-similarity however, the trajectories retain these properties at all scales including arbitrary small ones. Thus, despite the fact that the diffusion equation can be derived from the feasible random walk with a finite velocity by passage to large scales, the way back is closed: the small-scale behavior of the particles holds its Brownian pattern which looks here even more unnatural than on large scales. For these reasons, the diffusion model and their modifications based on the diffusivity concept are not really in a position to give proper particle path lengths and their escape time distributions.

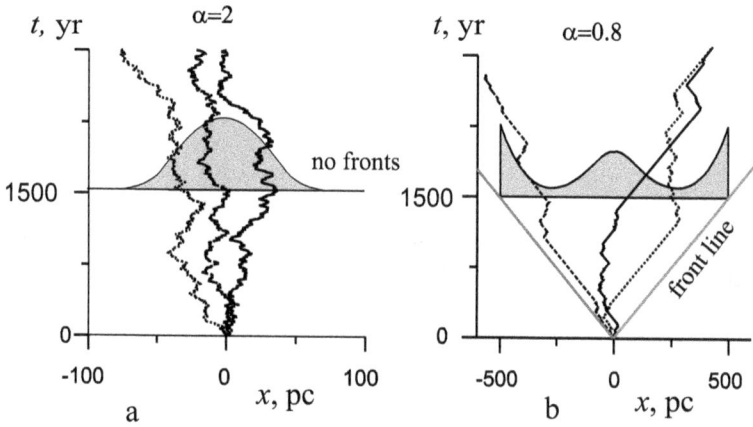

Fig. 7.1 Typical trajectories of random walk with constant velocity for different α. The 'front line' corresponds to the speed of light.

Speed limit effect

It is well known that even hydrodynamic turbulence deprives the Brownian motion status of a reliable mathematical model, and one could only wonder how long the diffusion ideas can live in the theory of CR transport. Nevertheless, one can say, the classical diffusion model is gradually losing its position. The process has been partially traced in reviews [Uchaikin (2013b, 2015)]. One can consider these reviews as a relatively young branch of transport theory: *relativistic Brownian motion*. The idea to get a relativistic analogue of the plain diffusion equation by generalizing the link of the latter with the quantum Schrödinger equation was first realized in [Gaveau *et al.* (1984)]. Applying the similar transformation to the Dirac equation, the authors obtained the result in the form of the telegraph equation representing the relativistic Brownian motion with a finite velocity (limited by the speed of light) and including exponentially distributed free paths. A key item of the approach is replacement of the nonrelativistic first Fick law

$$\mathbf{j}(\mathbf{x}, t) = -D \ \text{grad} \ n(\mathbf{x}, t), \tag{7.1}$$

by its 'relativistic' counterpart containing a retarded operator,

$$\mathbf{j}(\mathbf{x}, t) = - \int_0^t v^2 e^{-v^2 \tau / D} \text{grad} \ n(\mathbf{x}, t - \tau) d\tau, \quad D = lv/3. \tag{7.2}$$

Combining it with the continuity equation yields the telegraph equation derived for one-dimensional transport problems half century earlier [Fock

(1930)]:

$$\frac{\partial n}{\partial t} + \theta \frac{\partial^2 n}{\partial t^2} - D \frac{\partial^2 n}{\partial x^2} = S(x,t), \qquad (7.3)$$

where $S(x,t)$ is a source term ($S(x,t) = \delta(x)\delta(t)$ for instantaneous point source). More formal mathematical generalization of classical Brownian motion to its relativistic counterpart is performed by Dunkel and Hänggi [Dunkel and Hänggi (2009)].

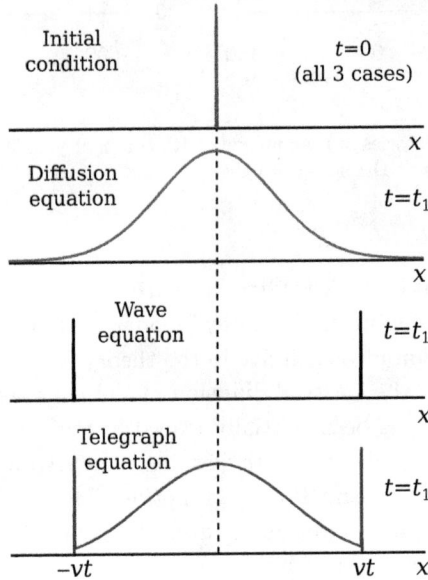

Fig. 7.2   Nature of 'relativistic' solution to the telegraph equation according to [Masoliver and Weiss (1996)].

The authors have developed a few versions of this model, we discuss one of them here. Considering a relativistic particle traveling from the 4-event $x_0 = (t_0, x_0)$ to $x = (t, x)$ on assumption that it can experience multiple scatterings on its way, and that the velocity is approximately constant between two successive scattering events, the authors write the relativistic action (per mass) in the form

$$A(x, x_0) = - \int_{t_0}^{t} \sqrt{[1 - [v(t')]^2} dt',$$

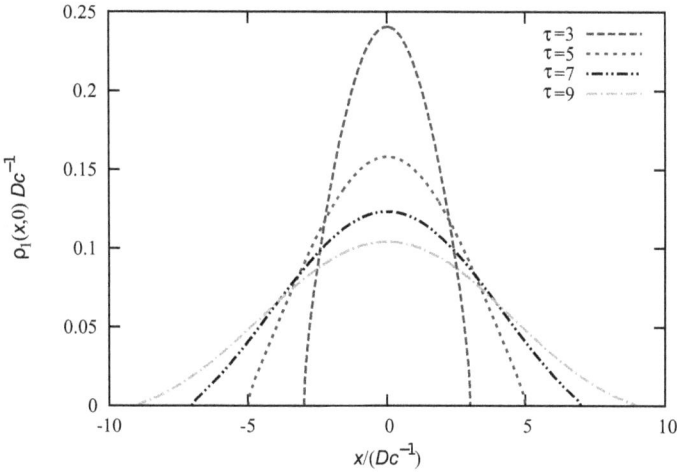

Fig. 7.3 One-dimensional relativistic [Dunkel *et al.* (2007)] propagator (7.4) at different times $\tau = tc^2/D$.

where the velocity $v(t')$ is a piecewise constant function. Evidently, this action takes a minimal value

$$a_-(\mathrm{x}, \mathrm{x}_0) = -\sqrt{(t - t_0)^2 - (\mathrm{x} - \mathrm{x}_0)^2},$$

if the particle does not collide, that is moves with a constant velocity. The maximum action value is realized when for particles moving at light speed, yielding $a_+ = 0$. As a result, the d-dimensional relativistic propagator reads

$$p_d(\mathrm{x}, \mathrm{x}_0) = \begin{cases} N\left\{\exp\left[-\frac{a_-(\mathrm{x}, \mathrm{x}_0)}{2D}\right]\right\}, & (\mathrm{x} - \mathrm{x}_0)^2 \le (t - t_0)^2; \\ 0, & \text{otherwise.} \end{cases} \tag{7.4}$$

Here N stands for normalizing constant and D is an arbitrary (positive) coefficient.

Graphs of the propagator (for the one-dimensional case) are shown in Fig. 7.3. Fig. 7.4 compares the mean square displacements in one-dimensional relativistic [Dunkel *et al.* (2007)] and nonrelativistic diffusion. The new propagator is bounded by the relativistic cone, but like its nonrelativistic counterpart describes propagation through a system of mutually independent scatterers, whereas real points of CR scattering are bounded by magnetic field lines along which the particles fly. This means that free path of CR particles should be distributed according to power type (similarly to other laws in turbulent media) rather than to exponential one (the

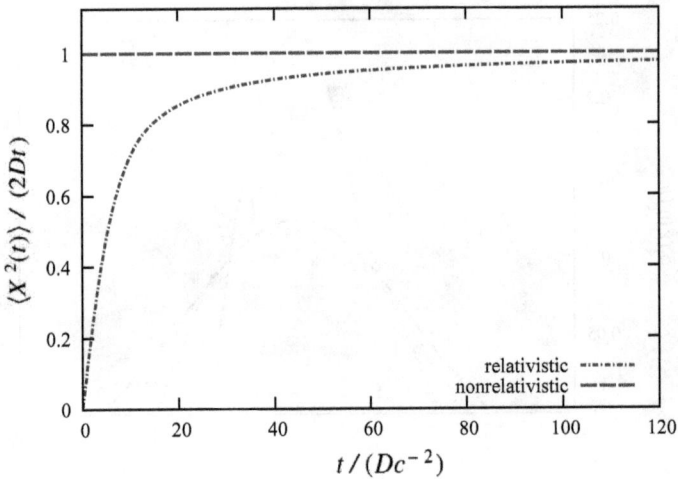

Fig. 7.4 The mean square displacements in one-dimensional relativistic [Dunkel *et al.* (2007)] and nonrelativistic diffusion, divided by $2Dt$.

latter relates to an ideal gas in the equilibrium state). For this reason, the right-hand side of Eq. (7.3) is replaced by some nonlocal (in space-time) operator, averaging it over turbulent fluctuations (see [Uchaikin (2013b)]). As a result, the exponential law changes to the power type Lévy distribution $p_\xi(x) \propto x^{-\alpha-1}$ with an exponent $\alpha \in (0, 2]$ connected with the fractal dimension of the ISM. All these improvements have drastically changed the picture making it closer to a real transport process (Fig. 7.1b). One can see now the front lines ($x = \pm vt$), free paths being parallel to one of the lines and having random lengths distributed according to the inverse power law. The latter is characterized by numerous small segments intermitted by long 'Lévy jumps'.

7.2 NoRD model

NoRD integral equations

As was mentioned above, neither diffusion equation nor the equation used by [Lagutin *et al.* (2001b)] satisfy the relativistic speed limit requirement. Galactic cosmic rays are mainly ultrarelativistic particles and a natural way to derive ultrarelativistic propagator is to relate delay times and flight lengths independent in classic CTRW-model by a proportionality factor equal to the speed of free motion [Uchaikin (1998b); Uchaikin and Zolotarev

(1999)]. Let us consider the generalized CTRW-process described by the following system of integral equations

$$\Phi(\mathbf{x}, \mathbf{\Omega}, t) = \int_0^t d\tau P_{\mathbf{\Omega}}(v\tau) F_1(\mathbf{x} - \mathbf{v}\tau, \mathbf{\Omega}, t - \tau) + w(\mathbf{\Omega}) \int_0^t d\tau F_0(\mathbf{x}, t - \tau) Q(\tau),$$
(7.5)

$$F_0(\mathbf{x}, t) = \int d\mathbf{\Omega} \int_0^t d\tau \; p_{\mathbf{\Omega}}(v\tau, \mathbf{\Omega}) \; F_1(\mathbf{x} - \mathbf{v}\tau, \mathbf{\Omega}, t - \tau) + \delta(\mathbf{x})\delta(t), \quad (7.6)$$

$$F_1(\mathbf{x}, \mathbf{\Omega}, t) = w(\mathbf{\Omega}) \int_0^t d\tau q(\tau) \, F_0(\mathbf{x}, t - \tau).$$
(7.7)

Here, $\Phi(\mathbf{x}, \mathbf{\Omega}, t)$ is the phase density, $p_{\mathbf{\Omega}}(x)$ is the pdf for random free paths between consecutive collisions in direction $\mathbf{\Omega}$, $P_{\mathbf{\Omega}}(x) = \int_x^\infty p(\xi)d\xi$ is the corresponding survival probability function, $q(t)$ and $Q(t)$ are pdf and survival probability function for waiting times in localized state in magnetic traps, $F_0(\mathbf{x}, t)$ and $F_1(\mathbf{x}, \mathbf{\Omega}, t)$ are densities of transitions from motion to trap and vice versa, $w(\mathbf{\Omega})$ is the angular distribution of flight direction for particle leaving a trap. It is assumed that the process starts from a localized state at the origin. The spatial density of particles can be found from Φ by integration over directions, $\psi(\mathbf{x}, t) \equiv f_0(\mathbf{x}, t) = \int d\mathbf{\Omega} \, \Phi(\mathbf{x}, \mathbf{\Omega}, t)$.

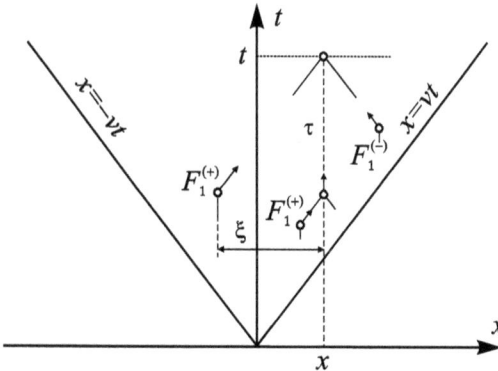

Fig. 7.5 The space-time diagram of random walk with finite constant velocity and trapping.

For better understanding what the equations mean, let us refer to the space-time diagram for one-dimensional case (Fig. 7.5). Every random trajectory is made up of a set of segments each of which is parallel to one of the straight lines $x = 0$, $x = vt$, $x = -vt$. The density $\psi(x, t)$ falls into the sum

$$\psi(x, t) = \psi^{(0)}(x, t) + \psi^{(+)}(x, t) + \psi^{(-)}(x, t).$$
(7.8)

Taking into account that $F_0(x,t)dxdt$ is the probability for the particle to fall into trap in the space-time region $dxdt$ and $Q(\tau)$ is the probability to wait here longer than τ, one can write

$$\psi^{(0)}(x,t) = \int_0^t d\tau Q(\tau) F_0(x, t - \tau). \tag{7.9}$$

This is the first term of the sum. Reasoning along similar lines for moving particle, one can obtain the others.

Applying the Fourier-Laplace transformation

$$\tilde{\psi}(\mathbf{k}, \lambda) = \int_0^\infty dt \int_{\mathbb{R}^3} d^3\mathbf{x}\; e^{-\lambda t + i\mathbf{k}\mathbf{x}} \psi(\mathbf{x}, t) \tag{7.10}$$

to the system (7.5)–(7.7) we obtain the following system of algebraic equations,

$$\tilde{\Phi}(\mathbf{k}, \boldsymbol{\Omega}, \lambda) = \tilde{P}_{\boldsymbol{\Omega}}\left(\frac{\lambda}{v} - i\mathbf{k}\boldsymbol{\Omega}\right)\tilde{F}_1(\mathbf{k}, \boldsymbol{\Omega}, \lambda) + w(\boldsymbol{\Omega})\tilde{Q}(\lambda)\tilde{F}_0(\mathbf{k}, \lambda), \tag{7.11}$$

$$\tilde{F}_0(\mathbf{k}, \lambda) = \int d\boldsymbol{\Omega}\; \tilde{p}_{\boldsymbol{\Omega}}\left(\frac{\lambda}{v} - i\mathbf{k}\boldsymbol{\Omega}, \boldsymbol{\Omega}\right)\tilde{F}_1(\mathbf{k}, \boldsymbol{\Omega}, \lambda) + 1, \tag{7.12}$$

$$\tilde{F}_1(\mathbf{k}, \boldsymbol{\Omega}, \lambda) = w(\boldsymbol{\Omega})\tilde{q}(\lambda)\tilde{F}_0(\mathbf{k}, \lambda). \tag{7.13}$$

Substituting the latter expression into two first equations, we obtain

$$\tilde{\Phi}(\mathbf{k}, \boldsymbol{\Omega}, \lambda) = \left\{w(\boldsymbol{\Omega})\tilde{P}_{\boldsymbol{\Omega}}\left(\frac{\lambda}{v} - i\mathbf{k}\boldsymbol{\Omega}\right)\tilde{q}(\lambda) + w(\boldsymbol{\Omega})\tilde{Q}(\lambda)\right\}\tilde{F}_0(\mathbf{k}, \lambda), \tag{7.14}$$

$$\tilde{F}_0(\mathbf{k}, \lambda) = \tilde{q}(\lambda)\left\langle \tilde{p}_{\boldsymbol{\Omega}}\left(\frac{\lambda}{v} - i\mathbf{k}\boldsymbol{\Omega}\right)\right\rangle \tilde{F}_0(\mathbf{k}, \lambda) + 1. \tag{7.15}$$

Here,

$$\left\langle \tilde{p}_{\boldsymbol{\Omega}}\left(\frac{\lambda}{v} - i\mathbf{k}\boldsymbol{\Omega}\right)\right\rangle = \int d\boldsymbol{\Omega}\; w(\boldsymbol{\Omega})\tilde{p}_{\boldsymbol{\Omega}}\left(\frac{\lambda}{v} - i\mathbf{k}\boldsymbol{\Omega}\right). \tag{7.16}$$

From Eq. (7.15) one can express $\tilde{F}_0(\mathbf{k}, \lambda)$,

$$\tilde{F}_0(\mathbf{k}, \lambda) = \left[1 - \tilde{q}(\lambda)\left\langle \tilde{p}_{\boldsymbol{\Omega}}\left(\frac{\lambda}{v} - i\mathbf{k}\boldsymbol{\Omega}\right)\right\rangle\right]^{-1} \tag{7.17}$$

and substitute it into the expression for $\tilde{\Phi}(\mathbf{k}, \boldsymbol{\Omega}, \lambda)$. As a result, we have

$$\left[1 - \tilde{q}(\lambda)\left\langle \tilde{p}_{\boldsymbol{\Omega}}\left(\frac{\lambda}{v} - i\mathbf{k}\boldsymbol{\Omega}\right)\right\rangle\right]\tilde{\Phi}(\mathbf{k}, \boldsymbol{\Omega}, \lambda) = w(\boldsymbol{\Omega})\tilde{P}_{\boldsymbol{\Omega}}\left(\frac{\lambda}{v} - i\mathbf{k}\boldsymbol{\Omega}\right)\tilde{q}(\lambda) + w(\boldsymbol{\Omega})\tilde{Q}(\lambda). \tag{7.18}$$

Equation (7.18) for phase density represents the image of equation of anisotropic (in general case) random walk of CRs. Integrating it over Ω, we obtain the expression for the Fourier-Laplace transform of the spatial density

$$\tilde{\psi}(\mathbf{k}, \lambda) = \frac{\left\langle \tilde{P}_\Omega \left(\frac{\lambda}{v} - i\mathbf{k}\Omega \right) \right\rangle \tilde{q}(\lambda) + \tilde{Q}(\lambda)}{1 - \tilde{q}(\lambda) \left\langle \tilde{p}_\Omega \left(\frac{\lambda}{v} - i\mathbf{k}\Omega \right) \right\rangle}, \qquad (7.19)$$

in case of isotropic scattering without trapping taking the simpler form:

$$\tilde{\psi}(\mathbf{k}, \lambda) = \frac{\frac{1}{v} \left\langle \frac{1 - \tilde{p}(\lambda/v - i\mathbf{k}\Omega)}{\lambda/v - i\mathbf{k}\Omega} \right\rangle}{\left\langle 1 - \tilde{p} \left(\frac{\lambda}{v} - i\mathbf{k}\Omega \right) \right\rangle}. \qquad (7.20)$$

Here, we used relation $\tilde{P}(\lambda) = (1 - \tilde{p}(\lambda))/\lambda$.

Assuming the power law distribution of free path lengths

$$p(r) = -\frac{d\mathbf{P}(R > r)}{dr} \sim \alpha r_0^\alpha r^{-\alpha-1}, \quad r \to \infty, \qquad (7.21)$$

we find in the vicinity of $\lambda = 0$

$$\tilde{p}(\lambda) = \int_0^\infty p(r) e^{-\lambda r} dr \sim \begin{cases} 1 - \mu_\alpha \lambda^\alpha, & 0 < \alpha < 1; \\ 1 - \langle R \rangle \lambda + \mu_\alpha \lambda^\alpha, & 1 < \alpha < 2; \\ 1 - \langle R \rangle \lambda + \langle R^2/2 \rangle \lambda^2, & \alpha > 2, \end{cases} \qquad (7.22)$$

and, as a consequence,

$$1 - \tilde{p}(\lambda/v - i\mathbf{k}\Omega) \sim \begin{cases} \mu_\alpha \langle (\lambda/v - i\mathbf{k}\Omega)^\alpha \rangle, & 0 < \alpha < 1; \\ \langle R \rangle \langle (\lambda/v - i\mathbf{k}\Omega) \rangle - \mu_\alpha \langle (\lambda/v - i\mathbf{k}\Omega)^\alpha \rangle, & 1 < \alpha < 2; \\ \langle R \rangle \langle (\lambda/v - i\mathbf{k}\Omega) \rangle - \langle R^2/2 \rangle \langle (\lambda/v - i\mathbf{k}\Omega)^2 \rangle, & \alpha > 2. \end{cases} \qquad (7.23)$$

Here, the angular brackets denote averaging over the isotropically distributed vector Ω and the power-law functions are treated, as usual, in the sense of the principal branch of the analytic function z^α in the plane with a cut along the positive semiaxis [Samko et al. (1993)]:

$$z^\alpha = |z|^\alpha e^{i\alpha \, \arg z}, \quad \lim_{\varepsilon \to 0} \arg z|_{z=s+i\varepsilon, \, s>0} = 0. \qquad (7.24)$$

In accordance with this choice,

$$(\lambda/v - i\mathbf{k}\Omega)^\alpha = [(\lambda/v)^2 + (\mathbf{k}\Omega)^2]^{\alpha/2} e^{-i\phi\alpha}, \qquad (7.25)$$

where $\tan \phi = \frac{\mathbf{k}\Omega}{\lambda/v}$. The functions $\tilde{P}(\lambda/v - i\mathbf{k}\Omega)$ corresponding to each of these intervals are found similarly.

Substitution of (7.23) into (7.20) and inverse transformation lead to the following asymptotic equations in natural variables:

$$\left\langle \left(\frac{\partial}{\partial t} + v\mathbf{\Omega}\nabla\right)^\alpha\right\rangle \psi(\mathbf{x}, t) = \frac{t^{-\alpha}}{\Gamma(1-\alpha)}\langle\delta(\mathbf{x} - v\mathbf{\Omega}t)\rangle, \quad |\mathbf{x}| \in (0, vt), \quad 0 < \alpha < 1; \quad (7.26)$$

$$\left[\frac{\partial}{\partial t} - \frac{\mu_\alpha v}{\langle l\rangle}\langle(v\mathbf{\Omega}\nabla)^\alpha\rangle\right]\psi(\mathbf{x}, t) = \delta(\mathbf{x})\delta(t), \quad |\mathbf{x}| \ll vt, \quad 1 < \alpha < 2; \quad (7.27)$$

$$\left(\frac{\partial}{\partial t} - \frac{v\langle l^2\rangle}{\langle l\rangle}\langle\mathbf{\Omega}^2\rangle\triangle\right)\psi(\mathbf{x}, t) = \delta(\mathbf{x})\delta(t), \quad |\mathbf{x}| \ll vt, \quad \alpha > 2. \quad (7.28)$$

The pseudo-differential operator in these equations can be considered as a fractional power of the material (substantial) derivative:

$$\left(\frac{\partial}{\partial t} + v\mathbf{\Omega}\nabla\right)^\alpha \psi(\mathbf{x}, t) = \left(\frac{\partial}{\partial t} + v\mathbf{\Omega}\nabla\right)\int_0^t \frac{\psi(\mathbf{x} - v\mathbf{\Omega}(t-\tau), \tau)}{\Gamma(1-\alpha)(t-\tau)^\alpha}d\tau. \quad (7.29)$$

In case of the spatial homogeneity of ψ the operator proceeds to the fractional Riemann-Liouville time derivative, and in the stationary case, it yields the fractional directional derivative. Averaging in Eq. (7.27) performed in terms of the Fourier transformation,

$$\langle(-i\mathbf{k}\mathbf{\Omega})^\alpha\rangle = \frac{1}{2}k^\alpha\int_{-1}^1(-i\mu)^\alpha d\mu = \frac{i}{2}k^\alpha\int_i^{-i}z^\alpha dz$$

$$= \frac{i}{2(\alpha+1)}k^\alpha\left(e^{-i(\alpha+1)\pi/2} - e^{i(\alpha+1)\pi/2}\right) = -\frac{|\cos(\alpha\pi/2)|}{\alpha+1}|k|^\alpha, \quad 1 < \alpha < 2,$$
$$(7.30)$$

leads to the fractional Laplacian of order $\alpha/2$. Equation (7.28) describes normal diffusion with the diffusion coefficient $D_2 = v\langle l^2\rangle/6\langle l\rangle$ for $w(\mathbf{\Omega}) = 1/4\pi$.

Let us list the obtained equations for isotropic case in table.

Table 7.1. Asymptotic equations for isotropic case.

$\alpha \in (0,1)$	$\int \frac{d\mathbf{\Omega}}{4\pi}\left(\frac{\partial}{\partial t} + v\mathbf{\Omega}\nabla\right)^\alpha \psi(\mathbf{x}, t) = \frac{t^{-\alpha}}{\Gamma(1-\alpha)}\frac{\delta(r-vt)}{4\pi r^2}$,	superdiff. II
$\alpha \in (1,2)$	$\left[\frac{\partial}{\partial t} + \frac{\mu_\alpha v}{\langle l\rangle}\frac{|\cos(\pi\alpha/2)|}{\alpha+1}(-\triangle)^{\alpha/2}\right]\psi(\mathbf{x}, t) = \delta(\mathbf{x})\delta(t)$	superdiff. I
$\alpha > 2$	$\left(\frac{\partial}{\partial t} - \frac{v\langle l^2\rangle}{6\langle l\rangle}\triangle\right)\psi(\mathbf{x}, t) = \delta(\mathbf{x})\delta(t)$	normal diff.

The solutions of these equations are not continuous functions of exponent α in the range of its values, they change jump-wise at $\alpha = 1$ and $\alpha = 2$.

For $\alpha < 2$, the coefficient D_α is not the diffusion coefficient and does not transform into it even if $\alpha \uparrow 2$, because D_α is determined by the asymptotic behavior of the range distribution at large arguments, while the classical diffusion coefficient is determined by variance, being finite for $\alpha > 2$, but infinite for any $\alpha < 2$, which means that as α approaches 2 from below, the variance limit is also infinite and cannot coincide with the classical value, we observe the discontinuity.

For $\alpha > 2$, the asymptotics $t \to \infty$ produces the normal diffusion with coefficient $D_2 = v\langle l^2 \rangle / 6\langle l \rangle$. In case of the exponential distribution of free paths, $p(r) = (1/\langle l \rangle) \exp\{-r/\langle l \rangle\}$, the mean square $\langle l^2 \rangle = 2\langle l \rangle^2$, and the diffusion coefficient takes the ordinary form $D_2 = v\langle l \rangle / 3$. For $\alpha < 2$, we separate the superdiffusion regime into two with different diffusion equations and propagators: the first refers to the case $1 < \alpha < 2$, and the second to the case $0 < \alpha < 1$.

Superdiffusion of the first kind

For superdiffusion of the first kind ($1 < \alpha < 2$), in asymptotic regime, far from the ballistic fronts, we can use the equation with fractional Laplacian from Table 7.1, where the coefficient is given by $D_\alpha = v\mu_\alpha |\cos(\alpha\pi/2)|/(\alpha + 1)\langle l \rangle$. From a physical point of view, the fractional Laplacian can be considered as a consequence of averaging procedure of the diffusion operator with random diffusion coefficient $\tilde{D}$,

$$\langle \nabla(\tilde{D}\nabla\psi(\mathbf{x}, t)) \rangle \mapsto -D_\alpha(-\triangle)^{\alpha/2}\langle \psi(\mathbf{x}, t) \rangle.$$

That is the equation used by [Lagutin *et al.* (2001b)] to explain the knee in the energy spectrum of cosmic rays. However, it should be noted that [Lagutin *et al.* (2001b)] following [Ragot and Kirk (1997)] considered this equation derived from the CTRW-model with instantaneous jumps. Here, we consider this equation as a particular regime of random walk of ultrarelativistic particles with a constant finite velocity. In contrast to [Lagutin *et al.* (2008)], we consider it as applicable only for $\alpha \in (1, 2)$, far from the ballistic front position and not close to the source. Moreover, given anomalous diffusion coefficient differs from that obtained in the CTRW-model with instantaneous flights. In the following papers of Lagutin and coauthors, first time derivative was replaced by a fractional operator,

$$\frac{\partial^\beta \psi(\mathbf{x}, t)}{\partial t^\beta} = -D_\alpha(-\triangle)^{\alpha/2}\psi(\mathbf{x}, t) + \delta(\mathbf{x})\delta^{(\beta)}(t), \ 0 < \alpha < 2, \ 0 < \beta \le 1.$$

$$(7.31)$$

Fractional order ($0 < \beta \le 1$) of the time derivative is interpreted as a sign of particle trapping for a random time distributed according to a power law. The solution of this equation is expressed through the *fractional stable distribution* (FSD) [Uchaikin (2003a)]. Like stable distributions, FSDs are characterized by heavy power tales leading to diverging variance for all $\alpha < 2$, but sufficiently differ from them near the source.

The Green function of (7.31) can be represented in the form (see e.g. [Uchaikin and Sibatov (2012b)]):

$$G_{\alpha,\beta}(\mathbf{x}, t) = \int_0^\infty G_{2,1}(\mathbf{x}, \tau) \, t^{-2\beta/\alpha} q_+(\tau t^{-2\beta/\alpha}; \alpha/2, \beta) \, d\tau, \qquad (7.32)$$

where

$$q_+(\tau; \alpha/2, \beta) = \int_0^\infty \xi^{2\beta/\alpha} g_+(\tau \xi^{2\beta/\alpha}; \alpha/2) g_+(\xi; \beta) \, d\xi \qquad (7.33)$$

is a one-sided fractional stable density [Kolokoltsov *et al.* (2001)] and $G_{2,1}(\mathbf{x}, \tau)$ is the Gaussian density. Here, $g_+(t; \alpha)$ is the one-sided Lévy stable density ($t > 0$, $0 < \alpha < 1$, $g_+(t; 1) = \delta(t - 1)$). This representation is convenient to compute the propagator by means of the Monte-Carlo algorithm,

$$G_{\alpha,\beta}(\mathbf{x}, t) = \int_0^\infty G_{2,1}(\mathbf{x}, S_+(\alpha/2)/[S_+(\beta)]^{-2\beta/\alpha}). \qquad (7.34)$$

The function $t^{-2\beta/\alpha} q_+(\tau t^{-2\beta/\alpha}; \alpha/2, \beta)$ has a sense of an *operational time distribution* at fixed *observational time* t, if we interpret these expressions in terms of conception of *subordinated processes* [Feller (1971)], or of distribution of travel time along different paths if we think in terms of *mixing-length theory*.

Superdiffusion of the second kind

One-dimensional propagator for superdiffusion described by Eq. (7.26) ($0 < \alpha < 1$) is expressed through the [Lamperti (1958)] distribution,

$$G(x, t) = \frac{2 \sin \pi\alpha}{\pi v t} \frac{(1 - x^2/v^2 t^2)^{\alpha-1}}{(1 - x/vt)^{2\alpha} + (1 + x/vt)^{2\alpha} + 2 (1 - x^2/v^2 t^2)^\alpha \cos \pi\alpha}. \qquad (7.35)$$

To derive the isotropic propagators in infinite spaces of higher dimensions, it is not necessary to solve the equation, it is enough to use the theorem linking the distribution by differentiating (see sec. 7.4 in [Uchaikin and Zolotarev (1999)]). In case $0 < \alpha < 1$, the result for radial density $f(\rho)$ ($\rho = r/vt$) has the following form [Magdziarz and Zorawik (2016)],

$$f(\rho) = \frac{8(\alpha+1)}{\alpha\pi} \sin(\alpha\pi)\rho(1-\rho^2)^{\alpha-1}$$

$$\times \frac{(1+\rho)^{2+2\alpha}(1+\alpha-\rho) - (1-\rho)^{2+2\alpha}(1+\alpha+\rho) - 2\rho(1-\rho^2)^{1+\alpha}\cos(\alpha\pi)}{[(1+\rho)^{2+2\alpha} + (1-\rho)^{2+2\alpha} + 2(1-\rho^2)^{1+\alpha}\cos(\alpha\pi)]^2}. \quad (7.36)$$

Note that asymptotic propagators (7.35) and (7.36) do not depend on anomalous diffusion coefficient and consequently on energy. It means that if asymptotic regime is achieved by an ensemble of CRs (with distributed energy) from a certain source, this ensemble at a certain point is characterized by spectrum coinciding with the injection spectrum, i.e., propagation does not change the energy distribution of particles.

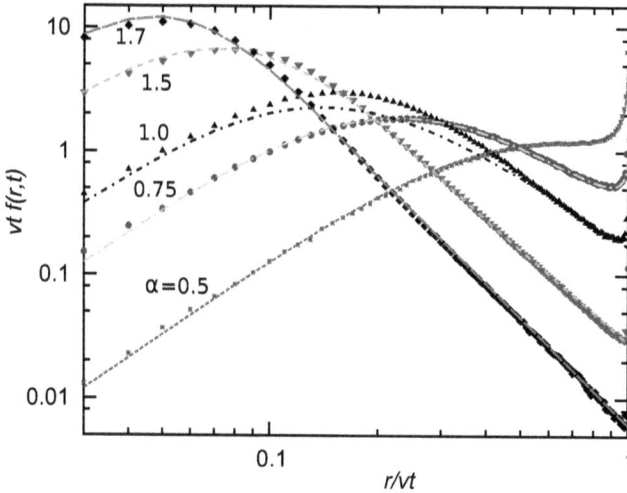

Fig. 7.6 Radial distributions for different α values ($v = 1$, $t = 1000$, $x_0 = 1$).

Figure 7.6 presents comparison of solutions (7.32) and (7.36) with the results of Monte-Carlo simulations. Expression (7.36) describes well the asymptotic distribution of particles for $0 < \alpha < 1$ in the whole region $|\mathbf{x}| < vt$. As expected, FSDs (7.32) describe sufficiently well the middle part, but do not reproduce fronts. In that case density in vicinity $|\mathbf{x}| = vt$ should be considered separately, e.g. by means of the Monte Carlo approach.

7.3   One-dimensional NoRD theory

Assuming that the time needed for particles to pass between magnetic field lines is small, the longitudinal motion can be considered as a continuous one although may reverse the direction at random instants of time. The one-dimensional symmetric random walk of a particle with a constant velocity v along z-axis in the absence of traps is completely characterized by the distribution density of its free paths. In a medium consisting of uniformly distributed independent scatterers, free paths are distributed according to the exponential law, and the process is governed by a second-order partial differential equation (the telegraph equation) [Uchaikin and Saenko (2000)]. The solution of this equation is expressed in terms of the modified Bessel functions and transforms into the normal (Gaussian) distribution in the asymptotic regime of large times. In case of an arbitrary density $p(z)$ with a finite second moment, the asymptotic part of the solution satisfies the telegraph equation. In [Shlesinger *et al.* (1986); Sokolov and Metzler (2003); Uchaikin and Sibatov (2004)], the authors investigated one-dimensional random walks with the asymptotically power-law distribution $p(\xi) \propto \xi^{-\alpha-1}$, $0 < \alpha < 2$, which are sometimes called fractal walks.

Consider a one-dimensional random walk of a particle with a finite speed v along the z-axis. Its pdf at the moment t (longitudinal propagator) is linked with the collision density $f(z,t)$ via relation

$$G(z,t) = \int_0^t [\gamma_1 f(z - v\tau, t - \tau) + \gamma_2 f(z + v\tau, t - \tau)]P(v\tau)d\tau, \quad (7.37)$$

where

$$P(z) = \int_z^\infty p(z')dz',$$

and $p(z)$ stands for the free path pdf. After each path, the particle goes in the positive direction with probability γ_1, and in the negative direction with probability $\gamma_2 = 1 - \gamma_1$. If the particle starts from the origin at moment $t = 0$, the integral equation for $f(z,t)$ takes the form [Uchaikin and Zolotarev (1999)]:

$$f(z,t) = \int_0^t [\gamma_1 f(z - v\tau, t - \tau) + \gamma_2 f(z + v\tau, t - \tau)]p(v\tau)d\tau \quad (7.38)$$

$$+ \delta(z)\delta(t).$$

In the case of an exponential path length distribution,

$$p(\xi) = \mu e^{-\mu\xi}, \quad P(\xi) = e^{-\mu\xi} = p(\xi)/\mu. \tag{7.39}$$

Pair of equations (7.37) and (7.38) can be combined into a unique integral equation, for the longitudinal density $G(z,t)$:

$$G(z,t) = v\mu \int_0^t \left[\gamma_1 G(z - v\tau, t - \tau) + \gamma_2 G(z + v\tau, t - \tau)\right] e^{-\mu v\tau} d\tau \tag{7.40}$$

$$+ \left[\gamma_1 \delta(z - vt) + \gamma_2 \delta(z + vt)\right] e^{-\mu vt}.$$

It is equivalent to the generalized telegraph equation, whose solution is of the form

$$G(z,t) = \tilde{G}(z,t) + G^0(z,t), \tag{7.41}$$

where

$$G^0(z,t) = e^{-\mu vt}[\gamma_1 \delta(z - vt) + \gamma_2 \delta(z + vt)] \tag{7.42}$$

means the density of unscattered particles, and

$$\tilde{G}(z,t)$$

$$= \frac{\mu}{2} e^{-\mu vt} \left\{ I_0\left(\mu\sqrt{(vt)^2 - z^2}\right) + \frac{vt + (\gamma_2 - \gamma_1)z}{\sqrt{(vt)^2 - z^2}} I_1\left(\mu\sqrt{(vt)^2 - z^2}\right) \right\} \tag{7.43}$$

represents the density of multiple scattered particles [Uchaikin and Saenko (2000)].

After the Fourier-Laplace transformation

$$\tilde{G}(k,\lambda) = \int_0^\infty dt \int_{-\infty}^{+\infty} dx \exp(ikx - \lambda t) G(x,t),$$

integral relations (7.37) and (7.38) become algebraic

$$\tilde{G}(k,\lambda) = \frac{1}{v} W(k,\lambda)\tilde{f}(k,\lambda), \quad \tilde{f}(k,\lambda) = 1 + w(k,\lambda)\tilde{f}(k,\lambda). \tag{7.44}$$

Here

$$w(k,\lambda) = \gamma_2\, \tilde{p}(\lambda/v - ik) + \gamma_1\, \tilde{p}(\lambda/v + ik) \tag{7.45}$$

and

$$W(k,\lambda) \tag{7.46}$$

$$= \frac{\lambda/v - ik\beta - \gamma_1(\lambda/v - ik)\tilde{p}(\lambda/v + ik) - \gamma_2(\lambda/v + ik)\tilde{p}(\lambda/v - ik)}{(\lambda/v)^2 + k^2},$$

where

$$\tilde{p}(\lambda) = \int_0^\infty e^{-\lambda z} p(z) dz$$

denotes the Laplace transform of the path length distribution. Skewness parameter $\beta = \gamma_1 - \gamma_2 = 2\gamma_1 - 1$.

Resolving relations (7.44) with regard to the propagator transform yields

$$\tilde{G}(k, \lambda) = \frac{W(k, \lambda)}{v[1 - w(k, \lambda)]}. \tag{7.47}$$

Substituting expressions (7.45) and (7.46) into (7.47), we obtain:

$$\tilde{G}(k, \lambda)$$

$$= \frac{\lambda/v - ik\beta - \gamma_1(\lambda/v - ik)\tilde{p}(\lambda/v + ik) - \gamma_2(\lambda/v + ik)\tilde{p}(\lambda/v - ik)}{v[k^2 + (\lambda/v)^2][1 - \gamma_2 \, \tilde{p}(\lambda/v - ik) - \gamma_1 \, \tilde{p}(\lambda/v + ik)]}. \tag{7.48}$$

Consider two cases of fractal walk. Let an asymptotic expansion ($\lambda \to 0$, i.e., $t \to \infty$) of the Laplace transform $\tilde{p}$ be of the form,

$$\tilde{p}(\lambda) = 1 - c\lambda^\alpha, \quad c = (A/\alpha)\Gamma(1 - \alpha), \quad 0 < \alpha < 1. \tag{7.49}$$

This case corresponds to the density with "heavy power law tail" $p(z) \sim \alpha z_0^\alpha z^{-\alpha-1}$, $0 < \alpha < 1$, and all moments of natural order diverge. The second case is characterized by a finite first moment ($1 < \alpha < 2$),

$$\tilde{p}(\lambda) = 1 - m_1\lambda + c_1\lambda^\alpha, \quad c_1 = \frac{(A/\alpha)\Gamma(2 - \alpha)}{\alpha - 1}, \quad 1 < \alpha < 2. \tag{7.50}$$

Substituting Eq. (7.49) into expression (7.48), we obtain for the first case

$$\tilde{G}(k, \lambda) = \frac{\gamma_2(\lambda/v - ik)^{\alpha-1} + \gamma_1(\lambda/v + ik)^{\alpha-1}}{v[\gamma_2(\lambda/v - ik)^\alpha + \gamma_1(\lambda/v + ik)^\alpha]}, \quad 0 < \alpha < 1. \tag{7.51}$$

In [Uchaikin and Sibatov (2004)], we inverted this transform in the symmetric case and expressed the result through elementary functions for all values of α of indicated interval. In the asymmetric case, we have

$$G(z, t) \tag{7.52}$$

$$= \frac{2\sin\pi\alpha}{\pi vt} \frac{\gamma_1\gamma_2 \left(1 - z^2/v^2t^2\right)^{\alpha-1}}{\gamma_1^2(1 - z/vt)^{2\alpha} + \gamma_2^2(1 + z/vt)^{2\alpha} + 2\gamma_1\gamma_2 \left(1 - z^2/v^2t^2\right)^\alpha \cos\pi\alpha},$$

$$0 < \alpha < 1.$$

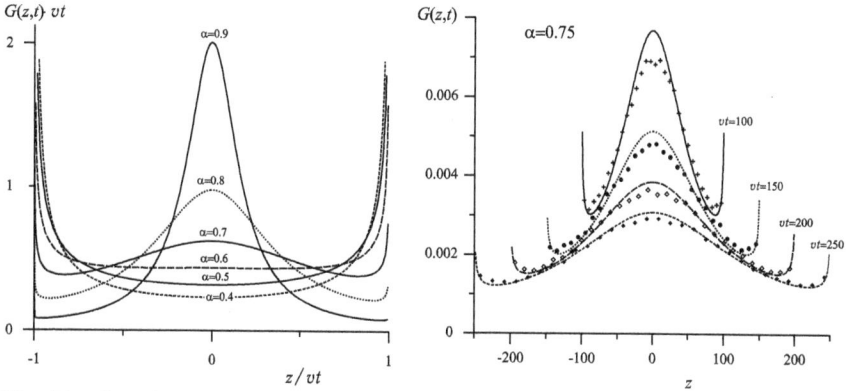

Fig. 7.7 One-dimensional symmetric propagators of the NoRD-model for $\alpha < 1$. Left panel: asymptotic densities in reduced coordinates for $\alpha = 0.4, 0.5, 0.6, 0.7, 0.8, 0.9$. Right panel: evolution of the asymptotic distribution density ($\alpha = 0.75$), points present the Monte Carlo simulation results.

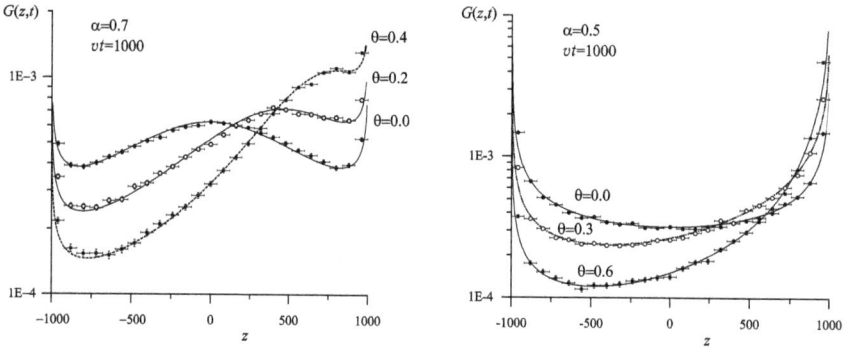

Fig. 7.8 One-dimensional propagators for $\alpha = 0.7$ and 0.5. Points are the result of Monte Carlo simulation.

Here $C = \gamma_1/\gamma_2$. Plots for these distributions for several values of α are presented in Figs. 7.7 and 7.8. The pdf of such type was obtained by Lamperti in frames of the mathematical theory of occupation times [Lamperti (1958)] and used in statistical physics of weakly nonergodic systems [Bel and Barkai (2006)].

Rewriting relation (7.51) in the form

$$v[\gamma_2(\lambda/v - ik)^\alpha + \gamma_1(\lambda/v + ik)^\alpha]\tilde{G}(k, \lambda) \qquad (7.53)$$

$$= \gamma_2(\lambda/v - ik)^{\alpha-1} + \gamma_1(\lambda/v + ik)^{\alpha-1}$$

and performing the inverse Fourier-Laplace transformation, we arrive at the equation with material derivatives of fractional order

$$\left[\gamma_2\left(\frac{\partial}{\partial t}+v\frac{\partial}{\partial x}\right)^\alpha+\gamma_1\left(\frac{\partial}{\partial t}-v\frac{\partial}{\partial x}\right)^\alpha\right]G(x,t)$$

$$=\left[\gamma_2\left(\frac{\partial}{\partial t}+v\frac{\partial}{\partial x}\right)^{\alpha-1}+\gamma_1\left(\frac{\partial}{\partial t}-v\frac{\partial}{\partial x}\right)^{\alpha-1}\right]\delta(x)\delta(t). \quad (7.54)$$

The latter equation can be rewritten in the form

$$\left[\gamma_2\left(\frac{\partial}{\partial t}+v\frac{\partial}{\partial x}\right)^\alpha+\gamma_1\left(\frac{\partial}{\partial t}-v\frac{\partial}{\partial x}\right)^\alpha\right]G(x,t)$$

$$=\frac{t^{-\alpha}}{\Gamma(1-\alpha)}[\gamma_2\delta(x+vt)+\gamma_1\delta(x-vt)]. \quad (7.55)$$

Multiplier $(\lambda\pm ivk)^\alpha$ in formulas derived above presents the Fourier-Laplace transform of the fractional material derivative [Sokolov and Metzler (2003)]:

$$(\lambda\mp ivk)^\alpha\tilde{f}(k,\lambda)=\int\limits_0^\infty dt\int\limits_{-\infty}^\infty dx e^{-\lambda t+ikx}\left(\frac{\partial}{\partial t}\pm v\frac{\partial}{\partial x}\right)^\alpha f(x,t),$$

that can be verified by rewriting the operator in the Riemann-Liouville form:

$$\left(\frac{\partial}{\partial t}\pm v\frac{\partial}{\partial x}\right)^\alpha f(x,t)=\frac{1}{\Gamma(1-\alpha)}\left(\frac{\partial}{\partial t}\pm v\frac{\partial}{\partial x}\right)\int\limits_0^t f(x-v(t-\tau),\tau)(t-\tau)^{-\alpha}d\tau,$$

$$0<\alpha<1,$$

and applying the Fourier-Laplace transformation.

Consider the second case: $1<\alpha<2$ (the process of the second kind). For asymptotic $\lambda\to 0$, $k\to 0$, $|\lambda/vk|\to 0$, the transform takes the form:

$$\tilde{p}(k,\lambda)=\frac{1}{\lambda+i\beta kv-(cv/m)\left[\gamma_1(ik)^\alpha+\gamma_2(-ik)^\alpha\right]}. \quad (7.56)$$

The inverse Laplace transformation leads to the characteristic function

$$\tilde{p}(k,t)=\exp\left(i\beta kvt-\frac{cvt}{m}\left[\gamma_1(ik)^\alpha+\gamma_2(-ik)^\alpha\right]\right),$$

which is related to the characteristic function

$$\tilde{g}(k;\alpha)=\exp\left(-|k|^\alpha[1-i\beta\tan(\pi\alpha/2)\,\mathrm{sign}(k)]\right)$$

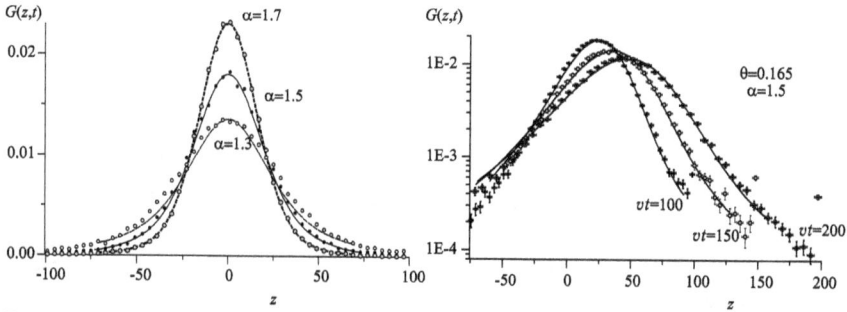

Fig. 7.9 One-dimensional propagators for $1 < \alpha < 2$. Points are the result of Monte Carlo simulation.

of the Lévy stable density $g(x; \alpha, \beta)$ by the following expression

$$\tilde{G}(k, t) = \exp(i\beta kvt) \, \tilde{g}((Kt)^{1/\alpha} k; \alpha; \beta).$$

Passage to the spatial coordinate leads to the propagator:

$$G(z, t) = (Kt)^{-1/\alpha} g\left((z - \beta vt)(Kt)^{-1/\alpha}; \alpha, \beta \right). \qquad (7.57)$$

Here

$$K = \frac{cv}{m} \sin \frac{\pi(\alpha - 1)}{2}.$$

These propagators are shown in Fig. 7.9.

Relation (7.56) can be rewritten in the form

$$\lambda \tilde{p}(k, \lambda) = -i\beta kv\tilde{p}(k, \lambda) + \frac{cv}{m} \left[\gamma_1 (ik)^\alpha + \gamma_2 (-ik)^\alpha \right] \tilde{p}(k, \lambda) + 1.$$

The inverse Fourier-Laplace transformation leads to the diffusion equation with a fractional operator:

$$\frac{\partial}{\partial t} p(z, t) = \beta v \frac{\partial}{\partial z} p(z, t) + \frac{cv}{m} \left[\gamma_1 \,_z D_\infty^\alpha + \gamma_2 \,_{-\infty} D_z^\alpha \right] p(z, t) + \delta(z)\delta(t).$$

The calculation results for different α are presented in Figs. 7.7–7.10 in comparison with data of the direct Monte Carlo simulation. Qualitative distinction of these distributions is explained by competition between two processes: diffusion expanding according to the law $\propto t^{1/\alpha}$ in the absence of restrictions, and ballistic motion bounding the particle position by segment $[-vt, vt]$. When $\alpha > 1$, the first process dominates at large times: segment $[-vt, vt]$, expanding rapidly, ceases to influence on diffusion. If $\alpha < 1$, the role of kinematic restriction grows and the distribution begins to concentrate near the boundaries of segment $[-vt, vt]$. Figure 7.10 shows variety of expansion dynamics for different values of α. It should be noted that the reduced asymptotic propagators for $0 < \alpha < 1$ are universal and do not depend on scale parameters of path length distributions.

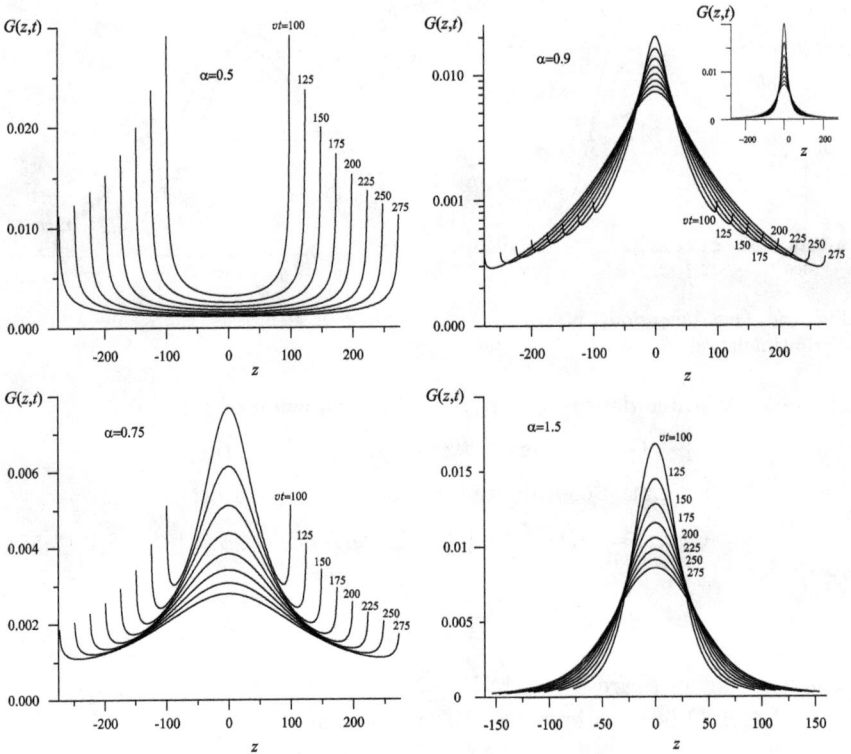

Fig. 7.10   Evolution of asymptotic NoRD-propagators for different α.

Some consequences of space-time correlations

The trajectory of a charged particle in the interstellar magnetic field, together with its derivative, is an extremely complex continuous curve in the phase space, which can be described in detail by an integer-order differential equation if the magnetic field and other characteristics of the medium affecting the particle motion are also specified in detail. Needless to say, we have no such information.

We return to the idea of passing from a continuous description to a coarse-grain description, this time focusing our attention on the trajectory shape. If we were dealing with a homogeneous medium, the partition elements would only be slightly different from each other, and, in the finite-element representation, we would again obtain an analog of the classical differential equation with the diffusion coefficient somewhat corrected due to the coarse graining. But the interstellar medium is not simply inhomogeneous in reality: its inhomogeneities have a multiscale character, as mentioned in the Introduction. This multiscale (fractal) property of the structure does not allow choosing the partition size for which the characteristics of elements could be assumed approximately identical. Selecting one such element, we find that the magnetic field in neighboring elements is an order of magnitude lower (and a particle freely goes away to a remote region) or higher (and a particle remains confined in one of the elements for a long time). The fractal structure assumes that this holds in a broad range of partition sizes. But while crossing the interface between neighboring elements occurs instantaneously, crossing a number of large-scale elements almost transparent to particles can no longer be assumed instantaneous, and the time spent for the passage should be taken into account. This is the difference between the NoRD model, taking the time spent by a particle to move from one trap to another into account, and the NoD model, in which such a transition is assumed instantaneous irrespective of its scale. If T is the time interval separating the instants of a particle arriving at some partition element and the next nonempty element, then in the NoRD model T consists of two terms: the residence time T_0 of a particle in the first element and the time T_1 spent to pass to the next element, whereas in the NoD model, this time is the residence time of a particle in a trap. Ranges and waiting times in both models are independent of each other, but the flight time in the NoRD model is proportional to the range (for a constant velocity) and gives rise to a $T - R$ correlation,

$$T = T_0 + R/v.$$

As shown in Section 2.5, a fractional order $\beta < 1$ of the time derivative means that residence times in traps are distributed with a density proportional to $t^{-\beta-1}$, $t \to \infty$, and the fractional order of the Laplacian suggests that ranges are distributed with a density proportional to $r^{-\alpha-1}$, $r \to \infty$. The unrelated terms $(\partial/\partial t)^\beta$ and $(-\Delta)^{\alpha/2}$ in the NoD equation operator mean the absence of correlations between T and R, and, as a consequence, a jumpwise (discontinuous) form of the trajectories, whereas the presence of the composite operator $\langle(\partial/\partial t + \mathbf{v}\nabla)^\alpha\rangle$ in the NoD equation indicates the presence of $R - T$ correlations.

Including $R - T$ correlations in the model

(1) transforms discontinuous trajectories into continuous ones, thereby eliminating nonphysical phantoms such as long instant flights and long rest times followed by instant gains of an infinite velocity;

(2) restricts the spatial position of a particle by a spherical region of the radius vt centered at the initial point, which allows reconciling the process with the relativistic concept, restricts the expansion law of the diffusion packet by a linear velocity, and returns the method of moments to the toolbox of computational techniques in the transfer theory (which is inapplicable in the NoD model due to the divergence of moments);

(3) for $\alpha < 1$, considerably changes the shape of the spatial distribution, giving rise to splashes near the ballistic boundary $r = vt$ and thereby transforming the usual bell-shaped diffusion packet to a W-shaped packet, and, for $\alpha < 1/2$, to the U-shaped packet.

The differences between the NoD and NoRD models are most strongly manifested for α, $\beta < 1$, when mathematical expectations of random variables are infinite and jumps of space-time trajectories can be seen at any scale. For $\alpha > 1$ and $\beta = 1$, the situation is different: as the size of the chosen space-time region increases, the relative role of jumps becomes less noticeable and becomes insignificant when the expectation values are greatly exceeded. This is well seen for a usual Brownian trajectory: despite the independence of the spatial and temporal parts of the differential operator, the trajectory is everywhere continuous (although not differentiable).

To elucidate the role of $R - T$ correlations in the model under study, we compare propagators in the NoRD and NoD models. In the NoRD model, we assume that independent traps are absent ($T_0 = 0$), and therefore T and $R = vt$ completely correlate, such that their distributions coincide up to the scale factor v (which we set equal to 1). In the NoD model, $v = \infty$ and random variables $T \equiv T_0$ and R are completely independent. We take the same distribution $\mathsf{P}(R > x) = \mathsf{P}(T > x) \propto x^{-\nu}$, $\nu > 0$, for them and note that in this case,

$$
\alpha = \begin{cases} \nu, \nu \leq 2; \\ 2, \nu > 2, \end{cases} \qquad \beta = \begin{cases} \nu, \nu \leq 1; \\ 1, \nu > 1. \end{cases}
$$

The principal difference between the propagators disappears only for $\nu > 2$. We can see from Fig. 18, representing calculations of one-dimensional and three-dimensional random walks in both models, that the $R - T$ correlations drastically change the process for $\nu \leq 1$: the propagators differ in their form, expansion law, and behavior near ballistic boundaries and near the radiation source (we return to these distributions in the study of one-dimensional random walks along field lines in Section 4.5). For $1 < \nu < 2$

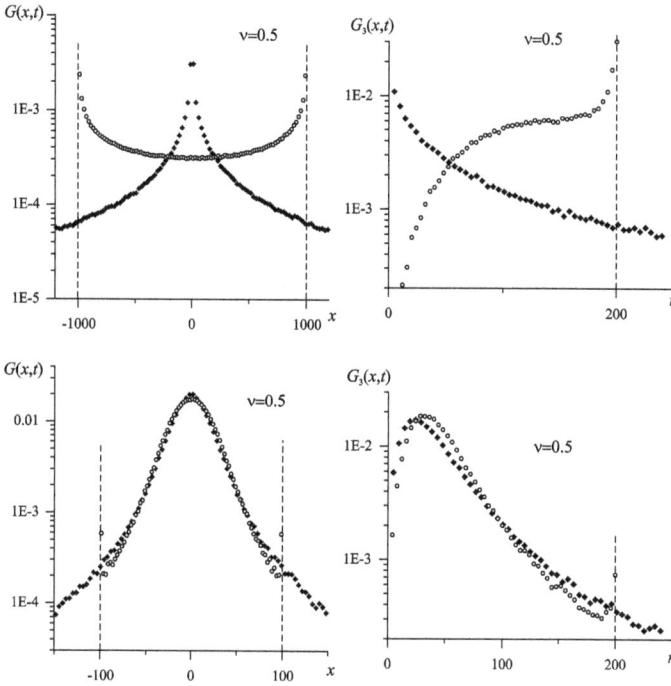

Fig. 7.11 Influence of range-time correlations on the form of (a) one-dimensional and (b) three-dimensional propagators. Filled symbols correspond to the process with independent R and T, and open symbols to the process with the linear dependence $R = vt$. The vertical straight lines $x = \pm vt$ show the boundaries of distributions.

(anomalous diffusion of the second kind), the differences are weaker, although quite noticeable: in one case, distributions are bounded, while in the other, they are not bounded; in the NoRD model, a front appears near the region $|\mathbf{x}| = vt$ and the densities are quantitatively different near the source.

7.4 On the energy spectrum

In this section, modeling of CR sources was performed under the following assumptions.

(1) All sources are isotropic, point and instantaneous with equal powers and energy spectra $S_k(E) \sim E^{-\gamma} \exp(-E/E_{\mathrm{max},k})$. Here γ is an exponent

of injection spectrum, and $E_{\max,k}$ characterizes maximum energies of k-type nuclei accelerated at supernova remnants.

(2) The sources are separated into two groups: relatively young supernovae (less than $1.5 \cdot 10^6$ years) whose space-time coordinates were taken individually (15 objects: Vela, Monogem Ring, Geminga and others), and 30000 sources with spatial distribution generated according to approach proposed in [Faucher-Giguere and Kaspi (2006)] and used in [Blasi and Amato (2012a)]. The radial distribution of discrete sources was simulated according to the density of pulsars (see [Case and Bhattacharya (1996)]),

$$ f(r) = \frac{\beta^4 e^{-\beta}}{12\pi R_\odot^2} \left(\frac{r}{R_\odot} \right)^2 \exp\left(-\beta \frac{r - R_\odot}{R_\odot} \right) $$

with $R_\odot = 8.5$ kpc and $\beta = 3.53$, the $|z|$-distribution was modeled by exponential function with $z_g = 200$ pc. To simulate the spiral structure, the algorithm proposed in [Faucher-Giguere and Kaspi (2006)] was used.

The example realization is presented in Fig. 7.12.

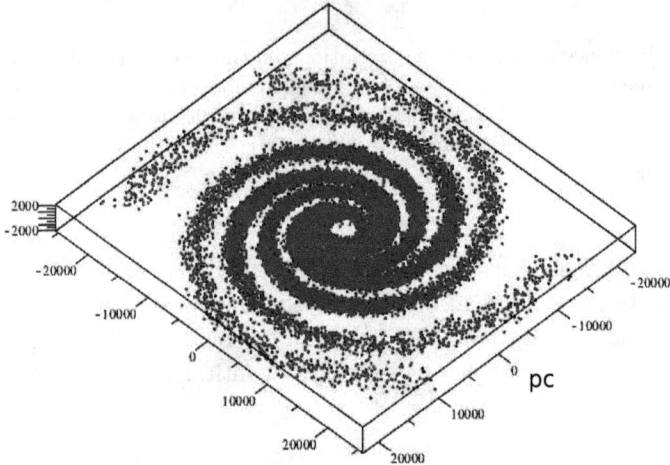

Fig. 7.12   Spatial arrangement of sources generated by Monte Carlo algorithm.

The area of CR diffusion is the cylinder of height $2H = 8$ kpc with radius $R_d = 20$ kpc surrounded by a nonreflective environment. The galactic disk being in the middle of the cylinder parallel to bases has the thickness

$2h = 600pc$. As in [Blasi and Amato (2012a)], we assume that the diffusion coefficient is independent of spatial and time variables, and determined by particle rigidity R,

$$D(E) = D_0(R/3\ GV)^\delta\ [\mathrm{pc^2 yr^{-1}}], \quad R > 3GV. \tag{7.58}$$

Results reported here are related to values $\delta = 0.6$ and $D_0 = 0.0729\ \mathrm{pc^2 yr^{-1}}$.

The simplest propagator of the LoD-model in the infinite homogeneous space has the Gaussian form

$$G_{\mathrm{LoD}}(\mathbf{x}, t, E; \mathbf{x}_s, t_s, E_0) = \frac{\delta(E - E_0)}{(4\pi D(E)\tau)^{3/2}} \exp\left(-\frac{(\mathbf{x} - \mathbf{x}_s)^2}{4D(E)\tau}\right); \quad \tau = t - t_s. \tag{7.59}$$

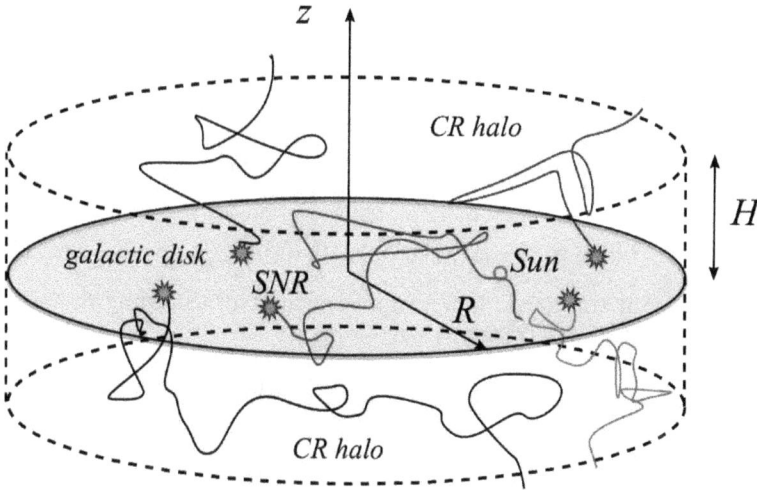

Fig. 7.13 Cylindrical model of a galactic domain for CR propagation.

To account for the halo boundedness (Fig. 7.13), solutions (7.59) should be modified: considering zero boundary conditions and neglecting by lateral surface of the cylinder, the image method can be used in local approach,

$$G_{\mathrm{LoD,H}}(\mathbf{x}, t, E; \mathbf{x}_s, t_s, E_0) = \frac{\delta(E - E_0)}{(4\pi D(E)\tau)^{3/2}} \exp\left(-\frac{(x - x_s)^2 + (y - y_s)^2}{4D(E)\tau}\right)$$

$$\times \sum_{j=-\infty}^{\infty} (-1)^j \exp\left(-\frac{(z - z_j)^2}{4D(E)\tau}\right). \tag{7.60}$$

Here $z_j = (-1)^j z_s + 2jH$ are z-coordinates of the image sources, $\{x_s, y_s, z_s\}$ are coordinates of an original source.

The NoD-propagator in an infinite medium is expressed through fractional stable Lévy-Feldheim density [Lagutin *et al.* (2001b)]. In case of $\beta = 1$ it can be presented in the form [Uchaikin and Sibatov (2012b)]

$$G_{\mathrm{NoD}}(\mathbf{x}, t, E; \mathbf{x}_0, t_0, E_0) \qquad (7.61)$$

$$= \int_0^\infty G_{\mathrm{LoD}}(\mathbf{x}, \theta, E; \mathbf{x}_0, 0, E_0) \frac{g_+(\theta(t - t_0)^{-2/\alpha}; \alpha/2)}{(t - t_0)^{2/\alpha}} d\theta.$$

Here $g_+(x, \nu)$ is a one-sided Lévy stable density (subordinator, $0 < \nu \le 1$) and G_{LoD} in (7.61) can be taken with the diffusion coefficient linked with D_α of the NoD-model by relation

$$D(E) = [D_\alpha(E)]^{2/\alpha}. \qquad (7.62)$$

If $D(E) \propto E^\delta$, then $D_\alpha(E) \propto E^{\delta_\alpha}$, where $\delta_\alpha = \delta\alpha/2$. For example, for $\delta = 0.6$ and $\alpha = 1.5$, we have $D_\alpha(E) \propto E^{0.45}$.

Formula (7.61) is suitable for Monte Carlo representation of the NoD-propagator

$$G_{\mathrm{NoD}} = \langle G_{\mathrm{LoD}}(\mathbf{x}, \Theta, E; \mathbf{x}_0, 0, E_0) \rangle_{\Theta_{\alpha/2}(t-t_0)}, \qquad (7.63)$$

where $\Theta_{\alpha/2}$ is a stable random variable generated according to (see e.g. [Uchaikin and Zolotarev (1999)])

$$\Theta_{\alpha/2}(t - t_0) = (t - t_0)^{2/\alpha} \frac{\sin(\alpha\pi U_1/2)[\sin((1 - \alpha/2)\pi U_1)]^{2/\alpha-1}}{[\sin(\pi U_1)]^{2/\alpha}[\ln U_2]^{2/\alpha-1}}.$$

Formulae (7.62) and (7.63) indicate the correspondence principle between NoD- and LoD-models (when $\alpha \to 2$) and allow us to use the standard diffusion coefficient to estimate CR characteristics in frames of the nonlocal models.

In calculations of energy spectrum in frames of the NoRD-model (for $\alpha \in (1, 2)$) we have used approximate expression for propagators

$$G_{\mathrm{NoRD}}(\mathbf{x}, t, E; \mathbf{x}_s, t_s, E_0)$$

$$= G_{\mathrm{NoD}}(\mathbf{x}, t, E; \mathbf{x}_s, t_s, E_0) \, 1\left(c\tau - |\mathbf{x} - \mathbf{x}_s|\right)$$

$$+ \delta(E - E_0)\delta(|\mathbf{x} - \mathbf{x}_s| - c\tau)\mathrm{Prob}(|\mathbf{x} - \mathbf{x}_s| > c\tau), \qquad (7.64)$$

where $1(x)$ is the Heaviside step function.

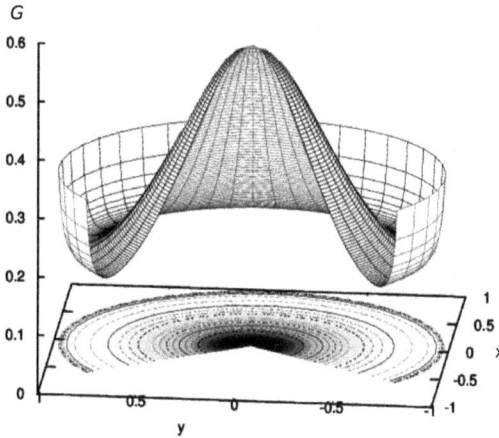

Fig. 7.14 Typical shape of the NoRD-propagator in two dimensions.

It should be noted that zero boundary conditions for nonlocal models are fulfilled approximately, if the mean scattering length of CRs much smaller than thickness of the halo. To account for restriction of the diffusion domain (halo), the generalized Marshak boundary conditions can be used. The solutions have been modified by means of the image method. In this approximation, we use $G_{\mathrm{LoD,H}}$ instead of G_{LoD} in expressions (7.61) and (7.63) using procedure (7.64).

We estimated energy spectra accounting for spatial and temporal distribution of supernova remnants and diffusion in halo in frames of four models (LoD, LoRD, NoD, NoRD). Note that such calculations for the local model were performed in many papers (see e.g. [Sveshnikova (2003); Blasi and Amato (2012a); Evoli *et al.* (2008)]) and for nonlocal model in [Lagutin *et al.* (2001b); Erlykin *et al.* (2003); Lagutin *et al.* (2015)]. All of them, however, utilize propagators ignoring the relativistic speed limit requirement. Relativistic nature of CRs sufficiently modify the propagators and spectra as well. NoRD-propagators are restricted to domain $|\mathbf{x} - \mathbf{x}_s| \le vt$, have ballistic splash at this boundary whereas in the middle part they are spreading superdiffusively (remind, that we restrict ourselves by values of superdiffusion parameter $\alpha \in (1,2]$). Spallation effects are taken into account according to Blasi and Amato [Blasi and Amato (2012a)] using formulae provided in [Hörandel *et al.* (2007)].

As is known, the standard assumptions about the origin of galactic CRs: power law spectra of particles injected by SNR, and diffusive propagation

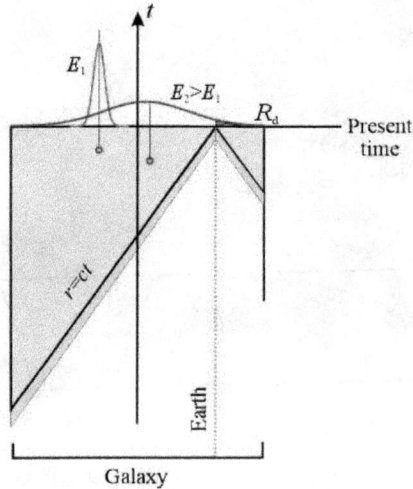

Fig. 7.15   To explain the steepening in equilibrium spectrum in frames of the LoRD and NoRD model.

throughout the Galaxy with a coefficient $D(E) \propto E^\delta$ leads to the spectrum $n(E) \propto N(E)/D(E) \propto E^{-\gamma+\delta}$ (from the SNR ensemble) reflecting the balance between injection and escape of CRs from the confinement volume. Accounting for relativistic restriction (in terms of relativistic propagators) steepens the spectrum at high energies, and the position of the spectrum break, its sharpness and slopes of asymptotes depend on $D_\alpha(E)$ and α [Sibatov and Uchaikin (2015)] (Figs. 7.16 and 7.17). The cause of this effect is universal and must be kept in mind in all calculations of propagation of high energy CRs. Explanation of the spectrum steepening at high energies in the NoRD model is schematically represented in Fig. 7.15. The grey field corresponds to the space-time area containing CR sources which do not take part in the formation of the spectrum due to the relativistic restriction. Their accounting in the ordinary diffusion model leads to a pure power law spectrum $\propto E^{-\gamma-\delta}$ under the assumption about the pure power law spectrum of injected particles. Taking the relativistic restriction into account in the NoRD model excludes automatically these sources and this exclusion affects predominantly on high energy part of CRs due to $D(E) \propto E^\delta$.

It is important to emphasize that, the NoRD model (compared to the NoD-version used by Lagutin *et al.*) operates with acceptable values of injection exponent γ (≈ 2) (in contrast with $\gamma > 2.7$, e.g. $\gamma = 2.85$, in

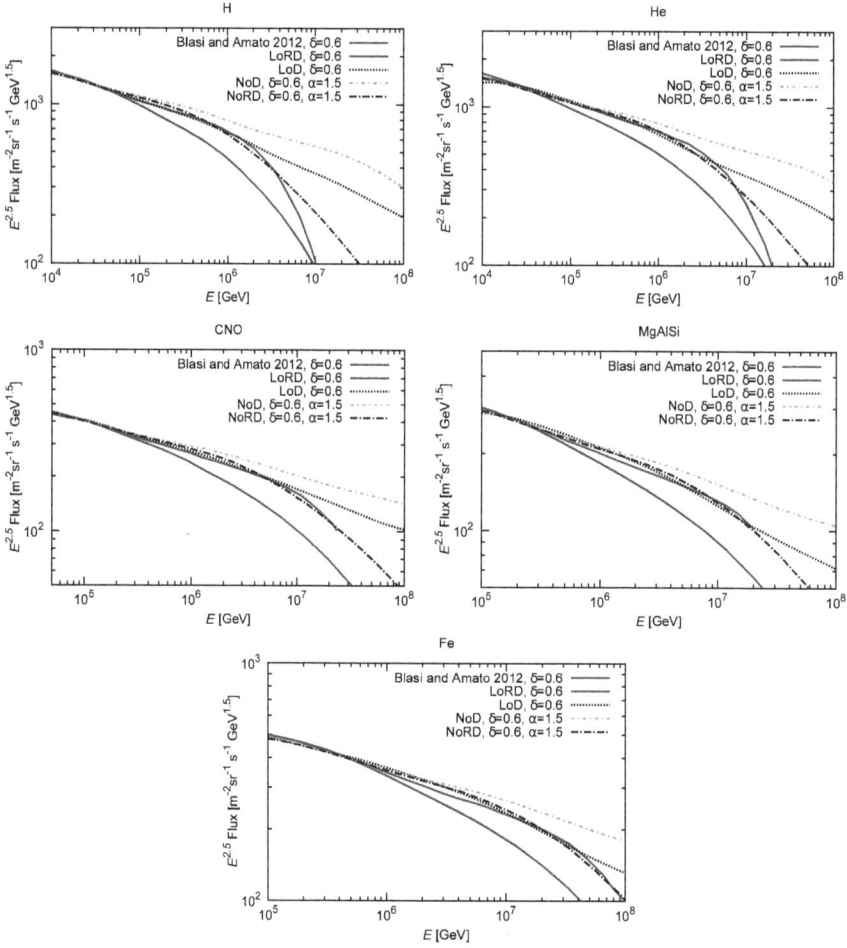

Fig. 7.16 Energy spectra for different groups of nuclei calculated in frames of four models of CR diffusion ($\delta = 0.6$, $\gamma = 2.07$).

recent calculations by [Lagutin *et al.* (2015)]).

Figures 7.16 and 7.17 demonstrate computational results for spectrum of protons, CNO-group and other groups of nuclei obtained in frames of LoD, LoRD, NoD and NoRD models. Relativistic speed limit requirement taken into account in LoRD and NoRD leads to steepening of spectra as mentioned above. The energy of the 'knee' E_{knee} obtained in NoRD is very close to that provided in [Blasi and Amato (2012a)] by exponential cut-off of injection spectrum, but we observe power law NoRD-spectrum above E_{knee}

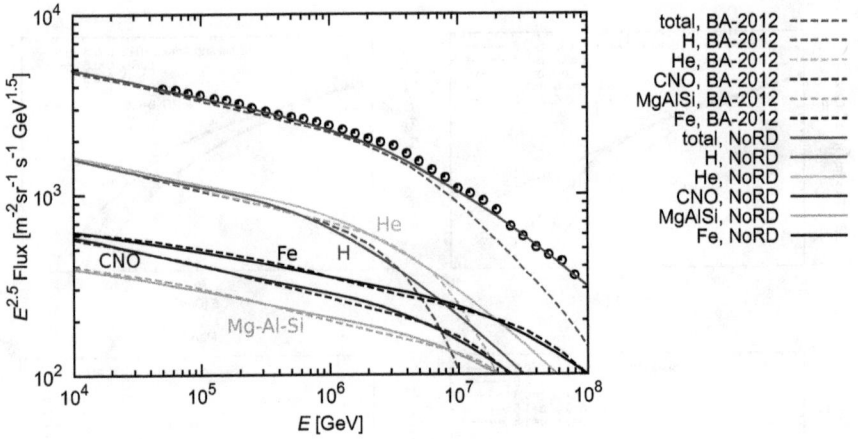

Fig. 7.17   Energy spectra in the NoRD model ($\alpha = 1.5$, $\delta = 0.6$, $\gamma = 2.07$) compared with results of [Blasi and Amato (2012a)].

in contrast to exponential decay obtained in [Blasi and Amato (2012a)]. Sure, this requires greater maximum energies of particles accelerated in sources. The exponentially falling tails for all nuclei contradict to observed power law spectrum above E_{knee}, if we consider it as produced by GCRs (without extragalactic component). From this point of view NoRD-model seems to be preferable.

7.5   Cosmic ray anisotropy

Irregularities in the magnetic fields in the Interstellar Medium (ISM) cause the direction of a CR observed at Earth to bear little relation to the direction to its actual source, except for CR of ultra high energies. The anisotropy coefficient of a flux in the standard diffusion theory, determined by the ratio of the current density $J_r = J$ to the particle flux density vN,

$$\delta(r,t) = 3\frac{J(r,t)}{vN(r,t)},$$ (7.65)

is described at a distance r from a point-like instantaneous source in a homogeneous infinite medium by the known expression [Hayakawa (1969)]

$$\delta(r,t) = \frac{3r}{2vt}.$$ (7.66)

We note that (i) the derivation of expressions (7.65) and (7.66) a priori assumes a weak anisotropy of the flux at the point under study and (ii) the diffusion approximation itself is applicable to random walks at rather large times, when the number of flights performed by a particle is large enough for the total displacement to reach the asymptotic regime.

In the case of anomalous diffusion, the derivation based on the usual Fick law is of course invalid; but the fact that anomalous diffusion equations preserve the self-similarity of their solutions is sufficient for finding the general expression relating the current and concentration, even without specifying the form of a self-similar solution. According to the physical meaning of J, we have

$$J(r,t) = \frac{1}{r^2} \frac{d}{dt} \int_r^\infty N(r,t) r^2 dr \qquad (7.67)$$

in a spherically symmetric problem.

Assuming the self-similarity of the propagator

$$N(\mathbf{x},t) = t^{-3H} \phi(rt^{-H}), \; r = |\mathbf{x}|,$$

and substituting expression

$$\int_r^\infty N(r,t) r^2 dr = \int_{rt^{-H}}^\infty \phi(z) z^2 dz$$

into (7.67), after differentiation with respect to t

$$\frac{\partial}{\partial t} \int_{rt^{-H}}^\infty \phi(z) z^2 dz = Hr^3 t^{-3H-1} \phi(rt^{-H}) = Hr^3 t^{-1} N(r,t)$$

we obtain the formula for flux anisotropy on the distance r from the point instantaneous source:

$$\delta(r,t) \equiv= 3 \frac{J(r,t)}{vN(r,t)} = 3H \frac{r}{vt}. \qquad (7.68)$$

In [Lagutin and Tyumentsev (2004)], the anisotropy was described by classical formula (7.65), in which the concentration N was replaced with the product kN_{near}, where k is the number of nearest sources and N_{near} is the concentration of one of them. However, it is obvious that in this case, the motion of particles never satisfies the continuity condition mentioned above. On the contrary, the concentration of particles moving at a given instant in the Lagutin-Tyimentsev model is zero, while the velocity of particles performing instant flights is infinite. This gives rise to an uncertainty in the product vN, which can lead to any result. Another consideration requiring a careful treatment of general expression (7.68) is that this expression is exact with respect to the fractional differential equation, but this equation itself is only an asymptotic form of the system of integral equations. Integral equations include the distributions of ranges and time intervals between flights, and therefore correctly describe random walks with specified characteristics, whereas the fractional differential equation contains some part of this information in the diffusion coefficient, and the characteristics of flight lengths can no longer be separated from waiting times. In considering this model less formally, we must admit that the concept of particles at rest in traps cannot correspond to reality. Simply put, we are dealing with regions which a particle can leave only with difficulty because of the small diffusion coefficient strongly entangling the trajectory along which the particle can continue its motion with the same velocity v (which for light particles (electrons and positrons) gives enhanced synchrotron radiation). In this case, we should consider real anisotropy at a point located in a trap (with N meaning the total concentration) and anisotropy at a point located outside the trap (with N being the concentration of particles not affected by traps).

Anisotropy inversion phenomenon near a boundary

The well-known fact is that there is a region at energies below 1 PeV which has an anisotropy phase opposite to expectation and a marked change of the phase in the Right Ascension (RA) plot starting from 1 PeV, where the amplitude of the anisotropy drops and then starts to rise again [Erlykin and Wolfendale (2015)].

In [Erlykin *et al.* (2015)] we answered positively on the fundamental question: can particles originated from a source in a particular direction be observed at Earth as coming from the opposite direction or generally from any different direction? In particular, can the 'flow' of particles from the Inner Galaxy give an anisotropy pointing to the opposite direction, i.e., to the Outer Galaxy?

We explained the anisotropy inversion as a result of stochastic reflection of the NoRD propagator front from dense medium. Monte Carlo simulation of this problem led us to the following conclusions. The anisotropy inversion does not take place in the case of stationary (time-independent) transport. The ordinary diffusion theory cannot catch this phenomenon because it is not in a position to describe the front splash: a diffusion packet is instantaneously spread around all space, breaking the relativistic principle. The NoRD reveals the reality of such anisotropy inversion mechanism: the most auspicious conditions for the phenomenon appear after the front splash passed through the boundary of two domains in the direction of the more dense one. Below, we present some numerical results confirming these conclusions.

The continuity equation

$$\frac{\partial n(x,t)}{\partial t} = -\frac{\partial j_x(x,t)}{\partial x}$$

shows that in the stationary case, when $n(x,t) = n(x)$, $j_x(x)$ doesn't depend on x and consequently has the same sign (positive to the right of the source and negative on the left). In other words, the flux is directed from the local source. This conclusion is valid for heterogeneous media too (if not, the particles should be accumulated in some region and the process will be time-dependent). In case of a time-dependent source, the ordinary diffusion model gives the similar result, because the diffusion equation is a parabolic one and has no wave-solutions which could reverse concentration gradient. However, a more realistic telegraph equation (7.3) is hyperbolic, it demonstrates such feature of the wave motion as the existence of a front splash and a retardation. But this equation relates to a uniform medium when free paths are distributed according to exponential law. The turbulent character is taken into account by its fractional counterpart.

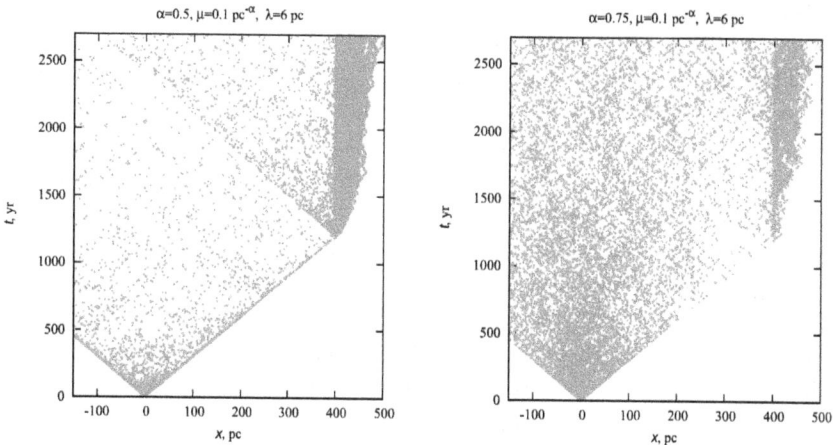

Fig. 7.18 Scattering diagram of random walks in a two-layer medium ($\alpha = 0.5$ and 0.75 for $x < 400$ pc and $\alpha = 1.75$ for $x > 400$ pc).

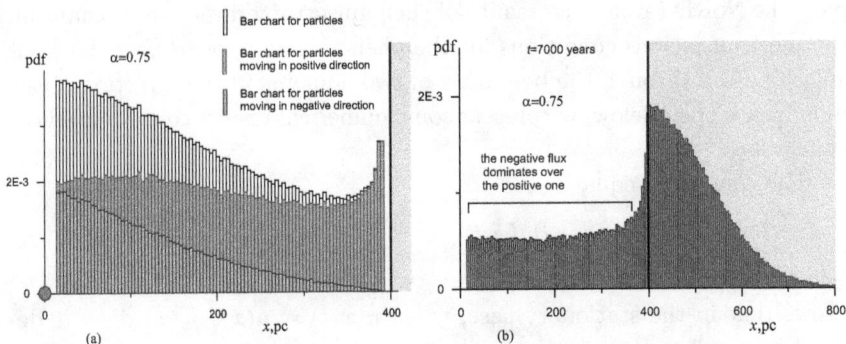

Fig. 7.19 Bar charts for particles moving in positive (dark gray bar chart) and negative (light gray bar chart) directions before (a) and after (b) hitting the boundary, $\alpha = 0.75$ for $x < 400$ and $\alpha = 1.75$ for $x > 400$.

We have solved this equation both analytically and numerically (by means of Monte Carlo method). The latter is especially effective in case of particles passing through a boundary separating two regions with different properties. As our calculations show, the fractional generalization also reveals the appearance of a weak short-term current inversion after the front flash. This weak phenomenon becomes more significant near the boundary between two domains with essentially different properties and long free paths in one of them (especially, when $\alpha < 1$). Some preliminary results obtained within the frames of one-dimensional (forward and backward) walk model, are presented in Figs. 7.18 – 7.22.

The results are obtained for the case when a point short-time source placed at the origin and emits particles moving along x-axis. We observe the forward and backward currents and obtain the anisotropy in the usual way. The inversion of anisotropy appears only when we combine a two-layer medium with a finite velocity of motion. In this case, we really observe negative isotropy after the short thick front layer of particles passed by, which give enough number of back-scattered particles close to the boundary. It is remarkable that the inversion of current is observed on the source side near the boundary, when the front of the incident packet is large enough. This phenomenon can be interpreted as a stochastic reflection of the packet front from the dense medium.

Calculated results of the time-dependent anisotropy are presented in Figs. 7.20 and 7.22 for different values of α and different detector positions. The source-boundary distance $a = 400$ pc. Random free paths are distributed according to the inverse power law with exponent α. The 'first-

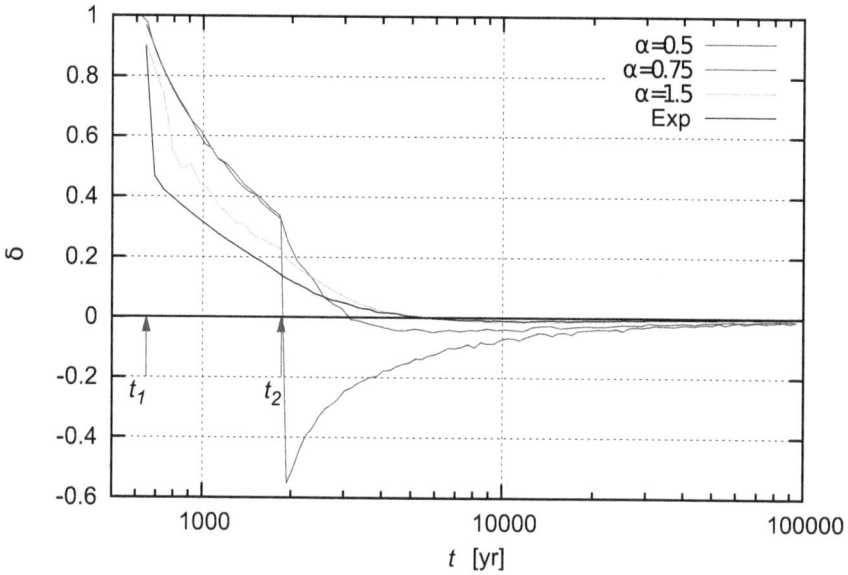

Fig. 7.20 The family of anisotropy time-dependence $\delta(t)$ for various values of α.

particles front' reaches the boundary, dives into the second medium, diffuses there, partially returns through the boundary and appears at the observer after $t_{\text{front}} = (a + (a - R))/v$. In case of $R = 200$ pc, $t_{\text{front}} \approx 1957$ yr.

We argue in [Erlykin *et al.* (2015)] that the presence of local regions of reduced diffusion coefficient (D) can seriously perturb the directions from which CR arrive from a specific source which generated CR for a limited period. For example, a close-by region of reduced D can "store" particles such that more source-particles come from the anti-source direction than the reverse. The location of the Solar System in the Local Bubble might satisfy this condition.

Attempting to give a more clear explanation of the anisotropy inversion phenomenon, we suggest to imagine what happens when a finite duration pulse of light falls on a perpendicular plane. If the latter is an absorber, the pulse is not reflected: it *sticks* in the absorber as the ordinary diffusion predicts. If the plane is of a mirror type, then the pulse is reflected with inverted form: the growing front part (moving now in negative direction) is superposed on the fading tail of the pulse (moving yet in positive direction), so the sum may become negative. This process is described by a plane kinematics. In reality, the particles penetrate through the plane into

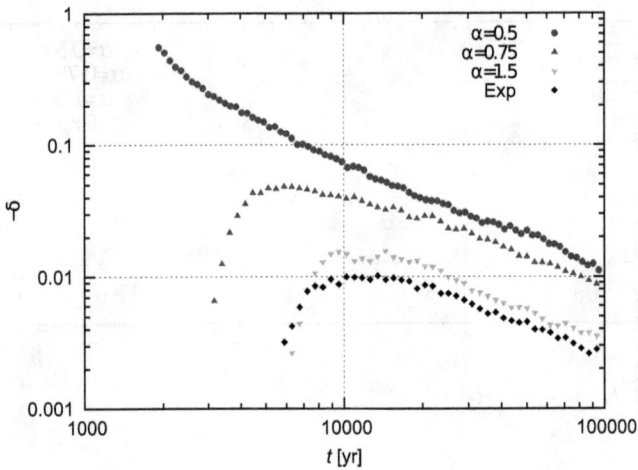

Fig. 7.21 Function $-\delta(t)$ in the log-log scale. The observation point is to the left of the boundary (that is, in 1-medium). Distances: source-observer $R = 200$ pc and source-boundary $a = 400$ pc. Random free paths are distributed according to a power law with exponents α in 1-medium and β in 2-medium ($\beta = 1.9$; mean path in 2-medium $\lambda_2 = 0.3$ pc). First-particles front reaches the detector at the time point $t_1 = R/v$ than it reaches boundary, dives into 2-medium, diffuses there, partially returns through the boundary and occurs the observer point after $t_2 = [a + (a - R)]/v$. Label 'Exp' corresponds to the case of exponential distribution of free path lengths.

small deep of the second medium and most of them are finally reflected. This is slightly more complex process described by the telegraph equation, so we meet some intermediate case, but our calculation showed that the phenomenon of anisotropy inversion continues to take place.

7.6   Milne problem for a fractal half-space

The backscattering problem is formulated like a specific problem of the transport theory called the Milne problem [Case and Zweifel (1967)]. In the simplest statement, it concerns the calculation of the backscattering flow of particles (albedo) from a half-space uniformly filled with an isotropically scattering medium (Fig. 7.23a).

Such a problem has important significance not only for astrophysical aims, when the property of planet atmospheres are investigated by using the back-scattered solar light measurements, but for the similar diagnostic aims in various fields of science and engineering, including meso- and nano-technologies.

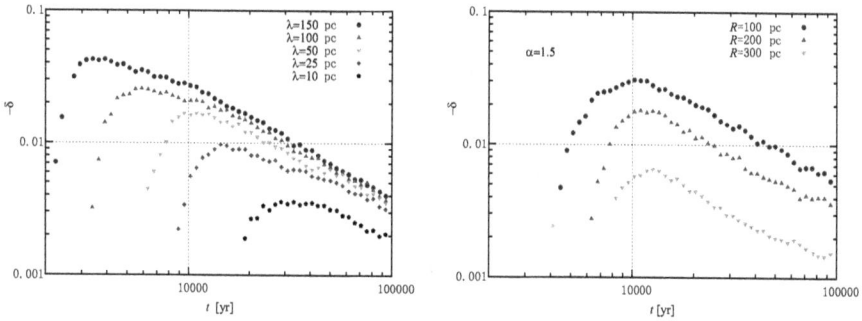

Fig. 7.22 Left panel: Anisotropy $\delta(t)$ for various values of mean path λ in the first layer ($\alpha = 1.5$). Right panel: $\delta(t)$ for various detector positions in the first layer (R is the source-detector distance). The observation point is to the left of the boundary (that is, in 1-medium). The source-boundary distance is equal to $a = 400$ pc. Random free paths are distributed according to a power law with exponents $\alpha = 1.5$ in 1-medium and $\beta = 1.9$ in 2-medium ($\beta = 1.9$; mean path in 2-medium $\lambda_2 = 0.3$ pc).

The backscattering of a flash (the radiation of a point isotropic instanta-neous source) on the boundary of a uniform medium has been considered in [Kuŝĉer and Zweifel (1965)] on the base of solution of the linear one-speed Boltzmann transport equation with isotropic indicatrix

$$\frac{\partial \psi}{\partial t} + \mu \frac{\partial \psi}{\partial x} = \frac{1}{2} \int_{-1}^{1} [\psi(x, \mu', t) - \psi(x, \mu, t)]d\mu, \quad x \geq 0, \mu \equiv \cos\theta \in [-1, 1].$$

$$(7.69)$$

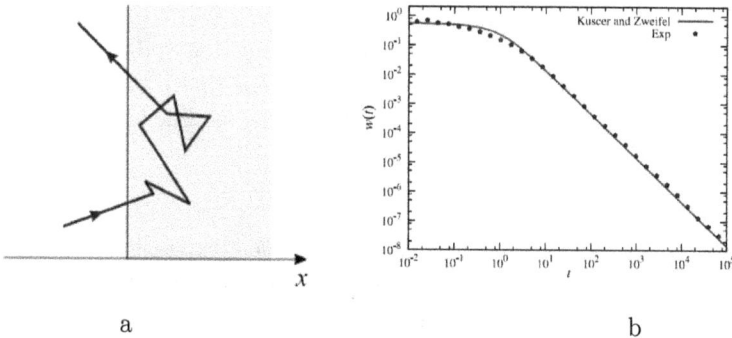

a   b

Fig. 7.23 a) Schematic picture of the particle backscattering; b) Escape time distribu-tion (7.70) for a uniform medium.

The authors use units in which the linear coefficient of scattering σ and the constant speed of particles v are equal to 1, so the integral

$$J(x,t) = \int_{-1}^{1} \psi(x,\mu,t)d\mu$$

means both the concentration and the flux of particles on the distance x from the boundary at time t whereas

$$j(x,t) = \int_{-1}^{1} \psi(x,\mu,t)\mu d\mu$$

has the sense of particles current. Under boundary-initial condition

$$\psi(0,\mu,0) = \begin{cases} 1/2, & -1 \geq \mu \geq; \\ 1 & \text{otherwise,} \end{cases}$$

the ratio

$$w(t) = \frac{|j_-(0,t)|}{j_+(0,0)}$$

of the current of particles returning from the scattering half-space

$$|j_-(0,t)| = \int_{-1}^{0} \psi(0,\mu,t)|\mu|d\mu$$

to the incident current

$$j_+(0,0) = \int_{0}^{1} \psi(0,\mu,0)\mu d\mu$$

can be interpreted as the probability density function (pdf) for the total time spent in the scattering half-space (the *escape time*) T. This function has been expressed in the cited article through the modified Bessel function $I_1(x)$ as follows

$$w(t) = t^{-1}e^{-t/2}I_1(t/2). \tag{7.70}$$

According to well-known small- and large-argument expansions of the function, Eq. (7.70) has the following asymptotics in these regions:

$$w(t) \sim \begin{cases} 1/4, & t \to 0; \\ \dfrac{t^{-3/2}}{\sqrt{\pi}}, & t \to \infty. \end{cases} \tag{7.71}$$

The first line relates to the process beginning whereas the second one describes the long-time behavior of the back current of particles at the boundary. The latter function can be considered as a solution to the correspondent diffusion problem being valid for all times $t > 0$. The function graphs are presented in Fig. 7.23 b.

The use of the ordinary Boltzmann equation with constant coefficients (7.69) assumes that the free paths of the particle between collisions are distributed according to exponential law,

$$\text{Prob}(R > x) = e^{-x}, \qquad x \in [0, x). \tag{7.72}$$

From probabilistic point of view, this means that scattering atoms are distributed in the half-space uniformly and independently of each other. This is an ideal gas structure generating the exponential free path distribution (7.71) and looking at large scales like a normal diffusion of Brownian type.

But many real physical structures obey neither first nor second conditions. Here, we concentrate our attention on the nonhomogeneous scattering media possessing the property of self-similarity (say, fractal or turbulent media).

To consider this case we should modify the problem by choosing a more general kind of jump distribution. For what follows, it will be more convenient to pass from positive random paths $R_1, R_2, R_3, \ldots$ to their x-projections coordinate of k-th scattering point is expressed by the sum

$$S_k = \Delta_1 + \Delta_2 + \Delta_3 + \ldots + \Delta_k.$$

We assume only that the random terms are mutually independent and symmetrically distributed. Evidently, all S_k are symmetrically distributed as well: for any positive c

$$\text{Prob}(S_k < -c) = \text{Prob}(S_k > c).$$

Let, now, N be a number of the last collision after which the particle escapes the scattering half-space. Obviously

$$p_n \equiv \text{Prob}(N = n) = \text{Prob}(S_1 > 0, S_2 > 0, \ldots, S_n > 0, S_{n+1} < 0).$$

According to *Sparre Andersen theorem* [Andersen (1953)], the generating function

$$G(u) = \sum_{n=1}^{\infty} u^n p_n, \qquad 0 \le u \le 1,$$

obeys the equation

$$\ln \frac{1}{1 - G(u)} = \sum_{n=1}^{\infty} \frac{u^n}{n} \text{Prob}(S_n > 0).$$

When one-step distributions are continuous and symmetric with respect to 0, then $\text{Prob}(S_n > 0) = 1/2$ and we have

$$\ln \frac{1}{1 - G(u)} = \ln \frac{1}{\sqrt{1 - u}}.$$

This yields the following expression for the generating function [Feller (1971)]:

$$G(u) = 1 - \sqrt{1 - u}.$$

Observe that the generating function and, as a consequence, the correspondent distribution itself doesn't depend on the exact form of the one-step displacement distribution and even on whether the moments of this displacement exist.

Using the Taylor expansion for a square root in the latter equation and the Gamma function representation yields

$$G(u) = \frac{1}{2} \sum_{n=1}^{\infty} \frac{\Gamma(n - 1/2)}{n!\Gamma(1/2)} u^n.$$

So, for the probability distribution p_n we have:

$$p_n = \frac{1}{2} \frac{\Gamma(n - 1/2)}{n!\Gamma(1/2)}, \quad n \in \mathbb{N}. \tag{7.73}$$

Now, denoting the pdf of random time T_k associated with displacement Δ_k by $g(t)$, we can represent pdf of sum of $T_1 + T_2 + \ldots + T_n$ as

$$W(t|n) = q^{*n}(t),$$

where the superscript $*$ means the Laplace convolution

$$q^{*0}(t) = \delta(t),$$

$$q^{*1}(t) = g(t),$$

$$q^{*2}(t) = \int_0^t q(t - t')q(t')dt'$$

and so on. To complete this calculation, we should average the conditional pdf over all possible values of the random variable k:

$$W(t) = \sum_{k=1}^{\infty} W(t|n) = \sum_{n=1}^{\infty} p_n q^{*n}(t).$$

Laplace transformation leads to the following expression

$$\hat{W}(\lambda) = \sum_{n=1}^{\infty} p_n \hat{q}^n(\lambda) = G(\hat{q}(\lambda)) = 1 - \sqrt{1 - \hat{q}(\lambda)}.$$

As noticed above, the exponential free path distribution seems wrong for the structures formed by correlated atom systems. An important example of such structures was investigated in the article [Uchaikin and Gusarov (1997)], where long-range correlations were introduced by means of the Lévy-flight process. As a result, the asymptotically fractal random point model was obtained, random free paths which turned out to be weak dependent and asymptotically distributed according to inverse power law [Uchaikin and Korobko (1999)]:

$$\text{Prob}(R > x) \propto x^{-\alpha}, \quad \alpha > 0, \quad x \to \infty. \tag{7.74}$$

Observe that the first moment of ξ exists only for $\alpha > 1$, and the variance only for $\alpha > 2$. Models of such kind are often used in astrophysics [Sánchez et al. (2005)], in solid state kinetics [Uchaikin and Sibatov (2013)].

We are in a position now to estimate long-time asymptotical behavior of back current of particles from the fractal half-space with different α, we take for $q(t)$ the one-sided Lévy-stable law [Uchaikin and Zolotarev (1999)], $q(t) = g_\alpha(t)$, $\alpha \in (0,1]$, defined by its Laplace characteristic function

$$\hat{g}_\alpha(\lambda) \equiv e^{-\lambda t} g_\alpha(t) dt = e^{\lambda^{-\alpha}}, \quad \alpha \in (0,1].$$

Thus,

$$1 - \hat{W}(\lambda) \sim \lambda^{\alpha/2} \text{ as } \lambda \to 0,$$

and according to Tauberian theorem [Feller (1971)]

$$W(t) \sim \frac{\alpha/2}{\Gamma(1 - \alpha/2)} t^{-\alpha/2-1} \text{ as } t \to \infty. \tag{7.75}$$

Observe that the case $\alpha = 1$ recovers the ordinary diffusion behavior.

As Fig. 7.24 shows, direct Monte Carlo simulations confirm validity of the approximations derived here.

Theoretical calculations confirmed by Monte Carlo simulation, have discovered two remarkable facts. First, the number events distribution in a half-space under the above-stated conditions do not depend on free path distribution between consecutive collisions. Second, the asymptotical behavior of flux pulse backscattered from the fractal semi-space falls according to power law with index connected to fractional dimensionality of the medium.

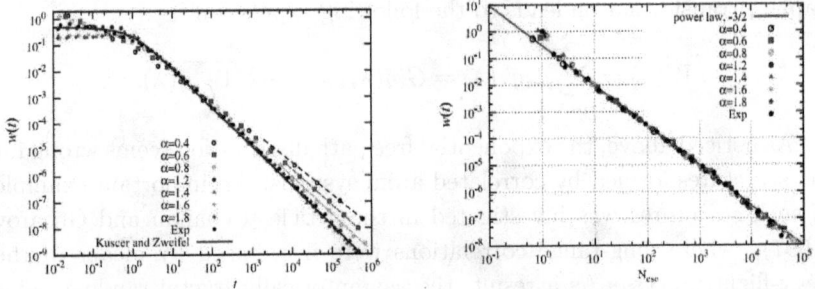

Fig. 7.24 Backscattering from a fractal medium: pdf of number of scattering events (left panel) and of escape time (right panel) for different α (dots show Monte Carlo results, lines represent the correspondent asymptotic solution (7.75).

Notice that Eq. (7.75) can be used in principle for the fractal type medium diagnostics by measurement of time-dependence of backscattering flash. The obtained result can be used in generalized models of acceleration of charged particles by shock waves and in description of the anisotropy inversion phenomenon.

Distributed reacceleration in a fractal half-space

The increment of the momentum in the multiplicative model is proportional (in the statistical case) to the absolute value of the momentum p' of the particle coming into interaction,

$$\Delta \mathbf{p} = p'\mathbf{Q}, \quad \int_{|\Delta \mathbf{p}|>p} w(\Delta \mathbf{p}; p')d\Delta \mathbf{p} \propto (p/p')^{-\gamma}, \quad p \to \infty. \tag{7.76}$$

On the assumption that the distribution of the random vector $\mathbf{Q}$ is independent of p', isotropic and has the form

$$W(\mathbf{q}; p')d\mathbf{q} = (1/2)V(q)dqd\cos(\mathbf{q}, \mathbf{p})$$

with

$$V(q) = \gamma q^{-\gamma-1}, \quad \gamma > 1, \tag{7.77}$$

the following equation approximates this process [Wandel *et al.* (1987)]:

$$\frac{\partial}{\partial t} f(p|t) = \sigma v \left\{ \int_1^\infty \gamma q^{-\gamma-1} f(p/q|t)dq/q - f(p|t) \right\} + f_0(p)\delta(t). \tag{7.78}$$

In what follows, we assume $f_0(p) = \delta(p-1)$. Let us stress that this equation is derived on assumption that the acceleration takes place at each collision and the random distance between consecutive collisions is distributed with an exponential pdf. Accounting the turbulent character of the interstellar magnetic field by means of replacement of exponential distribution of free path lengths by power law leads to a fractional version of Eq. (7.78) obtained in [Uchaikin (2010a)].

For the reasons discussed above, we represent the solution of the fractional version through the event number distribution $w(n)$ interpreting the latter as a continuous distribution,

$$f(p) = \int_0^\infty f(p|n)w(n)dn. \tag{7.79}$$

As one can understand from above, $f(p|n)$ is a pdf of random variable $\Pi_n = Q_1 Q_2 \ldots Q_n$,

$$f(p|n)dp = \text{Prob}(p < \Pi_n < p + dp) \tag{7.80}$$

being none other than the product of mutually independent random variables $Q_1, Q_2, \ldots Q_n$ distributed according to the common law (7.77). At first glance, it may seem that this random variable obeys the lognormal distribution at large n, but it is not the case: the distribution of Π_n follows the log-normal distribution only in the middle scale region, whereas its large deviations relating to cosmic rays preserve the inverse power tail up to a slowly varying factor (see investigation of this problem in [Sornette and Cont (1997)]). Our calculations confirm this fact (Fig. 7.25). As a result we obtain, that the albedo-particles momentum spectra from a fractal half-space accelerating the particles have the pdf of a power type with the index slightly greater than 1 independently of the fractal index α. This pdf can be represented by a universal curve (Fig. 7.26).

Finally, we investigated behavior of the process when the free path of the particles increases with p as p^δ and saw that in this case the domain of this universality narrows to $1 < \alpha < 2$ (Fig. 7.26).

As noted above, Eq. (7.78) describes the transport process as if it took place in a homogeneous medium where the random free paths between acceleration events have exponential distribution. In order to take into account the turbulent inhomogeneous we should pass from exponential pdf to power law in the momentum equation (7.78) as well, that is replace the first-order time-derivative by its fractional-order analog. Seeking its solution in the form of (7.79), we calculate the conditional pdf (7.80) and number event distribution $w(n)$ we found that:

(1) In case of identical pdf of each free path, the distribution $w(n)$ is practically independent of the fractal index $\alpha \in (0, 2)$.

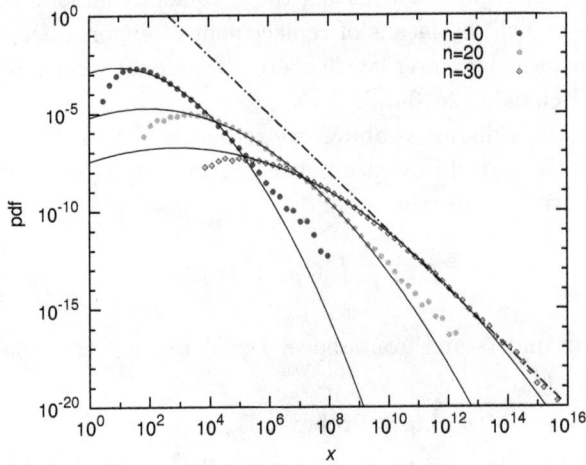

Fig. 7.25   Distribution of the product of n random multipliers distributed according to the power law, $p_\xi(x) = \alpha x_0 x^{-\alpha-1}$, $\alpha = 1.5$.

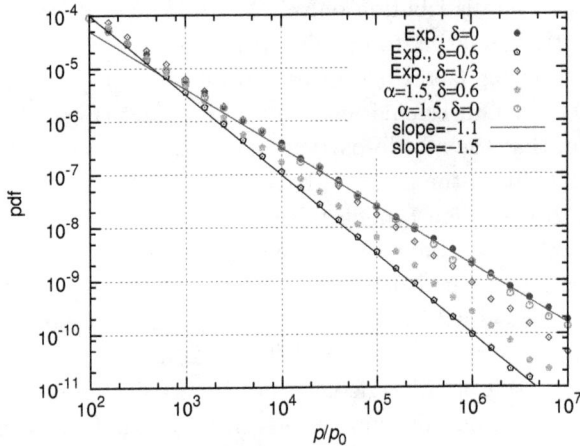

Fig. 7.26   Momentum distributions in case of the exponential and power distributions of free path lengths, and $\delta = 1/3$ and $\delta = 0.6$.

(2) Under this condition, the momentum distribution is represented by the universal power-type curve with an index, a little exceeding 1.

(3) In case when free path length increases proportionally to p^δ, the domain of this universality narrows to $1 < \alpha < 2$.

Chapter 8

Cosmological scales

This chapter set us on the largest scales in the world – cosmological scales. Section 8.1 begins with discussing three basic principles of the standard cosmology in the light of such intriguing concept as fractal. The next section (8.2) describes links between correlation functions, power spectra, fractals and mesofractals as sets of nods of random trajectories. Section 8.3 introduces fractional version of the Ornstein-Zernike equation for binary correlation function. Section 8.4 is devoted to the pencil-beam statistics with application to such kinetic processes as small-angle multiple scattering of particles and waves, and temporal broadening of a radio-pulse. In Section 8.5, we acquaint the reader to application of the fractional calculus to the problem of the large-scale structure of the Universe formation: the Press-Schechter formula is generalized to its fractional counterpart by means of fractional Brownian motion.

8.1 Two kinds of large-scale statistics

One of the important problems of cosmology connected to the large scale structure of the Universe is searching the most efficient ways of contraction of the information obtained from observations. Historically, the first way consisted in description of isolated clumps, clumps of clumps and so on by means of some individual parameters. Nowadays, we possess a very large amount of information on large scale structure of the Universe and this makes the statistical approach to be very efficient for its analysis. One can point at the *cell count method* [Zwicky (1957)], the *distance distribution method* [Pilkington and Scott (1965)], the *correlation function method* [Peebles (1980); Bertschinger (1998)], the *power spectrum analysis* [Webster and Ryle (1976)], the *topological analysis* [Melott (1990)], the *wavelet analysis* [Martínez et al. (1993)], the *percolation analysis* [Klypin and Shandarin

(1983)], the *scaling analysis* [Borgani (1995)], the *fractal analysis* [Coleman and Pietronero (1992)] and others.

For solving different astrophysical and cosmophysical problems, different levels of the information conciseness are required. Angular galaxy samples, three-dimensional samples, cluster samples, galaxy redshift surveys are the least contracted information. The use of more contracted information becomes possible on the base of preliminary assumptions. Some of them are generally accepted as principles of the standard cosmology [Weinberg (1972); Mandelbrot (1979); Peebles (1993)]. They are

(1) the Cosmological Principle: *There are no privileged points in the Universe,*
(2) the Ergodic Principle: *Ensemble averages equal spatial averages taken over one realization of the random field,*
(3) the Symmetry Principle: assumption about *homogeneity and isotropy (in the statistical sense) on large scale.*

The first principle claiming statistical equivalence of all *space points* with respect to their environment gives a basis for the *unconditional statistics*. In other words, it is of no importance if an origin of coordinates is occupied by some galaxy or not: averaging over the whole statistical ensemble of the random medium yield the same result. The second principle allows us to model this ensemble by a set of distant regions of the Universe with their own frame of coordinates. We will denote the averaged number of galaxies and other characteristics by using angle brackets.

The large-scale structure of the universe could be studied by measurements of coordinates and redshifts of galaxies in certain regions. Radial distributions of galaxies from spectroscopic 2dF survey (2 degree Field Galaxy Redshift Survey) of Anglo-Australian observatory are presented in [Colless *et al.* (1999); Peacock *et al.* (2001); Norberg *et al.* (2001); Colless *et al.* (2001)]. As a matter of fact, the visible galaxy distribution in some region of distances demonstrates some special patterns: it contains voids, clusters, groups of clusters and so on alternating at all scales up to some possible limit distance. Inside this region, the number of galaxies $N(r)$ within a sphere of radius r centered on our Milky Way, is not proportional to r^3 as could be expected in the case of a homogeneous spatial distribution, but is a very irregularly increasing function of r. Resorting to averaging shows that results of such procedure crucially depend on which sort of points are chosen as origins of frames: geometrical points or *material points* (i.e., galaxy centers of mass). The latter approach gives a basis for another kind of statistics known as the *conditional statistics*.

The conditional statistics began to play an essential role after B. Mandelbrot introduced the fractal conception and gave it status the *fractal geometry* [Mandelbrot (1983)]. It became logical to consider only points of fractal (that is material points), ignoring all others. The fractal model demonstrates long-range correlations of inverse power type, the presence of clusters and voids similar to that are observed in large-scale structure of the Universe and for this reason was proposed for simulation of galaxy spatial distributions. Investigations of catalogues show however that the results at $r < r^*$ are crucially depend on which sort of point will be chosen for these aims: geometrical points or *material points* (in other words, galaxy centers of mass). The latter approach demands the conditional statistics. Justifying this statement, Mandelbrot underlined that the Universe appears statistically the same to every observer *situated on a galaxy* but not in a region of void. More specifically, in a reference frame with origin O, the distribution of matter is independent of O under the sole condition that O must be a material point. Under this condition the mean number of galaxies

$$\langle \overset{\circ}{N}(r) \rangle \propto r^D,$$

where $\circ$ denotes the conditional kind of characteristics and D is between 1 and 2. Observe that if O is not a material point and r is fixed, then a sphere of radius r centered on O is empty with probability equal to one [Mandelbrot (1979)].

The situation described above relates to the region bounded by some radius r^* (the homogenization radius). As regards to much larger distances, most cosmologists believe in a homogeneous distribution, although estimations of the crossover distance are very dispersed: from 4 Mpc [Mandelbrot (1979)] to 100 Mpc or more [Mittal and Lohiya (2001)] depending on the galaxy catalogue and on the way of its statistical processing. However, some authors argue that the fractal distribution of visible matter has no cross-over to homogeneity and the observed scaling takes place at all distances. Without assigning a special central position to an observer, such a scaling can be explained only by assuming that galaxies are distributed on points belonging to a *fractal set with a fractal dimension D*, briefly called *fractal*.

The reader can compare arguments pros and cons taking in his hands two articles with very expressible titles "Why the Universe is not a fractal" [Martinez and Jones (1990)] and "Why the Universe is fractal" [King (2012)].

8.2   Correlation functions and power spectra

The correlation analysis of the galaxy distribution is a widely approach to describe statistical properties of spatial distribution of visible matter in the Universe. During the '70s, the availability of extended angular galaxy samples made possible the realization of accurate correlation analysis, mainly pursued by Peebles and his collaborators [Peebles (1980)]. The analysis begun with angular correlations showing inverse power type behavior with a break at large angular separations. On providing a method for deprojecting angular data, the angular function was converted into the spatial 2-point function, $\xi(r)$. These calculations show that in a wide region of distances ξ can be described by means of inverse power law [Peebles (1980); Coles and Lucchin (2003)]:

$$\xi(r) = (r/r_0)^{-\gamma}, \quad 0,06 < r \leq 60 \text{ Mpc},$$

where

$$\gamma = 1.77 \pm 0.04$$

and r_0 is the *correlation length* varying in a wide region from sample to sample.

The power law of spatial correlations reflects the cluster character of observed matter distribution. Each galaxy is nothing but a huge cluster of stars, therefore, a cluster of galaxies is the cluster of clusters of stars. This way of thinking may be continued, and one may assume that galaxy clusters form in their turns clusters, called *superclusters*. Astronomical investigations have not only confirmed this supposition but discovered that the reach cluster correlation function can also be fitted by the power law expression

$$\xi_c(r) = (r/r_{0c})^{-\gamma},$$

with $r_{0c} \simeq 25h^{-1}$ Mpc and $\gamma \simeq 1.8$, in the distant range $5 < r < 150h^{-1}$ Mpc (here h is the Hubble constant). This function exhibits the same slope as the galaxy function, but with a remarkably larger correlation length (see, for details, [Borgani (1995)]).

Observe that if the random position of the galaxies were independent of each other (i.e., formed the Poissonian ensemble), then the correlation function would be equal to zero. Often, the *structure function*

$$s(r) = 1 + \xi(r) \tag{8.1}$$

is used instead of ξ. If $s > 1$ (ξ is positive), we have an excess probability over the Poissonian case and, therefore, clustering. If $s < 1$ (ξ is negative), we meet anticlustering when the probability to find a galaxy in a neighborhood of another one is less than in the Poissonian ensemble. The distance r_0 at which $\xi = 1$ is called the *correlation length* and this implies that there should be no appreciable overdensities (clusters) and underdensities (voids) extending over distances essentially larger than r_0.

Fractals and mesofractals

Nowadays, majority of cosmologists believe that the Universe is not self-similar structure at all scales and looks like a homogeneous medium on large scales. One of the important arguments in favor of such conclusion is the high level of isotropy of background radiation which can be considered as an evidence of the Universe isotropy. Isotropy of the Universe means that, at any event, an observer who is at rest in it, cannot statistically distinguish any space direction from another. It is widely believed that isotropy from all points of observation implies homogeneity and therefore an inhomogeneous Universe like a fractal could not be isotropic [Peebles (1993)]. B. Mandelbrot wrote in his article [Mandelbrot (1998)] "the mainstream exemplified in [Peebles (1980)] views the overall distribution through a 'hybrid' model that grafts local perturbations on homogeneity in the large-scale range. In this mental picture, fractality gained a modicum of acceptance in the mainstream as a possible representation of the local perturbations."

The possibilities of fractal distributions on the largest scales of structure can be admitted for a variety of reasons including existence of a substantial quantity of dark matter [Durrer and Labini (1998)]. The prevalence of fractal scaling behaviors in the clustering of galaxies may also be significant because it would intimate, perhaps, some new physical mechanisms which could be capable of generating fractality at large-scale structure. Nowadays, a few phenomenological approaches are developed based on some statistical models and useful for practical aims [Mandelbrot (1998); Mittal and Lohiya (2001, 2003); Uchaikin (2004b); Capozziello and Funkhouser (2009); Capozziello *et al.* (2009)]. Some of the unsolved problems are observed in the fractal logic. Considering fractal as a pure mathematical idea, not limited by some inner and outer boundaries dictated by natural structures and processes, you can feel as if "the ground is slipping from under your feet". Really, in order to mathematically describe some heterogeneous continuum, you do usually divide it into small pieces assuming that these *small pieces are simpler for analysis because they are almost homogeneous*: their masses are proportional to their volumes, so you can operate with mass density etc., but there is no sense in doing this with fractals: *the small pieces of a fractal as complex as its large pieces*, they are similar to those ("*as within, so without*")... You cannot divide so you cannot gather, i.e., integrate, you cannot integrate so you cannot differentiate... What can you do then? This is one more reason for passage from ideal fractals to hybrid fractals or, as it is for shortness offered in [Uchaikin (2004a)], *mesofractals*.

Power spectrum

One of the most efficient tools for analyzing of statistical properties of random fields is the Fourier transformation method. In case of an isotropic random field, Fourier image $P(\mathbf{k})$ of two-point correlation function $\xi(r)$ is called the *power spectrum*,

$$P(k) = \int \xi(r) e^{i\mathbf{k}\mathbf{x}} d\mathbf{x}, \quad r = |\mathbf{x}|, \quad k = |\mathbf{k}|.$$

Using the explicit expression for the integral over angular variables,

$$\int\int e^{ikr\cos\theta} \sin\theta d\theta d\varphi = 4\pi \frac{\sin(kr)}{kr},$$

one can represent the power spectrum by means of a single integral, specifically

$$P(k) = \frac{4\pi}{k} \int\limits_0^\infty \xi(r) r \sin(kr) dr.$$

There exist several approximations of power spectrum. The simplest of them is

$$P(k) = A|k|^{-\alpha}, \quad A = \text{const.} \tag{8.2}$$

It corresponds to the correlation function with $\gamma = 3 - \alpha$ and

$$r_0^\gamma = (A/2\pi^2)\Gamma(2-\alpha)\sin(\alpha\pi/2).$$

One can invent a few approximations for the observation data, but one of them leads us to a very useful and significant analogy. It is of the form

$$P(k) = A/[e^{(bk)^\alpha} - c], \tag{8.3}$$

where $b > 0$ and $0 \leq c \leq 1$ can be chosen in an appropriate way.

Ornstein-Zernike equation

The point is that the inverse Fourier transformation of (8.3) yields the integral equation

$$\xi(r) = (A/b^3)p(r/b) + (c/b^3) \int \xi(|\mathbf{x} - \mathbf{x}'|)p(r'/b)d\mathbf{x}', \qquad (8.4)$$

where

$$p(\mathbf{x}) = q(\mathbf{x}; \alpha) \equiv (2\pi)^{-3} \int e^{-i\mathbf{k}\mathbf{x} - k^{\alpha}} d\mathbf{k}.$$

Equation (8.4) is the Ornstein-Zernike equation. It is derived for such random media as physical liquids and gases consisting from atoms and molecules and exhibits the relationship between statistical mechanics and the concepts under consideration (see, for example, [Balescu (1975)]). On the other hand, this equation leads directly to the random walk model as a tool for construction of random point distribution with given correlation function ξ.

The random walk model

As follows from (8.4), the function

$$g(\mathbf{x}) = A^{-1}\xi(b|\mathbf{x}|)$$

obeys the equation

$$g(\mathbf{x}) = p(\mathbf{x}) + c \int g(\mathbf{x} - \mathbf{x}')p(\mathbf{x}')d\mathbf{x}', \qquad (8.5)$$

where

$$p(\mathbf{x}) \geq 0, \quad \int p(\mathbf{x})d\mathbf{x} = 1, \quad 0 \leq c \leq 1.$$

Its solution can be interpreted as a density of collisions of some particle starting its movement from the origin $\mathbf{X}_0 = 0$ and performing the first collision at the point $\mathbf{X}_1 \in d\mathbf{x}'$ with probability $p(\mathbf{x}')d\mathbf{x}'$. Here it stops with the probability $1 - c$ or performs the next jump into $\mathbf{X}_2 \in d\mathbf{x}''$ with the probability $cp(\mathbf{x}'' - \mathbf{x}')d\mathbf{x}''$ and so forth.

The integral

$$\int g(\mathbf{x})d\mathbf{x} = 1/(1 - c), \quad c < 1, \qquad (8.6)$$

gives the mean number of all collisions of the particle including the final one, and the function $g(\mathbf{x})$ itself can be expressed by the Neumann series expansion

$$g(\mathbf{x}) = \sum_{j=1}^{\infty} c^{j-1}p^{*j}(\mathbf{x}),$$

where

$$p^{\star 1}(\mathbf{x}) \equiv p(\mathbf{x})$$

and

$$p^{\star(j+1)}(\mathbf{x}) = \int p^{\star j}(\mathbf{x} - \mathbf{x}')p(\mathbf{x}')d\mathbf{x}'$$

is the j-fold convolution of the distribution density $p(\mathbf{x})$.

The algorithm of random walk simulation looks as follows: the first galaxy is placed at some given point, say at the origin of coordinates, the radius-vector of the second galaxy is chosen from the isotropic inverse power distribution, the position of every next galaxy relative to the previous one is sampling from the same distribution. The obtained infinite sequence $\{\mathbf{X}_0 = 0, \mathbf{X}_1, \mathbf{X}_2, \mathbf{X}_3, ...\}$ can be considered as a set of collision points of a walker moving through some hypothetical medium in straight lines between collisions and changing its direction in collision events isotropically.

The random walks model with transition probability of inverse power kind of tail (the *Rayleigh-Lévy walks*, see [Martinez and Saar (2001)] for simulating galaxy statistics was introduced by B. Mandelbrot [Mandelbrot (1975)]. This model displayed in natural way such typical features of the distribution as clusters and voids and was used by its author to define the concept of *lacunarity* [Mandelbrot (1983)]. Further investigation of this model was performed in [Uchaikin and Gusarov (1997); Uchaikin *et al.* (1997)] on the base of Lévy-Feldheim stable laws chosen as transition probability densities. Some consequences from this model were discussed in review articles [Uchaikin (2004b,a)]. Typical realizations obtained with the use of this method on a plane are shown in Fig. 8.1.

It is worth noting that simulation of spatial distribution of galaxies is not a unique case of using the model of random walks in cosmology. The most known result obtained within the framework of this model (more specifically, of the Brownian motion model) is the Eddington-Weinberg formula expressing the link between the object size (say, the radius) R and the number N of constituents:

$$R(N) = R(1)\sqrt{N}.$$

One more example relates to so-called *excursion set formalism* used for derivation of the Press-Schechter mass function. This derivation is based on the ordinary Brownian motion model, although it looks not quite satisfactory [Schuecker *et al.* (2001)].

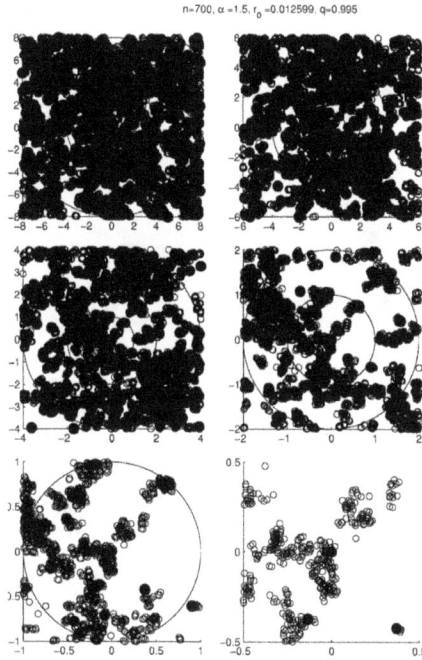

Fig. 8.1 Results of simulation [Uchaikin (2004a)] of the mesofractal embedded in a plane at different scales ($\alpha = 1.5$, number of trajectories $n = 700$, scale parameter $r_0 = 0.005$, survival probability $c = q = 0.995$).

8.3 Fractional Ornstein-Zernike equation

Rewriting the equality (8.2) for the Fourier image of the correlation function

$$(k^2)^{\alpha/2} P(k) = A \qquad (8.7)$$

and returning to originals on both sides yields the equation with fractional Laplacian

$$(-\Delta)^{\alpha/2} \xi(\mathbf{x}) = A\delta(\mathbf{x}).$$

It is very instructive to compare the result with the Ornstein-Zernike theory of critical phenomena (see, for instance, [Balescu (1975)]). This theory is based on three statements. The first of them is the relation between temperature-dependent isothermal compressibility $\chi(t)$ and fluctuations of the number of molecules N in a large (in comparison with the correlation length) volume V of the system,

$$\chi(T) = -\frac{\langle N \rangle}{nV^2}\left(\frac{\partial V}{\partial P}\right)_T = \frac{\delta^2}{nk_{\mathrm{B}}T},$$

where $n = \langle N \rangle / V$ is the mean concentration of molecules, and $\delta^2 = (\langle N^2 \rangle - \langle N \rangle^2)/\langle N \rangle$ is the ratio of the variance of number N to the Poisson variance with the same concentration. This formula means that the systems with high compressibility, e.g. diluted gases, are characterized by larger fluctuations of particle number than solids. Expressing these fluctuations through the integral of the correlation function over the whole space

$$\delta^2 = 1 + n \int \xi(\mathbf{r}, T) d\mathbf{r}$$

yields the formula

$$n k_B T \chi(T) = 1 + n \int \xi(\mathbf{r}, T) d\mathbf{r}$$

connecting the correlation function with temperature T.

It is known that for all stable systems $(\partial V / \partial P)_T < 0$, in a vicinity of a critical point of phase transition T_c the derivative of volume on pressure, $(\partial P / \partial V)_T$, turns into zero and $\chi(T) \to \infty$, $T \to T_c$. This is an evidence of extremely high fluctuations of density in a critical region that generate the phenomena of critical opalescence, light scattering, plasma oscillations and others. The arising divergence of the integral of the correlation function

$$\int \xi(\mathbf{r}; T_c) d\mathbf{r} = \infty$$

points to an infinite correlation radius caused by the presence of a long tail of the correlation function $C(\mathbf{r}; T_c)$. Theoretical analysis shows that this tail is of the power law type,

$$\xi(\mathbf{r}; T_c) \propto r^{-(d-2)}, \quad r \to \infty,$$

where d is a dimension of the space. At the same time, numerical simulations and some experiments show that

$$\xi(\mathbf{r}, T_c) \propto r^{-(d-2+\eta)},$$

where η is a positive number smaller than one.

The second thesis of the Ornstein-Zernike phenomenology is expressed by the integral equation allowing to pass on from one function $\xi(\mathbf{r}; T)$ to another function $c(\mathbf{r}; T)$ more convenient for analyzing critical phenomena

$$\xi(\mathbf{r}_{12}, T) = c(\mathbf{r}_{12}, T) + n \int \xi(\mathbf{r}_{23}, T) c(\mathbf{r}_{13}, T) d\mathbf{r}_3.$$

This equation divides the total correlation function $\xi(\mathbf{r}, T)$ into two terms: the direct term $c(\mathbf{r}, T)$ describing interaction $1 \to 2$ with the nearest environment and therefore short-range, and the nondirect term expressed by an integral. The latter term is related to the interactions $1 \to 3 \to 2$, $1 \to 3 \to 4 \to 2$, ... , $1 \to 3 \to 4 \to \cdots \to n \to 2$ and therefore defines correlations at large distances. Both correlation functions are dimensionless. In a critical point, the role of short distances is small. Disregarding them and normalizing $c(\mathbf{r}, T)$ in a proper way, one can consider it as a three-dimensional probability density function, and the equation may be interpreted as the equation of random walks.

The third thesis of the Ornstein-Zernike theory is that the function $\tilde{c}(\mathbf{k}, T)$ is assumed to be an even analytical function of $k = |\mathbf{k}|$ in a vicinity of $k = 0$ for all temperatures, including T_c:

$$\tilde{c}(\mathbf{k}, T) - \tilde{c}(0, T) \propto k^2, \quad k \to 0.$$

This fact provides finiteness of the variance of a "path length" and a diffusion (Brownian) character of the correlation function asymptotic

$$\xi^{as}(\mathbf{r}, T_c) = A r^{-1},$$

satisfying the stationary equation of normal diffusion,

$$-\Delta \, \xi^{as}(\mathbf{r}, T_c) = 4\pi A \delta(\mathbf{r}).$$

In the standard theory, the expansion up to the second member is used. It is presented in the form

$$\tilde{c}(\mathbf{k}, T) = \tilde{c}(0, T) - (R^2/n)k^2,$$

where $R(T)$ is the Debye length decay. In this case, the structural function takes the form

$$a_{\mathbf{k}} = \frac{1}{1 - nc(0, T) + R^2 k^2} = \frac{R^{-2}}{l^{-2} + k^2},$$

where l is the critical correlation length depending on the density and the temperature: it becomes infinite, when $T \to T_c$. In this approximation, the inverse Fourier transformation gives the following result for the total correlation function:

$$\xi(\mathbf{r}, T) \propto \begin{cases} (R^2 r)^{-1} \exp(-r/l), & T < T_c, \\ r^{-1}, & T = T_c. \end{cases}$$

One should note, however, that in case of a two-dimensional system, the approach under consideration leads to an absurd result: the correlation function grows logarithmically. Moreover, numerical simulations and a number of experiments with magnetic systems show that in a critical point the correlation function of d-dimensional system is better approximated by the formula proposed by Fisher,

$$\xi(\mathbf{r}, T_c) \propto r^{-(d-2+\eta)}, \tag{8.8}$$

where η is a positive number smaller than 1. This formula practically weakens the third thesis of the theory, replacing it by the following requirement

$$\tilde{c}(\mathbf{k}, T) - \tilde{c}(0, T) \propto k^\alpha, \ 0 < \alpha < 2 \quad k \to 0.$$

As one can see from [Uchaikin and Gusarov (1997)], in this case, the asymptotical part of the solution is characterized by the Fourier transform $\tilde{\xi}^{as}(\mathbf{k}, T_c)$ satisfying the equation

$$|\mathbf{k}|^\alpha \tilde{\xi}^{as}(|\mathbf{k}|, T_c) = A(\alpha).$$

It presents the Fourier image of the equation with the Laplace operator of fractional order

$$(-\Delta)^{\alpha/2} \xi^{as}(\mathbf{r}, T_c) = A(\alpha)\delta(\mathbf{r}).$$

In framework of the fractional model of critical state of medium at $T \to T_c$, the analyticity of the direct correlation function transform is violated. As a consequence, the Poisson equation describing the asymptotical ξ^{as} reduces to the fractional Poisson equation. This is an evidence of the fractal structure of random excitations of medium at the phase transition.

We have the reason to suppose that the mesofractal model described above can be used for the description of the observable statistical properties of the Universe. In any case the following statement seems to be highly plausible: *fractal cosmology should only be stochastic one*. It follows simply from the main attribute of a fractal, namely, from its self-similarity. If the fractal is stochastic at some scale then it should be stochastic at all scales. In other words there is not a scale at which the Universe could be described in terms of determined continuous medium.

However, it is impossible not to admit that the fractal model of the Universe is a very extreme kind of possible models requiring revision of not only the method of usual analysis of observation data but the very Cosmological Principle.

8.4 Pencil beam processes

Pencil beam statistics

In reality, galaxy and cluster catalogs contain samples from approximately conic volumes with relatively broad opening angles, up to 90°. There is however a different set of observations that refers catalogs with a very narrow opening angle, about 1°. From these observations, with some corrections due to luminosity effects and corrections for conic geometry, one obtains the behavior of the local density along a tiny but very long cylinder, so-called pencil beam survey. The observed distribution corresponds to the intersection of the full three-dimensional galaxy distribution with the one-dimensional cylinder: its width is small while its linear extent (depth) is very large.

In this respect the law of codimension additivity has been applied for finding the pencil beam distribution [Coleman and Pietronero (1992)]:

$$D_{A \cap B} = D_A + D_B - D,$$

where D_A and D_B are dimensions of two objects A and B embedded in a D-dimensional space. If A is the galaxy distribution with $D_A = 1.4$ and B is a pencil beam with $D_B = 1$ then

$$D_{A \cap B} = 1.4 + 1 - 3 = -0.6.$$

The negative dimension did not confuse the authors although their explanation does not seem to be clear.

To check this conclusion we performed numerical experiment in frame of model under consideration. The following results are obtained:

(i) the probability distribution for the random distance from one galaxy to the nearest one in the cylinder has a long power tail

$$q(\xi) \sim C\xi^{-\beta}, 0 < \beta < 1$$

with the exponent β depending on the fractal dimension but not coinciding with it;

(ii) the distribution of the next gap along this cylinder weakly depends on the previous one and can approximately be considered as independent of it.

These axioms generate stochastic fractal with dimension ω embedded in one-dimensional space and looks more convincing than a set with a negative dimension.

According to the generalized limit theorem, the probability that n points occur on the length x of the cylinder

$$P\{N(x) = n\} = w(n; x)$$

$$\sim G_+ \left([(n+1)C\Gamma(1-\beta)]^{-1/\beta} x; \beta \right) - G_+ \left([nC\Gamma(1-\beta)]^{-1/\beta} x; \beta \right),$$

where

$$G_+(x;\beta) = \int_0^x g_+(x';\beta)\,dx'$$

and

$$g_+(x;\beta) = (2\pi i)^{-1} \int_L \exp(\lambda x - \lambda^\beta)\,d\lambda.$$

After a simple algebra we obtain for $x \to \infty$

$$w(n;x) \sim \frac{x}{\beta n}\left[nC\Gamma(1-\beta)\right]^{-1/\beta} g_+\left(\left[nC\Gamma(1-\beta)\right]^{-1/\beta} x;\beta\right).$$

Using known expressions for the moments of negative orders

$$\int_0^\infty g_+(x;\beta)\,x^{-\nu}\,dx = \frac{\Gamma(1+\nu/\beta)}{\Gamma(1+\nu)},$$

we find positive moments of random number of events in interval $(0,x)$:

$$\langle N^k(x)\rangle = \frac{k!\,x^{k\beta}}{\left[C\Gamma(1-\beta)\right]^k \Gamma(1+k\beta)}.$$

Now, one can see that $\langle N(x)\rangle \propto x^\beta$, and moments of random variables $\zeta = N/\langle N(x)\rangle$ do not depend on x. As noticed above, this is an intrinsic property of a stochastic fractal. This self-similarity of fluctuations means that self-averaging is absent: there is no such a depth when statistical fluctuations can be neglected and averaging some continuous function $f(N)$ cannot be reduced to averaging the argument,

$$\langle f(N(x))\rangle \neq f(\langle N(x)\rangle), \quad x \to \infty$$

in contradistinction to a regular distribution. Moreover, there are observed gaps on all scales in contrast with the Poissonian case when the large-scale distribution looks like a homogeneous continuum.

8.5   The large-scale structure formation

Excursion set formalism[1]

Accurate prediction of the mass function of dark matter (DM) halo is an important problem in cosmology and astrophysics, because this function

---

[1]This section was written with participation of Dmitry Bezbat'ko.

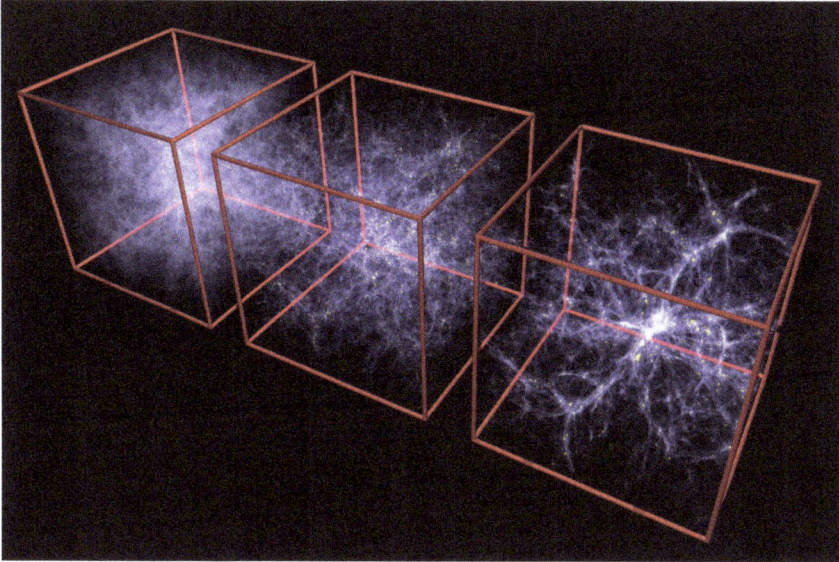

Fig. 8.2 Galaxy cluster formation. Supercomputer model of the formation of galaxies and clusters of galaxies from the contraction of a homogeneous cloud (left). Over time, the dark matter in the cloud coalesces due to mutual gravitational attraction. This leads to the formation of dense pockets in which stars (yellow) can form. Dark matter is a form of matter that does not emit radiation, making it hard to detect. It is only detectable by its gravitational effects. Models such as this can help to estimate the amount of dark matter in the universe by seeing whether they produce results that look like the real universe. Credit: Volker Springel/Max-Planck Institute for Astrophysics/Science Photo Library.

characterizes distribution of DM, formation and evolution of galaxies, and it is sensitive to cosmological parameters. In spite of great success in computing technology stimulating such numerical methods as the N-body simulation [Springel (2005)] (Fig. 8.2), the phenomenological approach still remains an important tool in these investigations.

A simple analytical description for the evolution of gravitational structure in a hierarchical universe was proposed by Press and Schechter [Press and Schechter (1974)]. In their theory, fluctuations $\delta(\mathbf{x})$ in density $\rho(\mathbf{x})$ of some random field (e.g, mass density) are considered, $\delta(\mathbf{x}) = [\rho(\mathbf{x}) - \langle\rho\rangle]/\langle\rho\rangle$. Here, $\delta(\mathbf{x})$ is the density contrast, $\langle\rho\rangle$ is the mean mass density in the universe and $\mathbf{x}$ is a comoving coordinate. Initially, the fluctuations are assumed to be small compared with unity, Gaussian and evolves linearly.

It is assumed that halos are formed in a smoothed field of contrast density $\delta(\mathbf{x}, R)$ when δ exceeds a certain threshold value $\delta > \delta_c$.

Smoothing of the density contrast on a particular scale, containing mass M, is determined via integral

$$\delta(\mathbf{x}; M) \equiv \delta_M = \int d\mathbf{x}' W_M(\mathbf{x} - \mathbf{x}')\delta(\mathbf{x}'). \qquad (8.9)$$

The window function (or filter function) $W_M(\mathbf{x} - \mathbf{x}')$ has units of inverse volume by dimensional arguments, so it is useful to think of a window as having a particular volume $V_M = M/\langle\rho\rangle$. The 'top hat' function and the Gaussian function are the most popular kernels used in a filter (see details in [Press and Schechter (1974)]).

The smoothed density contrast (8.9) is a function of the underlying total mass $M(R) \propto \overline{\rho}R^3$ included by the smoothing window; the M effectively represents the scale R. Instead of the mass M, the variance of the smoothed density contrast

$$S(M) \equiv \sigma_M^2 = \langle|\delta_M|^2\rangle = \int \langle|\widetilde{\delta}(\mathbf{k})|^2\rangle\widetilde{W}_M^2(\mathbf{k})d\mathbf{k}$$

can be used, so the smoothed density contrast is represented as

$$\delta_M = \int \widetilde{W}_M(\mathbf{k})\widetilde{\delta}(\mathbf{k})d\mathbf{k}.$$

The random function $\delta(S)$ can be interpreted as a random walk trajectory ($\delta \in \mathbb{R}$ is its coordinate, and $S \geq 0$ has the sense of time). If the δ in a region with mass M exceeds some critical value δ_c, the mass will collapse and by virialized in the future as to form a halo. Critical value δ_c depends on collapse parameters and filter function. In the case of spherical collapse and the so-called 'sharp-k' filter, $\delta_c = 1.686$ [Press and Schechter (1974)].

Further development of the Press-Schechter theory is the "excursion set formalism" introduced by [Bond *et al.* (1991)], which reformulated the problem in terms of the first passage of a threshold barrier. In particular, the halo mass distribution can be described by formula

$$\frac{dn}{dM} = 2SP(S)\frac{\langle\rho\rangle}{M^2}\left|\frac{d\ln\sqrt{S}}{d\ln M}\right| \qquad (8.10)$$

where $P(S)$ is the first passage time (FPT) density, providing the multiplicity function $f(\sigma) = 2\sigma^2 P(\sigma^2)$.

For sharp $\mathbf{k}$-space filtering function, the increment $\delta(S + dS) - \delta(S)$ comes from a new set of Fourier modes in a thin shell of $\mathbf{k}$-space being added to the sphere of radius k. The assumption that the increments are independent of $\delta(S)$, identically distributed on equal elements δS and have a finite second moment, lead to the Brownian motion process. Assuming that $\delta(S)$ is the Brownian motion and using the exact expression for the first time pdf p_{FT} known as the Lévy-Smirnov distribution:

$$p_{FT}(S, \delta_c) = -\frac{\partial}{\partial S} \int_{-\infty}^{\delta_c} Q(\delta, S) d\delta = \frac{\delta_c}{\sqrt{2\pi} S^{3/2}} \exp\left(-\frac{\delta_c^2}{2S}\right), \qquad (8.11)$$

one could obtain the Press-Schechter formula. Indeed, the comoving number density of halo of mass at epoch z is

$$\frac{dn}{dM} = p_{FT}(S, \delta_c) \frac{\bar{\rho}}{M} \left|\frac{dS}{dM}\right| = f_{PS}\left(\frac{\delta_c}{\sigma}\right) \frac{\bar{\rho}}{M^2} \left|\frac{d\ln\sigma}{d\ln M}\right|, \qquad (8.12)$$

where

$$f_{PS}(x) = \sqrt{\frac{2}{\pi}} \frac{\delta_c}{\sigma} e^{-\delta_c^2/2\sigma^2} \qquad (8.13)$$

is the well-known Press-Schechter mass function [Press and Schechter (1974)]. The ES-formalism had allowed to solve the 'cloud-in-cloud' problem, but had not provided a satisfactory agreement with N-body simulations.

The next step was the introduction of the so-called ellipsoidal collapse, which accounts for the nonsphericity of the emerging halos. In terms of the first passage process this leads to an evading barrier [Sheth *et al.* (2001); Sheth and Tormen (2002)]. This approach has made possible to significantly improve the agreement of calculations with numerical simulation. Further development includes substitution of constant or evading thresholds by a barrier performing Brownian motion [Maggiore and Riotto (2010a,b)] and consideration of Non-Gaussian initial fluctuations [Maggiore and Riotto (2010c)].

Independence of increments from any previous steps is an essential limitation of the normal Brownian model. This means that the formation of halos at small scales is not correlated with the density fluctuations smoothed at large scales. In other words, the process of halo formation develops independently of its environment. However, this property takes place only in case of the narrow filter window. The choice of more wide window generates correlation between the increment of $\delta(S)$ and $\delta(S)$ itself, so one should go beyond the ordinary Brownian model. In order to involve the correlations, [Pan (2007); Pan *et al.* (2008)] suggested to replace the Brownian motion in the excursion set formalism by the fractional Brownian motion (fBm), but they used incorrect FPT density obtained by means of the image method, inapplicable to fBm [Sanders and Ambjörnsson (2012); Jeon *et al.* (2014)]. Here, we recalculate the halo mass function using extensive stochastic simulations of fBm and known approximations for FPT pdf of fBm and show that the obtained solutions reverse conclusions of [Pan (2007); Pan *et al.* (2008)].

Fractional Brownian motion and the halo mass function

The fBm is a popular model of anomalous diffusion in complex media [Sokolov (2012); Ernst *et al.* (2012)]. It is defined by a single parameter, the Hurst exponent $H \in (0,1)$, and is a continuous-time Gaussian process $B_H(t)$ on $[0,T]$ starting at zero and having zero expectation for all times in $[0,T]$. The increments of fBm are correlated (except the case $H = 1/2$) and obtained from Gaussian noise via fractional integral [Mandelbrot and Van Ness (1968)],

$$B_H(t) = \frac{1}{\Gamma(H+1/2)} \int_{-\infty}^{\infty} \left[(t-t')_+^{H-1/2} - (-t')_+^{H-1/2} \right] dB(t')$$

producing an auto-correlation function

$$\langle B_H(t_1) B_H(t_2) \rangle = C \left[|t_1|^{2H} + |t_2|^{2H} - |t_1 - t_2|^{2H} \right],$$

where C is a constant.

When $H = 1/2$, fBm becomes the ordinary Bm. The case $1/2 < H < 1$ relates to persistent or fractional superdiffusion (enhanced diffusion), the process with $H < 1/2$ describes antipersistent or fractional subdiffusion.

In ES-formalism we have to calculate distributions of FPT of a random trajectory $\delta(S)$. One of the methods is the images method (IM) that is applicable for Markovian trajectories when the geometry of the problem is quite simple [Jeon *et al.* (2014)]. As [Jeon *et al.* (2014)] notice, it is still valid for subordinated Bm with traps characterized by power law distributions of waiting times with a diverging mean. But, it is incorrect for deeply non-Markovian processes such as fBm. IM demonstrates major deviations in the scaling behavior from the simulation for both persistent and antipersistent cases. The predicted asymptotic behavior of FPT density $P(S) \propto S^{-H-1}$ is opposite to the correct scaling $P(S) \propto S^{H-2}$. For this reason, we have recalculated multiplicity functions and reconsider conclusions, obtained in [Pan (2007); Pan *et al.* (2008)].

Spherical collapse

For the spherical collapse, ES-formalism deals with first passage of the constant threshold $\delta_c = 1.686$. As was mentioned above, the method of images is inapplicable in the case of fBm, and we use appropriate approximation used in [Sanders and Ambjörnsson (2012)]:

$$P(\delta_c, S|0) = \frac{2H\beta}{\Gamma\left(\frac{1-H}{2H\beta}\right)} (\gamma\delta_c^{2\beta})^{\frac{1-H}{2H\beta}} S^{H-2} \exp\left(-\gamma\left(\frac{\delta_c^2}{S^{2H}}\right)^{\beta}\right) \tag{8.14}$$

where β and γ are constant parameters having to be fitted for a certain H value. This approximation gives the following mass function

$$\frac{dn}{dM} dM = \frac{\langle\rho\rangle}{M^2} f_H(\sigma) \left|\frac{d\ln\sigma}{d\ln M}\right| dM \tag{8.15}$$

which can be expressed in terms of the multiplicity function $f_H(\sigma) = 2\sigma^2 P(\delta_c, \sigma^2|0)$. It has the form

$$f_H(\sigma) = \frac{4H\beta}{\Gamma\left(\frac{1-H}{2H\beta}\right)} (\gamma\delta_c^{2\beta})^{\frac{1-H}{2H\beta}} \sigma^{2(H-1)} \exp\left(-\gamma\left(\frac{\delta_c^2}{\sigma^{4H}}\right)^{\beta}\right). \tag{8.16}$$

Taking $H = 1/2$, $\beta = 1$, $\gamma = 1/2$, we arrive at the Press-Schechter formula (8.13).

Note that [Pan (2007); Pan *et al.* (2008)] used the solution obtained by means of the image method, that has the form

$$f_H^{\mathrm{IM}}(\sigma) = 2H\sqrt{\frac{2}{\pi}} \frac{\delta_c}{\sigma^{2H}} \exp\left(-\frac{\delta_c^2}{2\sigma^{4H}}\right). \tag{8.17}$$

By fitting the simulation results with curve defined by (8.14) we have determined β and γ for different H values. In all calculations $\delta_c = 1.686$, number of steps $n = 10^3$, number of trajectories $N_{\mathrm{traj}} = 5\cdot10^4$. Simulations are performed in R with usage of package 'dvfBm'.

Figure 8.3 demonstrates trajectories of fBm ($H = 0.7$) approaching the barrier $\delta_c = 1.686$ and corresponding histogram of FPT pdf approximated the generalized Lévy-Smirnov density (8.14) with $\beta = 1.73$ and $\gamma = 0.127$ (solid line) which differs sufficiently from the image method solution (dashed line).

Multiplicity functions obtained through approximation (8.14) of simulation data on FPT density are in contrast to the result of [Pan (2007)] and the behavior at small masses (large sigma) is directly opposite. In the persistent case ($H > 1/2$), we observe an excess of halos with small masses and fewer halos with large masses than predicted by the Press-Schechter theory. In the antipersistent case $H < 1/2$, the situation inverses.

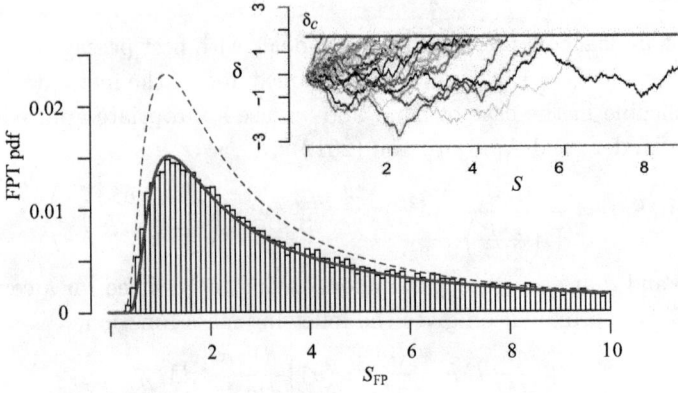

Fig. 8.3   Histogram of FPT of fBm ($H = 0.7$, $\delta_c = 1.686$) compared to the generalized Lévy-Smirnov density (8.14) with $\beta = 1.73$ and $\gamma = 0.127$ (solid line) and to the image method solution (dashed line). Inset: Some trajectories of fBm reached barrier δ_c.

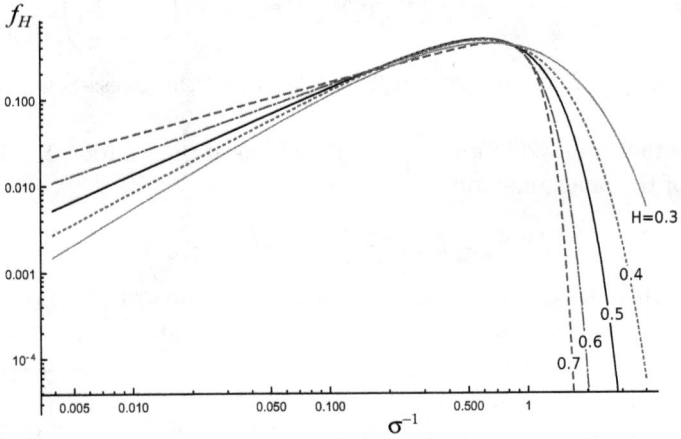

Fig. 8.4   The multiplicity functions (8.16) of ES-formalism for fBm with different Hurst exponent H.

Chapter 9

Conclusion: Invitation to fractional cosmology

9.1 Fractional methodology

Mark D. Roberts writes in [Roberts (2009)], that two distinct methods of converting ordinary cosmology into "fractional derivative cosmology" could be applied. One of them is based on the *last Step Modification* (LSM) in which Einstein's field equations for a given geometric configuration are replaced with fractional order field equations. Following another method called the *First Step Modification* (FSM), one starts by constructing fractional derivative geometry before derivation of the field equations. Unfortunately, none of them is free from logical contradictions. So, demonstrating the LSM-approach on the Newtonian gravity equation, the author replaces D'Alembert operator acting on a spherically symmetric function by the operator $D^{2p} + (2/r)D^p$ with a fractional order p, but this replacement is invalid when $p \neq 1$ because $D^{2p} + (2/r)D^p$ is not a p-power of the D'Alembert operator. The author notes as well, that this approach probably only gives consistent description in rectilinear coordinates; "the problem with the first step approach is that the whole of geometry has to be rethought through, even things like linear coordinate transformations have to be replaced by fractional ones, perhaps quadratic forms by fractional forms and so on; *Intermediate Step Approaches* (ISA) seem to have the disadvantages of both of the above approaches." The cited article is ended with author's opinion that for consistency one should begin the fractional cosmology development with the FSM-approach and this leads to the subject of fractional derivative geometry. "What fractional derivative geometry should look like is unclear", – the author concludes.

During last decade, we observe the intensively developing version of the fractional cosmology on the basis of a variational principle for the action of a fractional order (Fractional Action-Like Variational Approach, FALVA) [El-Nabulsi (2007a); El-Nabulsi and Torres (2008); El-Nabulsi (2012, 2013)] in combination with ISA. In the framework of this concept, the time-integral of action $A_L[q_i]$ of the Lagrangian $L(q_i(t), \dot{q}_i(t), t)$ is taken as a fractional-order α integral

$$A_L[q_i] = \frac{1}{\Gamma(\alpha)} \int_{t_0}^{t} L(t', q_i(t'), \dot{q}_i(t'), t')(t - t')^{\alpha-1} dt', \qquad (9.1)$$

and functions $q_i(t')$ obey the Euler-Lagrange equations

$$\frac{\partial L}{\partial q_i} - \frac{d}{dt'}\left(\frac{\partial L}{\partial \dot{q}_i}\right) = \frac{1 - \alpha}{t - t'}\frac{\partial L}{\partial \dot{q}_i} \equiv F^i, \quad i = 1, 2, \ldots, n, \quad t' \in (t_0, t),$$

where dots mark first derivatives with respect to t'. The function $F^i(t')$ is interpreted as a decaying force of 'friction', that is the general expression for non-conservative force [El-Nabulsi (2007b)]. The author states as well that $F_i \to 0$ as $t' \to \infty$ and provides some examples of application of FALVA to Riemannian geometry and perturbed cosmological models. Every time, he does not apply FALVA directly to the action, but tries to take into account influence of the fractional order action (9.1) on the Friedmann equations through perturbed (and time-dependent) classical gravitational constant. Some other authors (see, for instance, [Frederico and Torres (2006)]) treat time t' in (9.1) as an intrinsic (proper) time, considering t is the observer time. In [Shchigolev (2010)], the reader can find some details of critical analysis of this direction in development of fractional derivative cosmology.

Based on this fractional approach, [Jamil *et al.* (2012)] obtained a varying gravitational coupling constant and varying cosmological 'constant'. They also modeled dark energy and obtained several important cosmological parameters. [Debnath *et al.* (2013)] reconstructed the scalar potentials and scalar fields (quintessence, phantom, tachyon, k-essence, Dirac-Born-Infeld-essence, hessence, dilaton field, and Yang-Mills field) in frames of the FrAC theory. They express a scale factor in emergent, logamediate, and intermediate scenarios, leading to different regimes of expansion of the universe.

Fractal concept and fractional calculus are also used in the theory of quantum gravity [Calcagni (2010a); Calcagni *et al.* (2013)]. So standard quantum field theory indicates that on small scales, space-time is characterized by anomalous properties, which cannot be described by the standard geometry. Attributing the multifractality property to space-time, [Calcagni (2010a,b)] develops a new cosmological model. Fractal behavior is taken into account in the definition of action. The model satisfies important requirements: it is Lorentz-invariant, renormalizable. Calcagni [Calcagni (2010b)] starts with the action $S = \int \mathcal{L}(\phi, \partial_\mu \phi) d\rho(x)$, where $\mathcal{L} = -\frac{1}{2} \partial_\mu \phi \partial^\mu \phi - V(\phi)$ is the Lagrange function density and $d\rho(x) = \Pi_{\mu=0}^{D-1} f_\mu(x) dx^\mu$ is the isotropic measure.

In [Zeng *et al.* (2015); El-Nabulsi (2016)], a fractional calculus approach is used to characterize the accelerating expansion of the universe. The authors [Zeng *et al.* (2015)] claim that the accelerating expansion obeys a fractional α-exponential function and calculated values of the Hubble constant for several redshifts.

Different cosmological models could be testified by observations of CMB temperature anisotropies [Hu and Dodelson (2002)]. In Ref. [Tenreiro Machado *et al.* (2013)] present results of the 'fractional' Fourier analysis of the Wilkinson Microwave Anisotropy Probe (WMAP) data. They found that function $|H(i\omega)| = \left| k(i\omega)^{-\alpha} \left[1 + (i\omega/p)^\beta \right]^{-\gamma} \right|$, with adjustable parameters $\{k, p, \alpha, \beta, \gamma\}$, provides a good fit to the amplitude of the Fourier transform of CMB anisotropy intensity.

The cosmology based on fractional calculus is a developing field and some approaches and conclusions are controversial. Nevertheless, the results can be useful in interpretations of the fractal properties of quantum gravity, anisotropy of the cosmic microwave background (CMB), in description of interaction between dark energy and dark matter, accelerating expansion of the universe and in other problems. This direction facilitates the development of such branches of mathematics as fractional vector calculus [Tarasov (2010)], fractional differential geometry [Jumarie (2012); Vacaru (2012)], fractional calculus of variations (FCV) [Agrawal (2002); El-Nabulsi and Torres (2008); Malinowska and Torres (2012)], theory of space-time of fractional dimensions [Calcagni (2010a)] and others.

9.2 Turbulence

The matter in our universe forms a web of densely populated galaxy clusters and connecting filaments separated by vast voids. The rotating spiral galaxies look strikingly like fossil eddies. Both these arguments led to the primeval turbulence idea, which was first suggested by von Weizsäcker in the work [von Weizsäcker (1951)]. Discussed the possible role of turbulence in a broad range of astrophysical phenomena, von Weizsäcker wrote that "in an expanding universe gravitational instability would not be sufficient to form sub-systems, while turbulence could do it if its velocity v_t were large enough compared with the velocity of expansion v_{exp}" ([von Weizsäcker (1951)], p. 176). However, Russian theorist Gamow was not completely sure of the crucial role of turbulence in the evolution universe process and wrote: "... it is difficult to see, how such a motion could originate in a uniformity expanding homogeneous material." Peebles noted that in framework of this picture, it is hard to avoid the turbulence dissipation prematurely forming objects at $Z \sim Z_{dec}$ that would be denser than galaxies [Peebles (1980)]. Nevertheless, the turbulence problem remains actual in modern cosmology [Nakamichi and Morikawa (2009); Brevik *et al.* (2012, 2011a,b); Gibson (2001)].

Cosmic magnetic fields play an important role in governing the motion and evolution of stars and galaxies, not to mention such subtle matter as cosmic rays. Galaxies and galaxy clusters, crashing into one another, are pretty turbulent environments. The magnetic fields embedded within these objects should twist and stretch, too. Kinetics energy of charged particles taking part in these events is transformed into the energy of magnetic fields. It has been suggested that primordial magnetic fields might arise during the early cosmic phase transitions, in which magnetic fields could be generated when bubbles of the new vacuum collide, whence a ring of magnetic field may arise in the intersecting region [Kibble and Vilenkin (1995)].

As is noted in [Brandenburg *et al.* (1996)], the turbulent nature of the magnetic field may have some effects on the various phase transitions in the early universe. Also, the inherent shift of energy from small to large scales may be of interest in connection with the density fluctuations due to the magnetic energy. So, one can conclude, that the link between turbulence and fractional calculus may find a further development in cosmology.

9.3   Fractional burgulence

The word 'Burgulence' appeared in the title of the article [Frisch and Bec (2001)] is a contraction of 'Burgers' and 'turbulence' and covers a class of random processes obeying the nonlinear Burgers equation

$$\left[\frac{\partial \mathbf{u}}{\partial t} + (\mathbf{u} \cdot \nabla)\right] \mathbf{u} = \nu \Delta \mathbf{u}, \quad \mathbf{u} = -\nabla \psi. \qquad (9.2)$$

The randomness may be involved through a random initial condition $\mathbf{u}_0 = -\nabla \psi_0$ or a random driving force $\mathbf{f} = -\nabla F$ added to the right-hand side of (9.2).

Burgers' equation (9.2) has a lot in common with the Navier-Stockes equation: same type of nonlinearity, diffusion term, invariance and conservation properties (see article quoted). It was used by the Russian school [Zel'dovich (1970); Gurbatov and Saichev (1984); Gurbatov *et al.* (1989)] and other scientific groups (see, for example, [Vergassola *et al.* (1994)]). Motivation for using the Burgers equation can be explained on a simplified one-dimensional example which has neither background nor expansion. In this case, the gravitation potential at a point x is simply proportional to the difference of masses located right and left of x, so while the particles do not cross, these masses are conserved. Consequently, one can write down the equations

$$x = q + tu_0(q) + \frac{t^2}{2}g_0(q),$$

$$u(x,t) = u_(q) + tg_0(q),$$

$$\rho(x,t) = \rho_0(q)\left|\frac{\partial x}{\partial q}\right|^{-1},$$

where q, u, g and ρ are the particle position, velocity, acceleration and density respectively. Assuming that the initial velocity and acceleration are proportional,

$$g_o(q) = \lambda u_0(q),$$

and defining a new time $\tau = t + \lambda t^2/2$, we arrive at one-dimensional Burgers' equation without viscosity:

$$\frac{\partial u}{\partial \tau} + u\frac{\partial u}{\partial x} = 0.$$

On Burgers' equation, the so-called Zeldovich approximation was based, which was extended by Gurbatov and Saichev to the dynamical model, known in cosmology as the "adhesion approximation", Vergassolla et al. investigated links between this adhesion approximation and the Jeans-Vlasov-Poisson description of such matter.

In the light of what has been said in Chapter 2 of our book, the Laplace operator maps the Brownian type of motion, whereas its fractional power, whose exponent is related to the spectral index of the turbulent process, is more suitable for describing the turbulent motion. Therefore, it is natural to expect that equation

$$\left[\frac{\partial \mathbf{u}}{\partial t} + (\mathbf{u} \cdot \nabla)\right]\mathbf{u} = \nu\Delta^{\alpha/2}\mathbf{u}$$

can lead to new interesting results.

9.4 Visible subsystem as an open system

It was realized decades ago that the spatial clustering of observable galaxies need not precisely mirror the clustering of the bulk of the matter in the Universe. In its most general form, the galaxy density can be a non-local and stochastic function of the underlying dark matter density. This galaxy bias the relationship between the spatial distribution of galaxies and the underlying dark matter density field is a result of the varied physics of galaxy formation which can cause the spatial distribution of baryons to differ from that of dark matter. Stochasticity appears to have little effect on bias except for adding extra variance (e.g., [Scoccimarro (2000)]), and non-locality can be taken into account to first order by using smoothed densities over larger scales.

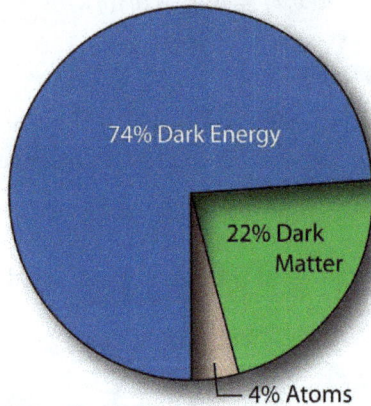

Fig. 9.1   Pie Chart of the content of the Universe according to the Wilkinson Microwave Anisotropy Probe data. Most of the universe's matter and energy are invisible. Image credit: NASA / WMAP Science Team.

Another, in some sense an inverse situation arises when only part of the system is accessible to observation, the rest part dynamics is unknown. We will call this part the *open system*[1] in contrast to the whole system assumed to be isolated from the rest world and called the *closed system*. On the basis of the simplest hypothesis that the subsystem environment

---

[1]We use this term in its narrow sense assuming that the observed subsystem can exchange with the rest subsystem by energy, moment, etc., but the number of particles remains constant.

(that is, the rest of the system called in this case the *thermal bath*) or the *thermostat* does not react to the evolution of the observed part, remaining in the same equilibrium state. On the basis of this hypothesis all the thermodynamics (which would be more appropriately termed thermostatic) is developed. Below, we describe a more subtle approach being free from such a "Procrustean" restriction and admitting that not only the environment affects the open system, but it itself is influenced by the system. According to the WMAP data, most of the universe's matter and energy are invisible (Fig. 9.1). The components of the visible matter (stars, diffuse gas, accreting black holes, etc.) are organized in a web consisting of filaments, and voids (Fig. 9.2). Embedded galaxies could be considered as the basic units of large scale structure of the Universe. Testing different scenarios of the evolution of dark matter and dark energy requires accurate description of related formation of structure of the visible matter [Vogelsberger *et al.* (2014); Genel *et al.* (2014)].

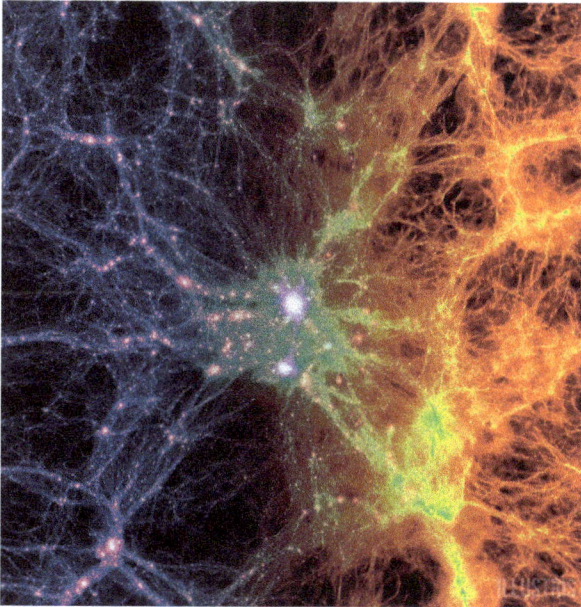

Fig. 9.2   Result of simulation [Vogelsberger *et al.* (2014); Genel *et al.* (2014)] of structure formation in the Universe (centered on the most massive cluster, 15 Mpc/h deep). Shows dark matter density (left) transitioning to gas density (right). Credit: Illustris Collaboration/Illustris Simulation.

In order to demonstrate a concrete example of such a situation, we come back for a moment to Sec.2.1, considered the velocity diffusion in a viscous incompressible. Suppose that the plate driving the liquid is affected by the force $F(t)$ per unit of its mass and directed along the y-axis, we arrive at the couple of equations

$$\frac{dV}{dt} = F(t) - 2\nu \left(\frac{\partial f(x,t)}{\partial x} \right)_{x=0},$$

$$\frac{\partial f(x,t)}{\partial t} = \nu \frac{\partial^2 f(x,t)}{\partial x^2},$$

describing the system "plate+liquid". Observe that the system of two equations contains derivatives of only integer orders. In case when only $V(t)$ is available for measurement, we eliminate another variable and obtain one equation with time-derivatives of orders 1 and 1/2:

$$\frac{dV}{dt} + 2\sqrt{\nu}V^{(1/2)}(t) = F(t).$$

Now, we give a more general consideration based on the Zwanzig-Mori formalism. We begin with the classical Liouville equation

$$\frac{\partial \rho}{\partial t} = -i\mathcal{L}\rho, \tag{9.3}$$

describing evolution of a closed (with fixed number of particles, their total energy, momentum, etc) system.

After mental splitting the system into observed (open) subsystem with phase coordinates $(q,p) \in \Gamma_1$ and hidden from observation (environment) with coordinates $(Q,P) \in \Gamma_2$, the Hamiltonian of total system splits into three terms

$$\mathcal{H}(q,p;Q,P) = \mathcal{H}_1(q,p) + \mathcal{H}_2(Q,P) + \mathcal{H}_{12}(q,p;Q,P)$$

where subscripts 1 and 2 relate to the open system and its environment correspondingly. Then, we will write the joint phase pdf ρ as $\rho(q,p,Q,P,t)$ and introduce the marginal densities related to each subsystem by

$$\sigma_1(q,p,t) = \int_{\Gamma_2} \rho(q,p,Q,P,t)dQdP$$

and

$$\sigma_2(Q,P,t) = \int_{\Gamma_1} \rho(q,p,Q,P,t)dqdp.$$

When the subsystems do not interact at $t > t_0$, their current correlations are determined only by initial states correlations. If the initial correlations are also absent,

$$\rho(q, p, Q, P, t_0) = \sigma_1(q, p, t_0)\sigma_2(Q, P, t_0),$$

then the solution of Eq. (9.3) marked for this case by superscript 0 becomes product of the marginal densities

$$\rho^0(q, p, Q, P, t) = \sigma_1^0(q, p, t)\sigma_2^0(Q, P, t),$$

each of which obeys its own equation with its own initial condition:

$$\frac{\partial \sigma_1^0}{\partial t} = -i\mathcal{L}_1\sigma_1^0(q, p, t), \quad \sigma_1^0(q, p, t_0) = \int_Y \rho(q, p, Q, P, t_0)dQdP;$$

$$\frac{\partial \sigma_2^0}{\partial t} = -i\mathcal{L}_2\sigma_2^0(Q, P, t), \quad \sigma_2^0(Q, P, t_0) = \int_X \rho(q, p, Q, P, t_0)dqdp.$$

Anyway, the future (at $t > t_0$) evolution of the joint system pdf ρ as well as of their parts under given $\rho(t_0)$ does not depend on its prehistory (at $t < t_0$).

Let us return now to the case of interacting subsystems assuming as before their initial states mutually independent. This case does also admit splitting the origin equation into two ones but they are not independent anymore but linked to each other though time-integral operators. This transformation is performed by using Zwanzig-Mori projection operators P_1, P_2, possessing the properties:

$$\mathsf{P}_1\mathsf{P}_1 = \mathsf{P}_1, \quad \mathsf{P}_2\mathsf{P}_2 = \mathsf{P}_2, \quad \mathsf{P}_1\mathsf{P}_2 = \mathsf{P}_2\mathsf{P}_1 = 0, \quad \mathsf{P}_1 + \mathsf{P}_2 = 1.$$

An explicit realization of operator P_1 can be defined via formula

$$\mathsf{P}_1(q, p, Q, P, t) = \left(\int_Y \rho(q, p, Q, P', t)dQdP'\right)w(Q, P) = \sigma_1(q, p, t)w(Q, P),$$

where non-negative function $w(Q, P)$ obeys the normalizing condition:

$$\int_Y w(Q, P)dQdP = 1.$$

Acting on both sides of the Liouville equation (9.3) one time by P_1 and another time by P_2 and introducing notations ρ_1 and ρ_2 for projections $\mathsf{P}_1\rho$ and $\mathsf{P}_2\rho$ respectively, we arrive at the system

$$\frac{\partial \rho_1}{\partial t} = -i\mathsf{P}_1\mathcal{L}\left[\rho_1(t) + \rho_2(t)\right], \tag{9.4}$$

$$\frac{\partial \rho_2}{\partial t} = -i P_2 \mathcal{L} \left[\rho_1(t) + \rho_2(t) \right]. \tag{9.5}$$

Observe that each of the projections depends on both variable sets q, p and Q, P,

$$\rho_1(t) = \rho_1(q, p, Q, P, t),$$

$$\rho_2(t) = \rho_2(q, p, Q, P, t),$$

and should be in agreement with a given initial condition for $\rho(t)$:

$$\rho_1(t_0) = \sigma_1(q, p, t_0) w(Q, P),$$

$$\rho_2(t_0) = \rho(q, p, Q, P, t_0) - \sigma_1(q, p, t_0) w(Q, P).$$

It is naturally from mechanical point of view to choose

$$w(Q, P) = \sigma_2(Q, P, t_0),$$

then

$$\rho_1(t) = \sigma_1(q, p, t) \sigma_2(Q, P, t_0)$$

and

$$\rho_2(t) = \rho(q, p, Q, P, t) - \sigma_1(q, p, t) \sigma_2(Q, P, t_0)$$

respectively. The ultimate aim of the open systems theory is investigation of OS pdf

$$\sigma_1(q, p, t) = \frac{\rho_1(q, p, Q, P, t)}{\sigma_2(Q, P, t_0)} \tag{9.6}$$

and some functionals of it.

The formal solution of Eq. (9.5) is of the form

$$\rho_2(t) = \mathcal{E}(t, t_0; \mathcal{Q}(\cdot)) \rho_2(t_0) + \int_{t_0}^{t} \mathcal{E}(t, \tau; \mathcal{Q}(\cdot)) \mathcal{Q}(\tau) \rho_1(\tau) d\tau, \tag{9.7}$$

where $\mathcal{Q}(t) = -i P_2 \mathcal{L}(t)$,

$$\mathcal{E}(t, t_0; \mathcal{Q}(\cdot)) \equiv \mathsf{T} \exp \left\{ \int_{t_0}^{t} d\tau \mathcal{Q}(\tau) \right\}$$

and T is the Dyson chronological operator (interaction between subsystems 1 and 2 causes the operator Q to be explicitly time-dependent, $Q = Q(t)$).

Substituting Eq. (9.7) into Eq. (9.5) leads to the integro-differential Zwanzig kinetic equation

$$\frac{\partial \rho_1(t)}{\partial t} = -i\mathsf{P}_1 \mathcal{L}(t)\rho_1(t) - i\mathsf{P}_1 \mathcal{E}(t, t_0; \mathcal{Q}(\cdot))\rho_2(t_0)$$

$$- i \int_{t_0}^{t} \mathsf{P}_1 \mathcal{L}(t)\mathcal{E}(t, \tau; \mathcal{Q}(\cdot))\mathcal{Q}\rho_1(\tau)d\tau. \tag{9.8}$$

$$\rho_1(t_0) = \rho_1(q, p, Q, P, t_0) = \sigma_1(q, p, t_0)\sigma_2(Q, P, t_0). \tag{9.9}$$

Remark that the variable Q, P is contained in initial phase density ρ_2 and in operator $\mathcal{Q}$. Thus, solving Eq. (9.8) under initial condition (9.9) and inserting the result into Eq. (9.6), we obtain evolution of the open system in its own phase space.

One can give a clear physical interpretation to each term in right hand side of Eq. (9.8). Its first term relates to the autonomous part of system's time evolution. The second one represents independent part of environment's ("heat bath's") time evolution making an instant impact on subsystem 1 at time t. First term often vanishes and second one can be made zero by judicious choice of initial conditions. The last (third) term takes into account the effect caused by subsystem 1 in environment 2 at time τ propagating into 2 and feeding back into system 1 at time t.

Recall that the Zwanzig equation (9.8) still remains exact but generally too complicated to be directly applied to practical problems. Several approximations and simplifications are usually made in order to weaken this difficulty. For instance, the initial condition is often chosen to satisfy

$$\rho_2(q, p, Q, P, t_0) = 0,$$

so the equation becomes slightly shorter:

$$\frac{\partial \rho_1(t)}{\partial t} = -i\mathsf{P}_1 \mathcal{L}(t)\rho_1(t) - i \int_{t_0}^{t} \mathsf{P}_1 \mathcal{L}(t)\mathcal{E}(t, \tau; \mathcal{Q}(\cdot))\mathcal{Q}\rho_1(\tau)d\tau. \tag{9.10}$$

Under more general assumptions, analogous transformations were presented in [Uchaikin (2014, 2016)]. In our opinion, these results gave the basis for hypotheses that the true cause of of fractional order operators is incompleteness of dynamic description of the system, due to the presence of hidden variables. Confirmation of this conclusion can be found in every chapter of this book: everywhere we easily can indicate a hidden part of the system variables, that affects the movement of the observed part.

9.5 Complexity and fractionality

As the reader could see above, fractional derivatives differ from the derivatives of integer orders by special properties that can easily reflect the distant spatial ("system fractality") and temporal ("system memory") correlations. Both of these properties, among others, are inherent in *complex systems*, i.e. the systems with internal structure and partially hidden (*latent*) dynamic variables. Moses [Moses (2000)] defined them as systems which are composed of many parts that interconnect in intricate ways. The dynamic complexity becomes apparent when cause and effect are divided by long time and space and for this reason are subtle but playing an essential role in explanation of phenomenon under observation. One of the ways of explaining such specific properties of complex systems lies through dividing time into operational and observational times [Feller (1971)]. Imposing the additional requirement of self-similarity inherent in all integer-order differential operators leads to a fractional differentiation operation (see discussion in [Uchaikin (2013a)], Vol. II, Chapter 15).

A complex system is normally composed of a group of subsystems, for which the degree and nature of the relationships is imperfectly known. In contrast to a simple mechanical system obeying the Newton laws, the complex dynamics is more difficult for prediction. But the real world is made up of complex systems whereas the simple systems and created by science.

Usually, experiments involve reducing the system to its parts and then studying those parts in a context formulated according to dynamics. However, this is not a universal approach. There are exist non-reducible systems, such as fractal: dividing it into fragments we obtain objects of the same complexity. Nothing becomes easier, a part of fractal is similar to the whole fractal. This property of fractals – their self-similarity – relates to the fractional calculus. But fractional operators are based on the continuity concept, whereas the fractal structure is discontinuous at all scales only by averaging over an ensemble of fractals, we can turn to continuity, hence the systems and processes must be random (stochastic). This is the idea, we started with which our book, and we want to finish it.

Bibliography

Abramenko, V., Carbone, V., Yurchyshyn, V., Goode, P., Stein, R., Lepreti, F., Capparelli, V., and Vecchio, A. (2011). Turbulent diffusion in the photosphere as derived from photospheric bright point motion, *Astrophysical Journal* **743**, 2, p. 133.

Agrawal, O. P. (2002). Formulation of Euler–Lagrange equations for fractional variational problems, *Journal of Mathematical Analysis and Applications* **272**, 1, pp. 368–379.

Almeida, J. S., Bonet, J., Viticchié, B., and Del Moro, D. (2010). Magnetic bright points in the quiet sun, *Astrophysical Journal Letters* **715**, 1, p. L26.

Andersen, E. S. (1953). On sums of symmetrically dependent random variables, *Scandinavian Actuarial Journal* **1953**, sup1, pp. 123–138.

Arkhincheev, V. (2010). Unified continuum description for sub-diffusion random walks on multi-dimensional comb model, *Physica A: Statistical Mechanics and its Applications* **389**, 1, pp. 1–6.

Arkhincheev, V. and Baskin, E. (1991). Anomalous diffusion and drift in a comb model of percolation clusters, *Sov. Phys. JETP* **73**, 1, pp. 161–165.

Arthur, A. D. and Le Roux, J. A. (2013). Particle acceleration at the heliospheric termination shock with a stochastic shock obliquity approach, *Astrophysical Journal Letters* **772**, 2, p. L26.

Artmann, S., Schlickeiser, R., Agueda, N., Krucker, S., and Lin, R. (2011). A diffusive description of the focused transport of solar energetic particles. intensity-and anisotropy-time profiles as a powerful diagnostic tool for interplanetary particle transport conditions, *Astronomy and Astrophysics* **535**, p. A92.

Aschwanden, M. J. (2012). The spatio-temporal evolution of solar flares observed with aia/sdo: Fractal diffusion, sub-diffusion, or logistic growth?, *Astrophysical Journal* **757**, 1, p. 94.

Axford, W. (1965). The modulation of galactic cosmic rays in the interplanetary medium, *Planetary and Space Science* **13**, 2, pp. 115–130.

Baeumer, B. and Meerschaert, M. M. (2010). Tempered stable lévy motion and transient super-diffusion, *Journal of Computational and Applied Mathematics* **233**, 10, pp. 2438–2448.

Bakunin, O. G. (2008). *Turbulence and diffusion: scaling versus equations* (Springer Science & Business Media).

Balescu, R. (1975). Equilibrium and nonequilibrium statistical mechanics, *NASA STI/Recon Technical Report A* **76**, p. 32809.

Balescu, R. (2000). Memory effects in plasma transport theory, *Plasma Physics and Controlled Fusion* **42**, 12B, p. B1.

Barenblatt, G. I. and Zel'dovich, Y. B. (1971). Intermediate asymptotics in mathematical physics, *Russian Mathematical Surveys* **26**, 2, p. 45.

Baryshev, Y. and Teerikorpi, P. (2002). *Discovery of cosmic fractals* (World Scientific).

Baskin, E. and Iomin, A. (2004). Superdiffusion on a comb structure, *Physical review letters* **93**, 12, p. 120603.

Bel, G. and Barkai, E. (2006). Random walk to a nonergodic equilibrium concept, *Physical Review E* **73**, 1, p. 016125.

Bell, A. (1978). The acceleration of cosmic rays in shock fronts–i, *Monthly Notices of the Royal Astronomical Society* **182**, 2, pp. 147–156.

Berezhko, E. and Krymskii, G. (1988). Acceleration of cosmic rays by shock waves, *Physics-Uspekhi* **31**, 1, pp. 27–51.

Berezhko, E. and Ksenofontov, L. (1999). Composition of cosmic rays accelerated in supernova remnants, *Journal of Experimental and Theoretical Physics* **89**, 3, pp. 391–403.

Berezinskii, V., Bulanov, S., Ginzburg, V., Dogel, V., and Ptuskin, V. (1984). The astrophysics of cosmic rays, *Moscow, Izdatel'stvo Nauka, 1984, 360 p. In Russian.*

Berger, T. E. *et al.* (1998). Measurements of solar magnetic element dispersal, *Astrophysical Journal* **506**, 1, p. 439.

Berk, H. and Roberts, K. (1967). Nonlinear study of Vlasov's equation for a special class of distribution functions, *Physics of Fluids (1958-1988)* **10**, 7, pp. 1595–1597.

Bernstein, I. B., Greene, J. M., and Kruskal, M. D. (1957). Exact nonlinear plasma oscillations, *Physical Review* **108**, 3, p. 546.

Bershadskii, A. G. (1990). Large-scale fractal structure in laboratory turbulence, astrophysics, and the ocean, *Physics-Uspekhi* **33**, 12, pp. 1073–1075.

Bertoin, J. (1998). *Lévy processes*, Vol. 121 (Cambridge University Press).

Bertschinger, E. (1998). Simulations of structure formation in the universe, *Annual Review of Astronomy and Astrophysics* **36**, 1, pp. 599–654.

Bian, N. and Browning, P. (2008). Particle acceleration in a model of a turbulent reconnecting plasma: a fractional diffusion approach, *Astrophysical Journal Letters* **687**, 2, p. L111.

Blackledge, J. (2007). Diffusion and fractional diffusion based models for multiple light scattering and image analysis, *ISAST Transaction of electronic and signal processing* **1**, pp. 1–23.

Blackledge, J. (2009). Diffusion and fractional diffusion based image processing, *EG UK Theory and Practice of Computer Graphics*, pp. 233–240.

Blackledge, J. and Blackledge, M. (2010). Fractional anisotropic diffusion for noise reduction in magnetic resonance images, *ISAST Transactions on Electronics and Signal Processing* **4**, pp. 44–57.

Blandford, R. and Ostriker, J. P. (1980). Supernova shock acceleration of cosmic rays in the Galaxy, *Astrophysical Journal* **237**, pp. 793–808.

Blasi, P. (2013). The origin of galactic cosmic rays, *The Astronomy and Astrophysics Review* **21**, 1, pp. 1–73.

Blasi, P. and Amato, E. (2012a). Diffusive propagation of cosmic rays from supernova remnants in the galaxy. I: spectrum and chemical composition, *Journal of Cosmology and Astroparticle Physics* **2012**, 01, p. 010.

Blasi, P. and Amato, E. (2012b). Diffusive propagation of cosmic rays from supernova remnants in the galaxy. II: anisotropy, *Journal of Cosmology and Astroparticle Physics* **2012**, 01, p. 011.

Bloemen, J., Dogiel, V., Dorman, V., and Ptuskin, V. (1993). Galactic diffusion and wind models of cosmic-ray transport. I-insight from CR composition studies and gamma-ray observations, *Astronomy and Astrophysics* **267**, pp. 372–387.

Bochner, S. (1949). Diffusion equation and stochastic processes, *Proceedings of the National Academy of Sciences* **35**, 7, pp. 368–370.

Bogdan, K. and Byczkowski, T. (2000). Potential theory of Schrödinger operator based on fractional Laplacian, *Probability and Mathematical Statistics – Wroclaw University* **20**, 2, pp. 293–335.

Boldyrev, S. (2005). On the spectrum of magnetohydrodynamic turbulence, *The Astrophysical Journal Letters* **626**, 1, p. L37.

Boldyrev, S. and Gwinn, C. R. (2003). Lévy model for interstellar scintillations, *Physical Review Letters* **91**, 13, p. 131101.

Bond, J., Cole, S., Efstathiou, G., and Kaiser, N. (1991). Excursion set mass functions for hierarchical Gaussian fluctuations, *Astrophysical Journal* **379**, pp. 440–460.

Borgani, S. (1995). Scaling in the universe, *Physics Reports* **251**, 1, pp. 1–152.

Borodin, A. N. and Salminen, P. (2012). *Handbook of Brownian motion-facts and formulae* (Birkhäuser).

Brandenburg, A., Enqvist, K., and Olesen, P. (1996). Large-scale magnetic fields from hydromagnetic turbulence in the very early universe, *Physical Review D* **54**, 2, p. 1291.

Brevik, I., Elizalde, E., Nojiri, S., and Odintsov, S. (2011a). Viscous little rip cosmology, *Physical Review D* **84**, 10, p. 103508.

Brevik, I., Gorbunova, O., Nojiri, S., and Odintsov, S. D. (2011b). On isotropic turbulence in the dark fluid universe, *The European Physical Journal C-Particles and Fields* **71**, 4, pp. 1–7.

Brevik, I., Myrzakulov, R., Nojiri, S., and Odintsov, S. (2012). Turbulence and little rip cosmology, *Physical Review D* **86**, 6, p. 063007.

Burnecki, K., Klafter, J., Magdziarz, M., and Weron, A. (2008). From solar flare time series to fractional dynamics, *Physica A* **387**, 2, pp. 1077–1087.

Cadavid, A., Lawrence, J., Ruzmaikin, A., Walton, S., and Tarbell, T. (1998). Spatiotemporal correlations and turbulent photospheric flows from soho/mdi velocity data, *Astrophysical Journal* **509**, 2, p. 918.

Cadavid, A., Lawrence, J., and Ruzmaikin, A. (1999). Anomalous diffusion of solar magnetic elements, *Astrophysical Journal* **521**, 2, p. 844.

Calcagni, G. (2010a). Fractal universe and quantum gravity, *Physical Review Letters* **104**, 25, p. 251301.

Calcagni, G. (2010b). Quantum field theory, gravity and cosmology in a fractal universe, *Journal of High Energy Physics* **2010**, 3, p. 120.

Calcagni, G., Eichhorn, A., and Saueressig, F. (2013). Probing the quantum nature of spacetime by diffusion, *Physical Review D* **87**, 12, p. 124028.

Candia, J., Epele, L. N., and Roulet, E. (2002). Cosmic ray photodisintegration and the knee of the spectrum, *Astroparticle Physics* **17**, 1, pp. 23–33.

Capozziello, S., De Filippis, E., and Salzano, V. (2009). Modelling clusters of galaxies by f (r) gravity, *Monthly Notices of the Royal Astronomical Society* **394**, 2, pp. 947–959.

Capozziello, S. and Funkhouser, S. (2009). Fractal large-scale structure from a stochastic scaling law model, *Modern Physics Letters A* **24**, 22, pp. 1743–1748.

Cartea, Á. and del Castillo-Negrete, D. (2007). Fluid limit of the continuous-time random walk with general Lévy jump distribution functions, *Physical Review E* **76**, 4, p. 041105.

Case, G. and Bhattacharya, D. (1996). Revisiting the galactic supernova remnant distribution. *Astronomy and Astrophysics Supplement Series* **120**, pp. 437–440.

Case, K. M. and Zweifel, P. F. (1967). *Linear transport theory* (Addison-Wesley).

Casse, F., Lemoine, M., and Pelletier, G. (2001). Transport of cosmic rays in chaotic magnetic fields, *Physical Review D* **65**, 2, p. 023002.

Chakrabarty, A. and Meerschaert, M. M. (2011). Tempered stable laws as random walk limits, *Statistics & Probability Letters* **81**, 8, pp. 989–997.

Chandran, B. D. (2003). Particle acceleration by slow modes in strong compressible magnetohydrodynamic turbulence, with application to solar flares, *Astrophysical Journal* **599**, 2, p. 1426.

Chavanis, P.-H. (2002). Statistical mechanics of two-dimensional vortices and stellar systems, in *Dynamics and thermodynamics of systems with long-range interactions* (Springer), pp. 208–289.

Chechkin, A. and Gonchar, V. Y. (2000). Linear relaxation processes governed by fractional symmetric kinetic equations, *Journal of Experimental and Theoretical Physics* **91**, 3, pp. 635–651.

Chechkin, A. *et al.* (2002). Fractional kinetics for relaxation and superdiffusion in a magnetic field, *Physics of Plasmas (1994-present)* **9**, 1, pp. 78–88.

Cho, J. and Lazarian, A. (2006). Particle acceleration by magnetohydrodynamic turbulence, *Astrophysical Journal* **638**, 2, p. 811.

Cho, J., Lazarian, A., and Vishniac, E. T. (2002). Simulations of magnetohydrodynamic turbulence in a strongly magnetized medium, *Astrophysical Journal* **564**, 1, p. 291.

Christon, S., Williams, D., Mitchell, D., Huang, C., and Frank, L. (1991). Spectral characteristics of plasma sheet ion and electron populations during disturbed geomagnetic conditions, *Journal of Geophysical Research: Space Physics* **96**, A1, pp. 1–22.

Chuvilgin, L. and Ptuskin, V. (1993). Anomalous diffusion of cosmic rays across the magnetic field, *Astronomy and Astrophysics* **279**, pp. 278–297.

Cocconi, G. (1951). On the origin of the cosmic radiation, *Physical Review* **83**, 6, p. 1193.

Coleman, P. H. and Pietronero, L. (1992). The fractal structure of the universe, *Physics Reports* **213**, 6, pp. 311–389.

Coles, P. and Lucchin, F. (2003). *Cosmology: The origin and evolution of cosmic structure* (John Wiley & Sons).

Colless, M. *et al.* (2001). The 2dF galaxy redshift survey: spectra and redshifts, *Monthly Notices of the Royal Astronomical Society* **328**, 4, pp. 1039–1063.

Colless, M. *et al.* (1999). First results from the 2dF galaxy redshift survey, *Philosophical Transactions – Royal Society of London, Series A, Mathematical Physical and Engineering Sciences* **357**, pp. 105–116.

Combes, F. (2000). Astrophysical fractals: Interstellar medium and galaxies, in *The Chaotic Universe* (World Scientific), pp. 143–172.

Cordes, J. M. and Lazio, T. J. W. (2001). Anomalous radio-wave scattering from interstellar plasma structures, *Astrophysical Journal* **549**, 2, p. 997.

Corrsin, S. (1959). Atmospheric diffusion and air pollution (Advances in Geophysics, Vol. 6, eds. F. N. Frenkiel, P. A. Sheppard), p. 187.

Corrsin, S. (1964). The isotropic turbulent mixer: Part II. Arbitrary Schmidt number, *AIChE Journal* **10**, 6, pp. 870–877.

Cowsik, R. (1986). Sporadic acceleration of cosmic rays, *Astronomy and Astrophysics* **155**, pp. 344–346.

De Vega, H., Sanchez, N., and Combes, F. (1998). The fractal structure of the universe: a new field theory approach, *Astrophysical Journal* **500**, 1, p. 8.

Debnath, U., Chattopadhyay, S., and Jamil, M. (2013). Fractional action cosmology: some dark energy models in emergent, logamediate, and intermediate scenarios of the universe, *Journal of Theoretical and Applied Physics* **7**, 1, p. 25.

del Castillo-Negrete, D., Carreras, B., and Lynch, V. (2003). Front dynamics in reaction-diffusion systems with Lévy flights: a fractional diffusion approach, *Physical Review Letters* **91**, 1, p. 018302.

del Castillo-Negrete, D., Carreras, B., and Lynch, V. (2004). Fractional diffusion in plasma turbulence, *Physics of Plasmas* **11**, 8, pp. 3854–3864.

del Castillo-Negrete, D., Carreras, B., and Lynch, V. (2005). Nondiffusive transport in plasma turbulence: a fractional diffusion approach, *Physical Review Letters* **94**, 6, p. 065003.

Dendy, R., Drury, L., and Bell, A. (1995). Nonlinear plasma effects in cosmic ray shock precursors, in *International Cosmic Ray Conference*, Vol. 3, p. 233.

Denisov, S., Horsthemke, W., and Hänggi, P. (2008). Steady-state Lévy flights in a confined domain, *Physical Review E* **77**, 6, p. 061112.

Dennis, T. J. and Chandran, B. D. (2005). Turbulent heating of galaxy-cluster plasmas, *Astrophysical Journal* **622**, 1, p. 205.

Dogiel, V., Colafrancesco, S., Ko, C., Kuo, P., Hwang, C., Ip, W., Birkinshaw, M., and Prokhorov, D. (2007). In-situ acceleration of subrelativistic electrons in the Coma halo and the halo's influence on the Sunyaev-Zeldovich effect, *Astronomy and Astrophysics* **461**, 2, pp. 433–443.

Dolginov, A. and Toptygin, I. (1966). Multiple scattering of particles in magnetic field with random inhomogeneities, *ZhETP* **51**, 6, p. 1771.

Dolginov, A. and Toptygin, I. (1967). Multiple scattering of particles in a magnetic field with random inhomogeneities, *Soviet Journal of Experimental and Theoretical Physics* **24**, p. 1195.

Doostmohammadi, S. and Fatemi, S. (2012). The characteristics of cosmic rays in a fractal medium, *ISRN High Energy Physics* **2012**.

Dorman, L. (1975). *Experimental and theoretical foundations of cosmic ray astrophysics*, Vol. 1 (Moscow, Izdatel'stvo Nauka, in Russian).

Dorman, L., Ghosh, A., and Ptuskin, V. (1985). Diffusion of the galactic cosmic rays in the vicinity of the solar system, *Astrophysics and Space Science* **109**, 1, pp. 87–97.

Dorman, L. and Kats, M. (1977). Cosmic ray kinetics in space, *Space Science Reviews* **20**, pp. 529–575.

Drury, L. O. (1983). An introduction to the theory of diffusive shock acceleration of energetic particles in tenuous plasmas, *Reports on Progress in Physics* **46**, 8, p. 973.

Du, Q., Gunzburger, M., Lehoucq, R. B., and Zhou, K. (2012). Analysis and approximation of nonlocal diffusion problems with volume constraints, *SIAM Review* **54**, 4, pp. 667–696.

Duffy, P., Kirk, J., Gallant, Y., and Dendy, R. (1995). Anomalous transport and particle acceleration at shocks, *arXiv preprint astro-ph/9509058*.

Dunkel, J. and Hänggi, P. (2009). Relativistic Brownian motion, *Physics Reports* **471**, 1, pp. 1–73.

Dunkel, J., Talkner, P., and Hänggi, P. (2007). Relativistic diffusion processes and random walk models, *Physical Review D* **75**, 4, p. 043001.

Durrer, R. and Labini, F. S. (1998). A fractal galaxy distribution in a homogeneous universe?, *arXiv preprint astro-ph/9804171*.

Eichler, D. (1980). Basic inconsistencies in models of interstellar cosmic-ray acceleration, *Astrophysical Journal* **237**, pp. 809–813.

Einstein, A. (1905). Über die von der molekularkinetischen theorie der Wärme geforderte Bewegung von in ruhenden Flüssigkeiten suspendierten teilchen, *Annalen der physik* **4**.

Einstein, A. (1956). *Investigations on the Theory of the Brownian Movement* (Courier Corporation).

El-Nabulsi, A. R. (2007a). Cosmology with a fractional action principle, *Romanian Reports in Physics* **59**, 3, pp. 763–771.

El-Nabulsi, R. (2007b). Differential geometry and modern cosmology with fractionaly differentiated Lagrangian function and fractional decaying force term, *Romanian Journal of Physics* **52**, 3/4, p. 467.

El-Nabulsi, R. A. (2012). Gravitons in fractional action cosmology, *International Journal of Theoretical Physics* **51**, 12, pp. 3978–3992.

El-Nabulsi, R. (2013). Nonstandard fractional exponential Lagrangians, fractional geodesic equation, complex general relativity, and discrete gravity, *Canadian Journal of Physics* **91**, 8, pp. 618–622.

El-Nabulsi, R. A. (2016). Implications of the Ornstein-Uhlenbeck-like fractional differential equation in cosmology, *Revista Mexicana de Física* **62**, 3, pp. 240–250.

El-Nabulsi, R. A. and Torres, D. F. (2008). Fractional actionlike variational problems, *Journal of Mathematical Physics* **49**, 5, p. 053521.

Elmegreen, B. G. (1998). Diffuse Hα in a fractal interstellar medium, *Publications of the Astronomical Society of Australia* **15**, 01, pp. 74–78.

Erlykin, A., Lagutin, A., and Wolfendale, A. W. (2003). Properties of the interstellar medium and the propagation of cosmic rays in the galaxy, *Astroparticle Physics* **19**, 3, pp. 351–362.

Erlykin, A., Sibatov, R., Uchaikin, V., and Wolfendale, A. (2015). A look at the cosmic ray anisotropy with the nonlocal relativistic transport approach, *Proc. of The 34th International Cosmic Ray Conference (ICRC'15)*.

Erlykin, A. and Wolfendale, A. (1997). A single source of cosmic rays in the range 10^{15}–10^{16} eV, *Journal of Physics G: Nuclear and Particle Physics* **23**, 8, p. 979.

Erlykin, A. and Wolfendale, A. (2001). Structure in the cosmic ray spectrum: an update, *Journal of Physics G: Nuclear and Particle Physics* **27**, 5, p. 1005.

Erlykin, A. D. and Wolfendale, A. W. (2013). Cosmic rays in the inner galaxy and the diffusion properties of the interstellar medium, *Astroparticle Physics* **42**, pp. 70–75.

Erlykin, A. and Wolfendale, A. (2015). The role of the galactic halo and the single source in the formation of the cosmic ray anisotropy, *Astroparticle Physics* **60**, pp. 86–91.

Ernst, D., Hellmann, M., Köhler, J., and Weiss, M. (2012). Fractional Brownian motion in crowded fluids, *Soft Matter* **8**, 18, pp. 4886–4889.

Everett, J. E., Zweibel, E. G., Benjamin, R. A., McCammon, D., Rocks, L., and Gallagher III, J. S. (2008). The Milky Ways kiloparsec-scale wind: a hybrid cosmic-ray and thermally driven outflow, *Astrophysical Journal* **674**, 1, p. 258.

Evoli, C., Gaggero, D., Grasso, D., and Maccione, L. (2008). Cosmic ray nuclei, antiprotons and gamma rays in the galaxy: a new diffusion model, *Journal of Cosmology and Astroparticle Physics* **2008**, 10, p. 018.

Falgarone, E., Phillips, T., and Walker, C. K. (1991). The edges of molecular clouds-fractal boundaries and density structure, *Astrophysical Journal* **378**, pp. 186–201.

Faucher-Giguere, C.-A. and Kaspi, V. M. (2006). Birth and evolution of isolated radio pulsars, *Astrophysical Journal* **643**, 1, p. 332.

Fedorov, Y. I., Kats, M., Kichatinov, L., and Stehlik, M. (1992). Cosmic-ray kinetics in a random anisotropic reflective non-invariant magnetic field, *Astronomy and Astrophysics* **260**, pp. 499–509.

Fedorov, Y. I. and Stehlik, M. (2008). Stochastic acceleration of cosmic rays in helical plasma turbulence, *in Proceedings of the 21st European Cosmic Ray Symposium (Kosice, 2008)*, p. 241244.

Feller, W. (1971). *An introduction to probability and its applications, Vol. II* (Wiley).

Fermi, E. (1949). On the origin of the cosmic radiation, *Physical Review* **75**, 8, p. 1169.

Fisher, J. C. (1951). Calculation of diffusion penetration curves for surface and grain boundary diffusion, *Journal of Applied Physics* **22**, 1, pp. 74–77.

Fock, V. A. (1926). Solution of one problem of light diffusion theory by finite differences method and its application to the diffusion of light, *Proceedings of State Optical Institute* **4**, 34, pp. 1–32.

Fock, V. (1930). Näherungsmethode zur lösung des quantenmechanischen mehrkörperproblems, *Zeitschrift für Physik* **61**, 1-2, pp. 126–148.

Forbush, S. E. (1954). World-wide cosmic ray variations, 1937–1952, *Journal of Geophysical Research* **59**, 4, pp. 525–542.

Forman, M., Jokipii, J., and Owens, A. (1974). Cosmic-ray streaming perpendicular to the mean magnetic field, *Astrophysical Journal* **192**, pp. 535–540.

Frederico, G. S. and Torres, D. F. (2006). Constants of motion for fractional action-like variational problems, *arXiv preprint math/0607472*.

Frisch, U. and Bec, J. (2001). Burgulence, in *New trends in turbulence Turbulence: nouveaux aspects* (Springer), pp. 341–383.

Gajda, J. and Magdziarz, M. (2010). Fractional Fokker-Planck equation with tempered α-stable waiting times: Langevin picture and computer simulation, *Physical Review E* **82**, 1, p. 011117.

Gaveau, B., Jacobson, T., Kac, M., and Schulman, L. (1984). Relativistic extension of the analogy between quantum mechanics and Brownian motion, *Physical Review Letters* **53**, 5, p. 419.

Genel, S., Vogelsberger, M., Springel, V., Sijacki, D., Nelson, D., Snyder, G., Rodriguez-Gomez, V., Torrey, P., and Hernquist, L. (2014). Introducing the illustris project: the evolution of galaxy populations across cosmic time, *Monthly Notices of the Royal Astronomical Society* **445**, 1, pp. 175–200.

Getmantsev, G. (1962). On the origin of cosmic radio-emission and cosmic rays, *Izvestia vuzov, Radiofizika [in Russian]* **5**, 1, pp. 172–174.

Getmantsev, G. (1963). On the isotropy of primary cosmic rays, *Soviet Astronomy* **6**, p. 477.

Giacalone, J. and Jokipii, J. R. (1999). The transport of cosmic rays across a turbulent magnetic field, *Astrophysical Journal* **520**, 1, p. 204.

Giannattasio, F. *et al.* (2013). Diffusion of solar magnetic elements up to supergranular spatial and temporal scales, *Astrophysical Journal Letters* **770**, 2, p. L36.

Gibson, C. H. (2001). Turbulence and mixing in the early universe, *Arxiv preprint astro-ph/0110012*.

Ginzburg, V. (1953). Origin of cosmic rays and radioastronomy, *Uspekhi Fizicheskikh Nauk* **51**.

Ginzburg, V. and Syrovatskii, S. (1964). *The origin of cosmic rays* (Macmillan).

Gleeson, L. and Axford, W. (1967). Cosmic rays in the interplanetary medium, *Astrophysical Journal* **149**, p. L115.

Goldreich, P. and Sridhar, S. (1995). Toward a theory of interstellar turbulence. 2: Strong alfvenic turbulence, *The Astrophysical Journal* **438**, pp. 763–775.

Goldstein, M. L., Roberts, D. A., and Matthaeus, W. (1995). Magnetohydrodynamic turbulence in the solar wind, *Annual Review of Astronomy and Astrophysics* **33**, pp. 283–326.

Guan, Q.-Y. and Ma, Z.-M. (2006). Reflected symmetric α-stable processes and regional fractional Laplacian, *Probability Theory and Related Fields* **134**, 4, pp. 649–694.

Gunzburger, M. and Lehoucq, R. B. (2010). A nonlocal vector calculus with application to nonlocal boundary value problems, *Multiscale Modeling & Simulation* **8**, 5, pp. 1581–1598.

Gurbatov, S. and Saichev, A. (1984). Probability distribution and spectra of potential hydrodynamic turbulence, *Radiophysics and Quantum Electronics* **27**, 4, pp. 303–313.

Gurbatov, S. N., Saichev, A., and Shandarin, S. (1989). The large-scale structure of the universe in the frame of the model equation of non-linear diffusion, *Monthly Notices of the Royal Astronomical Society* **236**, 2, pp. 385–402.

Hagenaar, H. *et al.* (1999). Dispersal of magnetic flux in the quiet solar photosphere, *Astrophysical Journal* **511**, 2, p. 932.

Hasegawa, A., Mima, K., and Duong-van, M. (1985). Plasma distribution function in a superthermal radiation field, *Physical Review Letters* **54**, 24, p. 2608.

Hasselmann, K. and Wibberenz, G. (1968). Scattering of charged particles by random electromagnetic fields, Tech. rep., Institut fuer Schiffbau, Hamburg.

Hayakawa, S. (1969). Cosmic ray physics. nuclear and astrophysical aspects, *Interscience Monographs and Texts in Physics and Astronomy*, Wiley-Interscience, **1**.

Heisenberg, W. (1948). Zur statistischen theorie der turbulenz, *Zeitschrift für Physik* **124**, 7-12, pp. 628–657.

Heisenberg, W. (1966). Die rolle der phänomenologischen theorien im system der theoretischen physik, *Preludes in Theoretical Physics in honor of VF Weisskopf* **1**, p. 166.

Heisenberg, W. (1985). On the theory of statistical and isotropic turbulence, in *Original Scientific Papers Wissenschaftliche Originalarbeiten* (Springer), pp. 115–119.

Hinze, J. (1959). *Turbulence: an introduction to its mechanism and theory* (McGraw-Hill).

Hnatich, M., Jurcisin, M., and Stehlik, M. (2001). Dynamo in helical MHD turbulence: quantum field theory approach, *Magnetohydrodynamics* **37**, pp. 80–86.

Hörandel, J. R. (2004). Models of the knee in the energy spectrum of cosmic rays, *Astroparticle Physics* **21**, 3, pp. 241–265.

Hörandel, J. R. (2006). A review of experimental results at the knee, in *Journal of Physics: Conference Series*, Vol. 47 (IOP Publishing), p. 41.

Hörandel, J. R., Kalmykov, N. N., and Timokhin, A. V. (2007). Propagation of super-high-energy cosmic rays in the galaxy, *Astroparticle Physics* **27**, 2, pp. 119–126.

Horbury, T. S. (1999). Waves and turbulence in the solar wind–an overview, in *Plasma Turbulence and Energetic Particles in Astrophysics*, pp. 115–134.

Horbury, T. and Balogh, A. (1997). Structure function measurements of the intermittent MHD turbulent cascade, *Nonlinear Processes in Geophysics* **4**, 3, pp. 185–199.

Hu, W. and Dodelson, S. (2002). Cosmic microwave background anisotropies, *Annual Review of Astronomy and Astrophysics* **40**, 1, pp. 171–216.

Hu, Y. and Kallianpur, G. (2000). Schrödinger equations with fractional Laplacians, *Applied Mathematics & Optimization* **42**, 3, pp. 281–290.

Iomin, A. (2011). Subdiffusion on a fractal comb, *Physical Review E* **83**, 5, p. 052106.

Isichenko, M. B. (1992). Percolation, statistical topography, and transport in random media, *Reviews of Modern Physics* **64**, 4, p. 961.

Jamil, M., Rashid, M. A., Momeni, D., Razina, O., and Esmakhanova, K. (2012). Fractional action cosmology with power law weight function, in *Journal of Physics: Conference Series*, Vol. 354 (IOP Publishing), p. 012008.

Jeng, M., Xu, S.-L.-Y., Hawkins, E., and Schwarz, J. (2010). On the nonlocality of the fractional Schrödinger equation, *Journal of Mathematical Physics* **51**, 6, p. 062102.

Jeon, J., Chechkin, A., and Metzler, R. (2014). First passage behavior of multidimensional fractional Brownian motion and application to reaction phenomena, *First-Passage Phenomena and Their Applications* **35**, p. 175.

Jokipii, J. R. (1966). Cosmic-ray propagation. I. charged particles in a random magnetic field, *Astrophysical Journal* **146**, p. 480.

Jokipii, J. (1967). Cosmic-ray propagation. II. Diffusion in the interplanetary magnetic field, *Astrophysical Journal* **149**, p. 405.

Jokipii, J. (1971). Propagation of cosmic rays in the solar wind, *Reviews of Geophysics* **9**, 1, pp. 27–87.

Jokipii, J. (1973a). The rate of separation of magnetic lines of force in a random magnetic field, *Astrophysical Journal* **183**, pp. 1029–1036.

Jokipii, J. R. (1973b). The rate of separation of magnetic lines of force in a random magnetic field, *Astrophysical Journal* **183**, pp. 1029–1036.

Jokipii, J. and Parker, E. (1969a). Cosmic-ray life and the stochastic nature of the galactic magnetic field, *The Astrophysical Journal* **155**, p. 799.

Jokipii, J. and Parker, E. (1969b). Stochastic aspects of magnetic lines of force with application to cosmic-ray propagation, *Astrophysical Journal* **155**, p. 777.

Jones, F. C. and Ellison, D. C. (1991). The plasma physics of shock acceleration, *Space Science Reviews* **58**, 1, pp. 259–346.

Jumarie, G. (2012). An approach to differential geometry of fractional order via modified Riemann-Liouville derivative, *Acta Mathematica Sinica* **28**, 9, pp. 1741–1768.

Kachelriess, M. (2008). Lecture notes on high energy cosmic rays, *arXiv preprint arXiv:0801.4376*.

Kadomtsev, B. and Pogutse, O. (1978). Plasma physics and controlled nuclear fusion research 1978. Proc. of the 7th Intern. Conf., IAEA, Innsbruck, Austria, August 23 - 30, 1978 vol. 1, in *Statistical description of transport in plasma, astro-and nuclear physics* (Vienna: Intern. Atomic Energy Agency), p. 649.

Kampert, K.-H. and Watson, A. A. (2012). Extensive air showers and ultra high-energy cosmic rays: a historical review, *The European Physical Journal H* **37**, 3, pp. 359–412.

Karakula, S. and Tkaczyk, W. (1993). The formation of the cosmic ray energy spectrum by a photon field, *Astroparticle Physics* **1**, 2, pp. 229–237.

Kaur, I. and Gust, W. (1989). *Handbook of grain and interphase boundary diffusion data*, Vol. 2 (Ziegler Press).

Kazanas, D. and Nicolaidis, A. (2003). Letter: Cosmic rays and large extra dimensions, *General Relativity and Gravitation* **35**, 6, pp. 1117–1123.

Kermani, H. A. and Fatemi, J. (2011). Cosmic ray propagation in a fractal galactic medium, *South African Journal of Science* **107**, 1-2, pp. 1–4.

Ketabi, N. and Fatemi, J. (2009). A simulation on the propagation of supernova cosmic particles in a fractal medium, *Transaction B: Mechanical Engineering, Sharif University of Technology* **16**, 3, pp. 269–272.

Kibble, T. W. B. and Vilenkin, A. (1995). Phase equilibration in bubble collisions, *Physical Review D* **52**, 2, p. 679.

Kichatinov, L. (1983). New mechanism for acceleration of cosmic particles in the presence of reflectively noninvariant turbulence, *JETP Lett. (Engl. Transl.); (United States)* **37**, 1.

Kichatinov, L. and Matyukhin, Y. (1981). Cosmic ray transport in anisotropic turbulent ISM, *Geomagnetizm and Aeronomy* **21**, 3, pp. 412–419.

King, C. (2012). Why the universe is fractal, *Prespacetime Journal* **3**, 3.

Klypin, A. and Shandarin, S. (1983). Three-dimensional numerical model of the formation of large-scale structure in the Universe, *Monthly Notices of the Royal Astronomical Society* **204**, 3, pp. 891–907.

Koay, J. and Macquart, J.-P. (2015). Scatter broadening of compact radio sources by the ionized intergalactic medium: prospects for detection with space VLBI and the square kilometre array, *Monthly Notices of the Royal Astronomical Society* **446**, 3, pp. 2370–2379.

Kolmogorov, A. N. (1941). The local structure of turbulence in incompressible viscous fluid for very large Reynolds numbers, **30**, 4, pp. 299–303.

Kolokoltsov, V., Korolev, V., and Uchaikin, V. (2001). Fractional stable distributions, *Journal of Mathematical Sciences* **105**, 6, pp. 2569–2576.

Koponen, I. (1995). Analytic approach to the problem of convergence of truncated Lévy flights towards the Gaussian stochastic process, *Physical Review E* **52**, 1, p. 1197.

Kovasznay, L. S. (1948). The spectrum of locally isotropic turbulence, *Physical Review* **73**, 9, p. 1115.

Kowal, G. and Lazarian, A. (2007). Scaling relations of compressible mhd turbulence, *Astrophysical Journal Letters* **666**, 2, p. L69.

Krause, F. and Rädler, K.-H. (2016). *Mean-field magnetohydrodynamics and dynamo theory* (Elsevier).

Krepysheva, N., Di Pietro, L., and Néel, M.-C. (2006). Space-fractional advection-diffusion and reflective boundary condition, *Physical Review E* **73**, 2, p. 021104.

Kritsuk, A. G., Norman, M. L., Padoan, P., and Wagner, R. (2007). The statistics of supersonic isothermal turbulence, *Astrophysical Journal* **665**, 1, p. 416.

Krymskii, G. (1964). Diffusion mechanism of diurnal cosmic-ray variations, *Geomagnetism Aeronomy (USSR)(English Transl.)* **4**.

Krymskii, G. (1977). A regular mechanism for the acceleration of charged particles on the front of a shock wave, in *Akademiia Nauk SSSR Doklady*, Vol. 234, pp. 1306–1308.

Kulakov, A. and Rumyantsev, A. (1994). Fractals and energy spectrum of high-energy cosmic particles, in *Physics-Doklady*, Vol. 39, pp. 337–338.

Kulikov, G. and Khristiansen, G. (1959). On the size spectrum of extensive air showers, *Soviet Physics JETP-USSR* **8**, 3, pp. 441–444.

Kuŝĉer, I. and Zweifel, P. (1965). Time-dependent one-speed albedo problem for a semi-infinite medium, *Journal of Mathematical Physics* **6**, 7, pp. 1125–1130.

Lagage, P. and Cesarsky, C. (1983). Cosmic-ray shock acceleration in the presence of self-excited waves, *Astronomy and Astrophysics* **118**, pp. 223–228.

Lagutin, A. (2010). Acceleration of cosmic rays on shock wave front: Fractional-differential approach, *Proceedings of 31st Russian Conference on Cosmic Rays [in Russian]. Moscow State University.*

Lagutin, A., Nikulin, Y. A., and Uchaikin, V. (2000). The 'knee' in the primary cosmic rays spectrum as consequence of fractal structure of the galactic magnetic field, *Preprint ASU-2000/4.*

Lagutin, A., Makarov, V., and Tyumentsev, A. (2001a). Proc. of the 27th Intern. Cosmic Ray Conf., 7 - 15 August, 2001, Hamburg, Germany vol. 5.

Lagutin, A., Nikulin, Y. A., and Uchaikin, V. (2001b). The 'knee' in the primary cosmic ray spectrum as consequence of the anomalous diffusion of the particles in the fractal interstellar medium, *Nuclear Physics B-Proceedings Supplements* **97**, 1, pp. 267–270.

Lagutin, A., Osadchiy, K., and Gerasimov, V. (2001c). Spectra of cosmic ray electrons and positrons in the galaxy, *Problems of Atomic Science and Technology.*

Lagutin, A., Strelnikov, D., and Tyumentsev, A. (2001d). Proc. of the 27th Intern. Cosmic Ray Conf., 7 - 15 August, 2001, Hamburg, Germany vol. 5.

Lagutin, A., Raikin, R., and Tyumentsev, A. (2004). Distribution of first path length in the galactic medium of fractal type, *Bulletin of Altay State University [in Russian].*

Lagutin, A. and Tyumentsev, A. (2004). Spectrum, mass composition and anizotropy of cosmic rays in the fractal galaxy, *Izv. Altai. Gos. Univ* **35**, 4, pp. 4–21.

Lagutin, A. and Tyumentsev, A. (2013). Monte Carlo simulations of anomalous diffusion of cosmic rays, in *Journal of Physics: Conference Series*, Vol. 409 (IOP Publishing), p. 012050.

Lagutin, A., Tyumentsev, A., Volkov, N., and Raikin, R. (2015). Spectra of cosmic-ray protons and nuclei from 1010 to 1020 eV within the galactic origin scenario of cosmic rays, *Bulletin of the Russian Academy of Sciences: Physics* **79**, 3, pp. 322–325.

Lagutin, A., Tyumentsev, A., and Yushkov, A. (2008). Energy spectrum and mass composition of primary cosmic rays around the 'knee' the framework of the model with two types of sources, *Nuclear Physics B-Proceedings Supplements* **175**, pp. 555–558.

Lagutin, A. and Uchaikin, V. (2001). *Proc. of the 27th International Cosmic Ray Conference, 7-15 August 2001, Hamburg, Germany, Vol. 5* , p. 1896.

Lagutin, A. and Uchaikin, V. (2003). Anomalous diffusion equation: Application to cosmic ray transport, *Nuclear Instruments and Methods in Physics Research Section B: Beam Interactions with Materials and Atoms* **201**, 1, pp. 212–216.

Lamperti, J. (1958). An occupation time theorem for a class of stochastic processes, *Transactions of the American Mathematical Society* **88**, 2, pp. 380–387.

Larson, R. B. (1981). Turbulence and star formation in molecular clouds, *Monthly Notices of the Royal Astronomical Society* **194**, 4, pp. 809–826.

Lawrence, J., Cadavid, A., Ruzmaikin, A., and Berger, T. (2001). Spatiotemporal scaling of solar surface flows, *Physical review letters* **86**, 26, p. 5894.

Lazarian, A. (2005). Astrophysical implications of turbulent reconnection: from cosmic rays to star formation, *arXiv preprint astro-ph/0505574*.

Lazarian, A. (2008). Obtaining spectra of turbulent velocity from observations, in *From the Outer Heliosphere to the Local Bubble* (Springer), pp. 357–385.

Lee, L. and Jokipii, J. R. (1975). Strong scintillations in astrophysics. II. A theory of temporal broadening of pulses, *Astrophysical Journal* **201**, pp. 532–543.

Lee, M. and Fisk, L. (1982). Shock acceleration of energetic particles in the heliosphere, *Space Science Reviews* **32**, 1, pp. 205–228.

Leighton, R. B. (1964). Transport of magnetic fields on the sun. *Astrophysical Journal* **140**, p. 1547.

Lerche, I. and Schlickeiser, R. (1982). On the transport and propagation of cosmic rays in galaxies. I. Solution of the steady-state transport equation for cosmic ray nucleons, momentum spectra and heating of the interstellar medium, *Monthly Notices of the Royal Astronomical Society* **201**, 4, pp. 1041–1072.

Levine, H. and MacCallum, C. (1960). Grain boundary and lattice diffusion in polycrystalline bodies, *Journal of Applied Physics* **31**, 3, pp. 595–599.

Lévy, P. (1965). *Processus stochastiques et mouvement brownien* (Paris).

Litvinenko, Y. E. and Effenberger, F. (2014). Analytical solutions of a fractional diffusion-advection equation for solar cosmic-ray transport, *Astrophysical Journal* **796**, 2, p. 125.

Litvinenko, Y. E., Effenberger, F., and Schlickeiser, R. (2015). The telegraph approximation for focused cosmic-ray transport in the presence of boundaries, *arXiv preprint arXiv:1505.05134*.

Liu, S., Petrosian, V., and Melia, F. (2004). Electron acceleration around the supermassive black hole at the galactic center, *Astrophysical Journal Letters* **611**, 2, p. L101.

Lubashevskii, I. and Zemlyanov, A. (1998). Continuum description of anomalous diffusion on a comb structure, *Journal of Experimental and Theoretical Physics* **87**, 4, pp. 700–713.

Lukacs, E. (1970). *Characteristics functions* (Griffin, London).

Magdziarz, M. and Zorawik, T. (2016). Explicit densities of multidimensional ballistic Lévy walks, *Physical Review E* **94**, 2, p. 022130.

Maggiore, M. and Riotto, A. (2010a). The halo mass function from excursion set theory. I. Gaussian fluctuations with non-Markovian dependence on the smoothing scale, *Astrophysical Journal* **711**, 2, p. 907.

Maggiore, M. and Riotto, A. (2010b). The halo mass function from excursion set theory. II. The diffusing barrier, *Astrophysical Journal* **717**, 1, p. 515.

Maggiore, M. and Riotto, A. (2010c). The halo mass function from excursion set theory. III. Non-Gaussian fluctuations, *Astrophysical Journal* **717**, 1, p. 526.

Malinowska, A. B. and Torres, D. F. (2012). *Introduction to the fractional calculus of variations* (World Scientific Publishing Co Inc).

Mandelbrot, B. B. (1975). Stochastic models for the Earth's relief, the shape and the fractal dimension of the coastlines, and the number-area rule for islands, *Proceedings of the National Academy of Sciences* **72**, 10, pp. 3825–3828.

Mandelbrot, B. B. (1977). *Fractals, Forms, Chance and Dimension* (Freeman, San Francisco).

Mandelbrot, B. B. and Van Ness, J. W. (1968). Fractional Brownian motions, fractional noises and applications, *SIAM Review* **10**, 4, pp. 422–437.

Mandelbrot, B. (1979). Correlations et texture dans un nouveau modele d'Univers hierarchise, base sur les ensembles tremas, *Comptes Rendus*, pp. 81–83.

Mandelbrot, B. (1983). *The fractal geometry of nature* (Macmillan).

Mandelbrot, B. (1998). Fractality, lacunarity, and the near-isotropic distribution of galaxies, in *Current Topics in Astrofundamental Physics: Primordial Cosmology* (Springer), pp. 583–601.

Mantegna, R. N. and Stanley, H. E. (1994). Stochastic process with ultraslow convergence to a Gaussian: the truncated Lévy flight, *Physical Review Letters* **73**, 22, p. 2946.

Marazzato, R. and Sparavigna, A. C. (2009). Astronomical image processing based on fractional calculus: the AstroFracTool, *arXiv:0910.4637*.

Marschalkó, G., Forgács-Dajka, E., and Petrovay, K. (2007). Molecular cloud abundances and anomalous diffusion, *Astronomische Nachrichten* **328**, 8, pp. 871–874.

Martinell, J., Del-Castillo-Negrete, D., Raga, A., and Williams, D. (2006). Non-local diffusion and the chemical structure of molecular clouds, *Monthly Notices of the Royal Astronomical Society* **372**, 1, pp. 213–218.

Martinez, V. J. and Jones, B. J. (1990). Why the universe is not a fractal, *Monthly Notices of the Royal Astronomical Society* **242**, 4, pp. 517–521.

Martínez, V. J., Paredes, S., and Saar, E. (1993). Wavelet analysis of the multifractal character of the galaxy distribution, *Monthly Notices of the Royal Astronomical Society* **260**, 2, pp. 365–375.

Martinez, V. J. and Saar, E. (2001). *Statistics of the galaxy distribution* (CRC Press).

Masoliver, J., Porrà, J. M., and Weiss, G. H. (1992). Solutions of the telegrapher's equation in the presence of traps, *Physical Review A* **45**, 4, p. 2222.

Masoliver, J., Porrà, J. M., and Weiss, G. H. (1993). Solution to the telegrapher's equation in the presence of reflecting and partly reflecting boundaries, *Physical Review E* **48**, 2, p. 939.

Masoliver, J. and Weiss, G. H. (1996). Finite-velocity diffusion, *European Journal of Physics* **17**, 4, p. 190.

Mathieu, B., Melchior, P., Oustaloup, A., and Ceyral, C. (2003). Fractional differentiation for edge detection, *Signal Processing* **83**, 11, pp. 2421–2432.

Matthaeus, W., Qin, G., Bieber, J., and Zank, G. (2003). Nonlinear collisionless perpendicular diffusion of charged particles, *Astrophysical Journal Letters* **590**, 1, p. L53.

McKinnon, M. (2014). The analytical solution to the temporal broadening of a Gaussian-shaped radio pulse by multipath scattering from a thin screen in the interstellar medium, *Publications of the Astronomical Society of the Pacific* **126**, 939, p. 476.

Melott, A. L. (1990). The topology of large-scale structure in the Universe, *Physics Reports* **193**, 1, pp. 1–39.

Metzler, R. and Klafter, J. (2000). The random walk's guide to anomalous diffusion: a fractional dynamics approach, *Physics reports* **339**, 1, pp. 1–77.

Meyer, P., Parker, E., and Simpson, J. (1956). Solar cosmic rays of february, 1956 and their propagation through interplanetary space, *Physical Review* **104**, 3, p. 768.

Miller, J. A., Guessoum, N., and Ramaty, R. (1990). Stochastic Fermi acceleration in solar flares, *Astrophysical Journal* **361**, pp. 701–708.

Milovanov, A. and Zelenyi, L. (2002). Nonequilibrium stationary states in the Earth's magnetotail: Stochastic acceleration processes and nonthermal distribution functions, *Advances in Space Research* **30**, 12, pp. 2667–2674.

Miroshnichenko, L. I. (2001). *Solar cosmic rays*, Vol. 2 (Springer).

Mittal, A. and Lohiya, D. (2001). From fractal cosmography to fractal cosmology, *arXiv preprint astro-ph/0104370*.

Mittal, A. and Lohiya, D. (2003). Fractal dust model of the universe based on Mandelbrot's conditional cosmological principle and general theory of relativity, *Fractals* **11**, 02, pp. 145–153.

Moffatt, H. K. (1978). *Field Generation in Electrically Conducting Fluids* (Cambridge University Press).

Monin, A. (1955). The equation of turbulent diffusion, *Doklady Akademii Nauk SSSR* **105**, p. 256.

Monin, A. and Yaglom, A. (1975). Statistical fluid mechanics: Mechanics of turbulence, vol. 2.

Moraal, H. (2013). Cosmic-ray modulation equations, *Space Science Reviews* **176**, 1, p. 299319.

Morrison, P., Olbert, S., and Rossi, B. (1954). The origin of cosmic rays, *Physical Review* **94**, 2, p. 440.

Morse, R. L. and Nielson, C. (1969). One-, two-, and three-dimensional numerical simulation of two-beam plasmas, *Physical Review Letters* **23**, 19, p. 1087.

Moses, J. (2000). Complexity and flexibility, *Quoted in Ideas on Complexity in Systems – Twenty Views, by JM Sussman*, **83**.

Nakamichi, A. and Morikawa, M. (2009). Cosmic dark turbulence, *Astronomy and Astrophysics* **498**, 2, pp. 357–359.

Nikolsky, S. and Romachin, V. (2000). Cosmic rays of energies in the range 10^3–10^5 TeV and higher, *Physics of Atomic Nuclei* **63**, 10, pp. 1799–1814.

Norberg, P. *et al.* (2001). The 2dF Galaxy Redshift Survey: luminosity dependence of galaxy clustering, *Monthly Notices of the Royal Astronomical Society* **328**, 1, pp. 64–70.

Obukhov, A. (1941). Spectral energy distribution in a turbulent flow, in *Doklady Akademii Nauk SSSR* **32**, pp. 22–24.

Osborne, J., Wdowczyk, J., and Wolfendal, A. (1976). Origin and propagation of cosmic rays in the range 100-1000 gev, *Journal of Physics A: Mathematical and General* **9**, 8, p. 1399.

Pallottini, A., Ferrara, A., and Evoli, C. (2013). Simulating intergalactic quasar scintillation, *Monthly Notices of the Royal Astronomical Society* **434**, 4, pp. 3293–3304.

Pan, J. (2007). Fractional Brownian motion and the halo mass function, *Monthly Notices of the Royal Astronomical Society: Letters* **374**, 1, pp. L6–L9.

Pan, J., Wang, Y., Chen, X., and Teodoro, L. (2008). Effects of correlation between merging steps on the global halo formation, *Monthly Notices of the Royal Astronomical Society* **389**, 1, pp. 461–468.

Panasyuk, M., Kalmykov, N., Kovtiukh, A., Kuznetsov, N., Kulikov, G., Kurt, V., Nymmik, R., and Roganova, T. (2006). *Radiation Conditions in Space [Radiatsionnye Usloviia v Kosmicheskom Prostranstve] (in Russian)* (Skobeltsyn Institute of Nuclear Physics, Lomonosov Moscow State University).

Pao, Y.-H. (1965). Structure of turbulent velocity and scalar fields at large wavenumbers, *Physics of Fluids (1958-1988)* **8**, 6, pp. 1063–1075.

Parker, E. (1958a). Cosmic-ray modulation by solar wind, *Physical Review* **110**, 6, p. 1445.

Parker, E. N. (1958b). Dynamics of the interplanetary gas and magnetic fields. *Astrophysical Journal* **128**, p. 664.

Parker, E. N. (1965). The passage of energetic charged particles through interplanetary space, *Planetary and Space Science* **13**, 1, pp. 9–49.

Peacock, J. A. *et al.* (2001). A measurement of the cosmological mass density from clustering in the 2dF Galaxy Redshift Survey, *Nature* **410**, 6825, pp. 169–173.

Peebles, P. J. E. (1980). *The large-scale structure of the universe* (Princeton University Press).

Peebles, P. J. E. (1993). *Principles of physical cosmology* (Princeton University Press).

Perri, S. and Zimbardo, G. (2007). Evidence of superdiffusive transport of electrons accelerated at interplanetary shocks, *Astrophysical Journal Letters* **671**, 2, p. L177.

Perri, S. and Zimbardo, G. (2008). Superdiffusive transport of electrons accelerated at corotating interaction regions, *Journal of Geophysical Research: Space Physics* **113**, A3.

Perri, S. and Zimbardo, G. (2009a). Ion and electron superdiffusive transport in the interplanetary space, *Advances in Space Research* **44**, 4, pp. 465–470.

Perri, S. and Zimbardo, G. (2009b). Ion superdiffusion at the solar wind termination shock, *Astrophysical Journal Letters* **693**, 2, p. L118.

Perri, S. and Zimbardo, G. (2015). Short acceleration times from superdiffusive shock acceleration in the heliosphere, *Astrophysical Journal* **815**, 1, p. 75.

Peters, B. (1961). Primary cosmic radiation and extensive air showers, *Il Nuovo Cimento (1955-1965)* **22**, 4, pp. 800–819.

Pilkington, J. D. and Scott, J. (1965). A survey of radio sources between declinations 20 and 40, *Memoirs of the Royal Astronomical Society* **69**, p. 183.

Pommois, P., Zimbardo, G., and Veltri, P. (2007). Anomalous, non-Gaussian transport of charged particles in anisotropic magnetic turbulence, *Physics of Plasmas (1994-present)* **14**, 1, p. 012311.

Popov, V. V. (2008). Analysis of possibilities of fishers model development, in *Solid State Phenomena*, Vol. 138 (Trans Tech Publ), pp. 133–144.

Potgieter, M. S. (2013). Solar modulation of cosmic rays, *Living Reviews in Solar Physics* **10**, 1, pp. 1–66.

Press, W. H. and Schechter, P. (1974). Formation of galaxies and clusters of galaxies by self-similar gravitational condensation, *Astrophysical Journal* **187**, pp. 425–438.

Ptuskin, V. and Chuvilgin, L. (1990). Preprint no 49, *Troitsk: IZMIRAN*.

Ptuskin, V., Rogovaya, S., Zirakashvili, V., Chuvilgin, L., Khristiansen, G., Klepach, E., and Kulikov, G. (1993). Diffusion and drift of very high energy cosmic rays in galactic magnetic fields, *Astronomy and Astrophysics* **268**, pp. 726–735.

Ptuskin, V. S. (2007). On the origin of galactic cosmic rays, *Physics-Uspekhi* **50**, 5, pp. 534–540.

Pucci, F., Malara, F., Perri, S., Zimbardo, G., Sorriso-Valvo, L., and Valentini, F. (2016). Energetic particle transport in the presence of magnetic turbulence: influence of spectral extension and intermittency, *Monthly Notices of the Royal Astronomical Society* **459**, 3, pp. 3395–3406.

Qin, G., Matthaeus, W., and Bieber, J. (2002a). Perpendicular transport of charged particles in composite model turbulence: Recovery of diffusion, *Astrophysical Journal Letters* **578**, 2, p. L117.

Qin, G., Matthaeus, W., and Bieber, J. (2002b). Subdiffusive transport of charged particles perpendicular to the large scale magnetic field, *Geophysical Research Letters* **29**, 4.

Rafeiro, H. and Samko, S. (2005). On multidimensional analogue of Marchaud formula for fractional Riesz-type derivatives in domains in r^n, *Fractional Calculus and Applied Analysis* **8**, 4, pp. 393–401.

Ragot, B. and Kirk, J. (1997). Anomalous transport of cosmic ray electrons, *arXiv preprint astro-ph/9708041*.

Raikin, R. and Lagutin, A. (2011). Changes in mass composition of primary cosmic rays above the knee: towards a model-independent evaluation, in *International Cosmic Ray Conference*, Vol. 1, p. 299.

Ramaty, R. and Lingenfelter, R. (1971). Isotopic composition of the primary cosmic radiation, in *Proc. of a Symp.*, Lyngby, Denmark, March.

Rechester, A. and Rosenbluth, M. (1978). Electron heat transport in a tokamak with destroyed magnetic surfaces, *Physical Review Letters* **40**, 1, p. 38.

Richardson, L. F. (1926). Atmospheric diffusion shown on a distance-neighbour graph, *Proceedings of the Royal Society of London. Series A, Containing Papers of a Mathematical and Physical Character* **110**, 756, pp. 709–737.

Roberts, D., Goldstein, M., and Klein, L. (1990). The amplitudes of interplanetary fluctuations: Stream structure, heliocentric distance, and frequency dependence, *Journal of Geophysical Research: Space Physics* **95**, A4, pp. 4203–4216.

Roberts, M. D. (2009). Fractional derivative cosmology, *arXiv preprint arXiv:0909.1171*.

Roberts, P. (1960). Research report no, *HSN-2* (New York Univ.).

Rosiński, J. (2007). Tempering stable processes, *Stochastic Processes and their Applications* **117**, 6, pp. 677–707.

Rytov, S., Kravtsov, Y. A., and Tatarskii, V. (1978). *Vvedenie v statisticheskuyu radiofiziku. Ch. 2. Sluchainye polya [in Russian] (Introduction to Statistical Radiophysics. Vol. 2. Random Fields)* (Moscow: Nauka).

Sabzikar, F., Meerschaert, M. M., and Chen, J. (2015). Tempered fractional calculus, *Journal of computational physics* **293**, pp. 14–28.

Saichev, A. I. and Zaslavsky, G. M. (1997). Fractional kinetic equations: solutions and applications, *Chaos: An Interdisciplinary Journal of Nonlinear Science* **7**, 4, pp. 753–764.

Sainz, R. M., González, M. M., and Ramos, A. A. (2011). Advection and dispersal of small magnetic elements in the very quiet sun, *Astronomy and Astrophysics* **531**, p. L9.

Samko, S. G., Kilbas, A. A., and Marichev, O. I. (1993). Fractional integrals and derivatives, *Theory and Applications, Gordon and Breach, Yverdon* **1993**.

Samorodnitsky, G. and Taqqu, M. S. (1994). *Stable non-Gaussian random processes: stochastic models with infinite variance*, Vol. 1 (CRC Press).

Sánchez, N., Alfaro, E. J., and Pérez, E. (2005). The fractal dimension of projected clouds, *Astrophysical Journal* **625**, 2, p. 849.

Sanders, L. and Ambjörnsson, T. (2012). First passage times for a tracer particle in single file diffusion and fractional Brownian motion, *Journal of Chemical Physics* **136**, 17, p. 05B605.

Sandev, T., Iomin, A., Kantz, H., Metzler, R., and Chechkin, A. (2016a). Comb model with slow and ultraslow diffusion, *Mathematical Modelling of Natural Phenomena* **11**, 3, pp. 18–33.

Sandev, T., Iomin, A., and Méndez, V. (2016b). Lévy processes on a generalized fractal comb, *Journal of Physics A: Mathematical and Theoretical* **49**, 35, p. 355001.

Scalo, J. (1990). Perception of interstellar structure: Facing complexity, in *Physical Processes in Fragmentation and Star Formation* (Springer), pp. 151–177.

Schlickeiser, R. (2002). *Cosmic ray astrophysics* (Springer Science & Business Media).

Schlickeiser, R. (2009). First-order distributed Fermi acceleration of cosmic ray hadrons in non-uniform magnetic fields, *Modern Physics Letters A* **24**, 19, pp. 1461–1472.

Schönfeld, J. (1962). Integral diffusivity, *Journal of Geophysical Research* **67**, 8, pp. 3187–3199.

Schuecker, P., Böhringer, H., Arzner, K., and Reiprich, T. (2001). Cosmic mass functions from Gaussian stochastic diffusion processes, *Astronomy and Astrophysics* **370**, 3, pp. 715–728.

Scoccimarro, R. (2000). The bispectrum: from theory to observations, *The Astrophysical Journal* **544**, 2, p. 597.

Seinfeld, J. H. (1986). Atmospheric Chemistry and Physics of Air Pollution.

Shakhov, B. and Stehlik, M. (2008). The α-effect and proton acceleration in the solar wind, *Kinematics and Physics of Celestial Bodies* **24**, 1, pp. 19–24.

Shalchi, A. (2010). A unified particle diffusion theory for cross-field scattering: subdiffusion, recovery of diffusion, and diffusion in three-dimensional turbulence, *Astrophysical Journal Letters* **720**, 2, p. L127.

Shalchi, A., Bieber, J., Matthaeus, W., and Qin, G. (2004). Nonlinear parallel and perpendicular diffusion of charged cosmic rays in weak turbulence, *Astrophysical Journal* **616**, 1, p. 617.

Shalchi, A. and Kourakis, I. (2007a). Analytical description of stochastic field-line wandering in magnetic turbulence, *Physics of Plasmas (1994-present)* **14**, 9, p. 092903.

Shalchi, A. and Kourakis, I. (2007b). A new theory for perpendicular transport of cosmic rays, *Astronomy and Astrophysics* **470**, 2, pp. 405–409.

Shalchi, A. and Kourakis, I. (2007c). Random walk of magnetic field-lines for different values of the energy range spectral index, *Physics of Plasmas (1994-present)* **14**, 11, p. 112901.

Shalchi, A., Le Roux, J., Webb, G., and Zank, G. (2009). Analytical description for field-line wandering in strong magnetic turbulence, *Physical Review E* **80**, 6, p. 066408.

Shchigolev, V. (2010). Cosmological models with fractional derivatives and fractional action functional, *arXiv:1011.3304v1*.

Sheth, R., Mo, H., and Tormen, G. (2001). Ellipsoidal collapse and an improved model for the number and spatial distribution of dark matter haloes, *Monthly Notices of the Royal Astronomical Society* **323**, 1, pp. 1–12.

Sheth, R. and Tormen, G. (1999). Large-scale bias and the peak background split, *Monthly Notices of the Royal Astronomical Society* **308**, 1, pp. 119–126.

Sheth, R. and Tormen, G. (2002). An excursion set model of hierarchical clustering: ellipsoidal collapse and the moving barrier, *Monthly Notices of the Royal Astronomical Society* **329**, 1, pp. 61–75.

Shlesinger, M. F., Klafter, J., and West, B. J. (1986). Lévy walks with applications to turbulence and chaos, *Physica A: Statistical Mechanics and its Applications* **140**, 1, pp. 212–218.

Sibatov, R. and Morozova, E. (2015). Multiple trapping on a comb structure as a model of electron transport in disordered nanostructured semiconductors, *Journal of Experimental and Theoretical Physics* **120**, 5, pp. 860–870.

Sibatov, R. and Svetukhin, V. (2017). Grain boundary diffusion in terms of the tempered fractional calculus, *Physics Letters A* **381**, 24, pp. 2021–2027.

Sibatov, R. and Uchaikin, V. (2010). Statistics of photocounts of blinking fluorescence of quantum dots, *Optics and Spectroscopy* **108**, 5, pp. 761–767.

Sibatov, R. T. and Uchaikin, V. V. (2011). Truncated lévy statistics for dispersive transport in disordered semiconductors, *Communications in Nonlinear Science and Numerical Simulation* **16**, 12, pp. 4564–4572.

Sibatov, R. and Uchaikin, V. (2015). On the energy spectrum of cosmic rays in the model of relativistic nonlocal diffusion, in *Proceedings of The 34th International Cosmic Ray Conference*, Vol. 30.

Sokolov, I. (2012). Models of anomalous diffusion in crowded environments, *Soft Matter* **8**, 35, pp. 9043–9052.

Sokolov, I. M. and Metzler, R. (2003). Towards deterministic equations for lévy walks: The fractional material derivative, *Physical Review E* **67**, 1, p. 010101.

Song, R. and Vondracek, Z. (2003). Potential theory of subordinate killed Brownian motion in a domain, *Probability Theory and Related Fields* **125**, 4, pp. 578–592.

Sornette, D. and Cont, R. (1997). Convergent multiplicative processes repelled from zero: power laws and truncated power laws, *Journal de Physique I* **7**, 3, pp. 431–444.

Sparavigna, A. C. and Milligan, P. (2009). Using fractional differentiation in astronomy, *arXiv preprint arXiv:0910.4243*.

Springel, V. (2005). The cosmological simulation code GADGET-2, *Monthly Notices of the Royal Astronomical Society* **364**, 4, pp. 1105–1134.

Stanev, T., Biermann, P. L., and Gaisser, T. K. (1993). Cosmic rays IV. the spectrum and chemical composition above 10^4 GeV, *arXiv preprint astro-ph/9303006*.

Stanislavsky, A., Weron, K., and Weron, A. (2008). Diffusion and relaxation controlled by tempered α-stable processes, *Physical Review E* **78**, 5, p. 051106.

Strömgren, B. (1948). On the density distribution and chemical composition of the interstellar gas, *Astrophysical Journal* **108**, p. 242.

Suzuoka, T. (1961). Lattice and grain boundary diffusion in polycrystals, *Transactions of the Japan Institute of Metals* **2**, 1, pp. 25–32.

Sveshnikova, L. G. (2003). The knee in the Galactic cosmic ray spectrum and variety in Supernovae, *Astronomy and Astrophysics* **409**, 3, pp. 799–807.

Swordy, S. (1995). Expectations for cosmic ray composition changes in the region 10^{14} to 10^{16} eV, in *International Cosmic Ray Conference*, Vol. 2, p. 697.

Tarasov, V. E. (2010). Fractional vector calculus, *Fractional Dynamics*, pp. 241–264.

Taylor, G. I. (1920). Diffusion by continuous movements, *Proc. London Math. Soc.* **20**, pp. 196–211.

Tchen, C.-M. (1954). Transport processes as foundations of the Heisenberg and Obukhoff theories of turbulence, *Physical Review* **93**, 1, p. 4.

Tchen, C. (1959). Diffusion of particles in turbulent flow, *Advances in Geophysics* **9**, pp. 165–174.

Tenreiro Machado, J., Stefanescu, P., Tintereanu, O., and Baleanu, D. (2013). Fractional calculus analysis of the cosmic microwave background, *Romanian Reports in Physics* **65**, 1, p. 316323.

Terletskii, Y. P. and Logunov, A. (1951). Energeticheskii spektr pervichnoi komponenty kosmicheskikh luchei, *Zhurnal Eksperimentalnoi i Teoreticheskoi Fiziki* **21**, 4, pp. 567–568.

Terletskii, Y. P. and Logunov, A. (1952). Funktsiya raspredeleniya kosmicheskikh chastits pervichnoi komponenty, *Zhurnal Eksperimentalnoi i Teoreticheskoi Fiziki* **23**, 6, pp. 682–685.

Toptygin, I. (1983). Cosmic rays in interplanetary magnetic fields, *Moscow, Izdatel'stvo Nauka, 1983, 304 p. In Russian.* **1**.

Tsytovich, V. N. (1977). Theory of turbulent plasma.

Tverskoy, B. (1968). Theory of turbulent acceleration of charged particles in a plasma, *Soviet Phys. JETP* **26**, 4, pp. 821–828.

Uchaikin, V. V. (1998a). Anomalous diffusion of particles with a finite free-motion velocity, *Theoretical and Mathematical Physics* **115**, 1, pp. 496–501.

Uchaikin, V. V. (1998b). Anomalous transport equations and their application to fractal walking, *Physica A: Statistical Mechanics and its Applications* **255**, 1, pp. 65–92.

Uchaikin, V. (1999). Subdiffusion and stable laws, *Journal of experimental and theoretical physics* **88**, 6, pp. 1155–1163.

Uchaikin, V. V. (2002). Multidimensional symmetric anomalous diffusion, *Chemical Physics* **284**, pp. 507–520.

chaikin, V. (2003a). Anomalous diffusion and fractional stable distributions, *Journal of Experimental and Theoretical Physics* **97**, 4, pp. 810–825.

Uchaikin, V. V. (2003b). Self-similar anomalous diffusion and Lévy-stable laws, *Physics-Uspekhi* **46**, 8, pp. 821–849.

Uchaikin, V. (2004a). The mesofractal universe driven by rayleigh-lévy walks, *General Relativity and Gravitation* **36**, 7, pp. 1689–1717.

Uchaikin, V. V. (2004b). If the universe were a Lévy-Mandelbrot fractal... *Gravitation and Cosmology* **10**, pp. 5–24.

Uchaikin, V. V. (2010a). Fractional models of cosmic ray acceleration in the Galaxy, *JETP letters* **92**, 4, pp. 200–205.

Uchaikin, V. V. (2010b). On the fractional derivative model of the transport of cosmic rays in the galaxy, *JETP letters* **91**, 3, pp. 105–109.

Uchaikin, V. V. (2013a). *Fractional derivatives for physicists and engineers* (Springer).

Uchaikin, V. V. (2013b). Fractional phenomenology of cosmic ray anomalous diffusion, *Physics-Uspekhi* **56**, 11, pp. 1074–1119.

Uchaikin, V. (2014). On fractional differential Liouville equation describing open systems dynamics. *Belgogrod State University Scientific Bulletin: Mathematics & Physics* **37**, 25.

Uchaikin, V. V. (2015). Nonlocal models of cosmic ray transport in the galaxy, *Journal of Applied Mathematics and Physics* **3**, 02, p. 187.

Uchaikin, V. V. (2016). On time-fractional representation of an open system response, *Fractional Calculus and Applied Analysis* **19**, 5, pp. 1306–1315.

Uchaikin, V. and Gusarov, G. (1997). Levy flight applied to random media problems, *Journal of Mathematical Physics* **38**, 5, pp. 2453–2464.

Uchaikin, V. and Korobko, D. (1998). *Scientific Notes of Ulyanovsk State University* **1**, 4, p. 3.

Uchaikin, V. and Korobko, D. (1999). Theory of multiple scattering in a fractal medium, *Technical Physics Letters* **25**, 6, pp. 435–437.

Uchaikin, V., Korobko, D., and Gismyatov, I. (1997). Modified Mandelbrot algorithm for stochastic analysis of a fractal-type distribution of galaxies, *Russian Physics Journal* **40**, 8, pp. 711–716.

Uchaikin, V. and Saenko, V. (2000). Telegraph equation in random walk problem, *Zhurnal Fyizichnikh Doslyidzhen'* **4**, 4, pp. 371–379.

Uchaikin, V. and Sibatov, R. (2004). One-dimensional fractal walk at a finite free motion velocity, *Technical Physics Letters* **30**, 4, pp. 316–318.

Uchaikin, V. and Sibatov, R. (2012a). On fractional differential models for cosmic ray diffusion, *Gravitation and Cosmology* **18**, 2, pp. 122–126.

Uchaikin, V. V. and Sibatov, R. T. (2012b). Anomalous kinetics of charge carriers in disordered solids: Fractional derivative approach, *International Journal of Modern Physics B* **26**, 31, p. 1230016.

Uchaikin, V. and Sibatov, R. (2013). Fractional kinetics in solids: Anomalous charge transport in semiconductors, dielectrics and nanosystems.

Uchaikin, V. V. and Zolotarev, V. M. (1999). *Chance and stability: Stable distributions and their applications* (Walter de Gruyter).

Uchaikin, V., Sibatov, R., and Byzykchi, A. (2014). Cosmic rays propagation along solar magnetic field lines: a fractional approach, *Communications in Applied and Industrial Mathematics* **6**, 1, pp. e–480.

Uchaikin, V. V., Cahoy, D. O., and Sibatov, R. T. (2008). Fractional processes: from poisson to branching one, *International Journal of Bifurcation and Chaos* **18**, 09, pp. 2717–2725.

Urch, I. (1977a). Charged particle diffusion in a turbulent magnetic field, *Astrophysics and Space Science* **48**, 1, pp. 231–236.

Urch, I. (1977b). Charged particle diffusion in the presence of Alfvén waves: The perpendicular particle flux, *Astrophysics and Space Science* **49**, 2, pp. 443–472.

Vacaru, S. I. (2012). Fractional dynamics from Einstein gravity, general solutions, and black holes, *International Journal of Theoretical Physics* **51**, 5, pp. 1338–1359.

Vallée, J. P. (1998). Observations of the magnetic fields inside and outside the solar system: From meteorites ($\sim$10 attoparsecs), asteroids, planets, stars, pulsars, masers, to protostellar cloudlets ($<$ 1 parsec), *Fundamentals of Cosmic Physics* **19**, pp. 319–422.

Vázquez, L. (2004). A fruitful interplay: from nonlocality to fractional calculus, in *Nonlinear Waves: Classical and Quantum Aspects* (Springer), pp. 129–133.

Veltri, P. and Zimbardo, G. (1993). Electron-whistler interaction at the earth's bow shock: 1. whistler instability, *Journal of Geophysical Research: Space Physics* **98**, A8, pp. 13325–13333.

Vergassola, M., Dubrulle, B., Frisch, U., and Noullez, A. (1994). Burgers' equa-